The Mechanical Properties of Matter

WILEY SERIES ON THE SCIENCE AND TECHNOLOGY OF MATERIALS

Advisory Editors: J. H. Hollomon, J. E. Burke, B. Chalmers, R. L. Sproull, A. V. Tobolsky

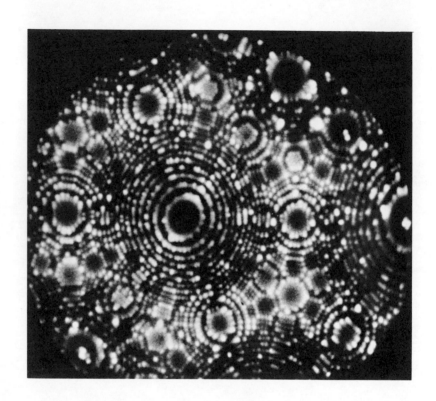

The Atomic Structure Of Tungsten

A spherical tip of crystalline tungsten, approximately 10^{-6} cm radius (i.e. 80 atomic spacings), photographed in the field-ion microscope by S. Ranganathan. Each bright spot is the image of a single atom of the surface. The concentric rings near the centre of the photograph are successive disk-shaped layers of the most densely packed atomic plane in the crystal, viewed from almost vertically above. Only the atoms along the edges of these disks emit brightly. Further out from the centre can be seen various small facets of other crystal planes that form the surface of the tip. The isolated very bright spots are probably oxygen atoms on the metal surface.

The Mechanical Properties of Matter

by
A. H. Cottrell

GOLDSMITHS' PROFESSOR
OF METALLURGY
UNIVERSITY OF CAMBRIDGE

John Wiley & Sons, Inc.
New York · London · Sydney
1964

PHYSICS

add

Library of Congress Catalog Card Number: 64-14262
Printed in the United States of America

HJPI

Preface

I have written this book for students of physical science and engineering who are at a fairly early stage in their university courses and are interested in how the mechanical properties of matter in bulk are determined by what the atoms are doing. The book thus deals mainly with what is usually called "General Properties of Matter", with short excursions into "Strength of Materials", "Physical Metallurgy" and "Materials Science". It differs from traditional books on the general properties of matter in several ways. First, I have left out mechanics and gravitation, which are more commonly dealt with as branches of applied mathematics these days. Instead, I have tried to give an explanation, in terms of atomic behaviour, of the mechanical properties of solids, liquids and gases; the traditional approach, by contrast, usually limits such explanations to the kinetic theory of gases and deals with solids and liquids in a severely empirical way. This atomistic approach has led me to discuss the structure of matter, particularly solids, at some length, but I cannot see how to understand the mechanical properties of matter in any other way. Finally, I have also interpreted the subject fairly broadly, so that at several places the text overlaps into neighbouring domains of physics, chemistry, metallurgy and engineering, in order to obtain an integrated view of the whole field as both a pure and an applied science.

I have tried to keep the text simple and introductory, and have assumed little background knowledge other than school science and some elementary mechanics and calculus. Two things have to be squarely faced, however. First, the partial differential equations which describe the variations of physical quantities through a medium. These, although an unavoidable *pons asinorum* on the road to elasticity and fluid mechanics, are really no more than the expression of Newton's laws and a little commonsense in a mathematical language appropriate to the problem. Second, the approximations and rough order-of-magnitude estimates which we have so often to make when trying to explain bulk properties

v

in terms of atoms. After the apparent rigour of classical science, these seem disturbingly crude, but they are the only way to gain a good physical insight into the behaviour of atoms in crowds without going through a mathematical jungle.

Finally, I would like to express my gratitude to several people. To Dr. J. H. Hollomon, from discussions with whom the idea for this book arose; to Professor B. A. Bilby and Dr. A. Kelly, who have both read the manuscript and commented most usefully upon it; and to Dr. D. G. Brandon and his research group for the field-ion microscope photograph in the frontispiece.

Department of Metallurgy   A. H. COTTRELL
University of Cambridge
June, 1963

Contents

Contents

chapter 1

Perfect Gases

*In this chapter we shall introduce the idea of atoms and show
how it has been outstandingly successful in giving a
quantitative explanation of the mechanical properties of the
simplest types of gases. In Section 1.2 we shall deduce the
pressure of a gas and other simple properties in terms of
atomic or molecular motion, and in Section 1.3 we shall go
on to consider the distribution of velocities among the gas
particles. This leads to various other topics such as the density
of the earth's atmosphere, distribution of energy among the
particles and their modes of motion, specific heats, random
thermal motion and adiabatic changes. Finally, in Section 1.9
we take account of the collisions which limit the mean free
paths of gas particles and show in the remaining sections how
this leads to a simple understanding of thermal conductivity,
viscosity and diffusion in gases.*

1.1 ATOMIC STRUCTURE OF MATTER

One of the oldest ideas in science is that the ordinary matter of the world
consists of large numbers of small atoms. It seemed impossible that a
piece of matter could be cut up repeatedly into smaller and smaller parts;
there must come an end to this when it was reduced to ultimate constituents. It was also felt that the sensations of heat, cold and the force of
the wind might be brought about by the impacts of innumerable small
particles flying about.

The atomic theory became popular in the 16th and 17th centuries, when
it was used to explain the structures and properties of solids. The dense-
ness and the hardness of solids suggested that atoms in them must be
packed together particularly closely. Really close packing has to be done

in a systematic orderly way, and this agreed with what was known then about crystals. Their flat faces, sharp edges and regular angles suggested that crystals are a form of matter in which the atoms are arranged in orderly patterns, such as may be produced by stacking large numbers of equal spheres together. Robert Hooke, in fact, showed that the various common shapes of alum crystals could all be reproduced from a pile of musket shot stacked in a single regular pattern. Other people tried to explain the plasticity and fracture of solids, the nature of alloyed metals and changes of crystal structure in terms of the ways in which atoms pack together and move about in crystals.

This early theory failed to take root, however, because it was speculative and lacked proof, even though it has since been proved broadly correct. Some of the best evidence for the atomic structure of matter has in fact

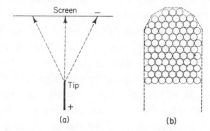

Fig. 1.1.   Field-ion microscopy.   (a) Arrangement of specimen and screen;   (b) schematic
section through tip, showing atomic structure.

come from the study of crystals during the present century. The ability of crystals to act as diffraction gratings for X-rays has proved beyond all doubt the regularity of atomic structure in them. In recent years even more direct methods have been developed, based on electron microscopy and other high-resolution microscopical techniques. In one such method, called *field-ion* microscopy, an extremely fine hemispherical tip of a specimen is examined, as shown in Fig. 1.1. This tip is positively charged electrically, so that lines of electric force radiate from it to a nearby fluorescent screen. A trace of gas such as helium is allowed to enter a vacuum chamber which surrounds the specimen. These atoms become positively charged when they touch the atoms of the tip and then accelerate down the lines of force to the screen, where they produce a visible image of the tip. The magnification and resolution of this image are high enough to show the individual atoms of the tip. The frontispiece shows an example. This is of course a flat picture of a hemispherical surface, so that there is some distortion; but the atoms and crystal facets of the surface are clearly visible.

The idea of atoms helped chemistry to make great progress in the 19th century. Striking regularities were observed in the proportions by which the chemical elements combine together. A table of unit chemical weights was drawn up, each element being given its own individual value in such a way that unit weight of one element always combines with one or a few unit weights of another. This suggested that each unit weight was proportional to the mass of a *unit particle* of the element and that chemical combination involves the joining together of such unit particles in definite and reproducible small numbers. It was also discovered that gases of the elements combine chemically in simple proportions by volume as well as by weight. The observed proportions suggested that equal volumes of gases contain equal numbers of particles at the same temperature and pressure (*Avogadro's hypothesis*). A complicating factor was that in most gases of elements the particles are not single atoms but *molecules* containing two or more atoms in chemical combination, e.g. O_2, O_3, H_2, N_2, Cl_2. Two tables were thus necessary, one for single atoms and one for molecules. These are of course the familiar tables of *atomic* and *molecular weights* of the elements.

The scales were chosen so that a single oxygen atom has a weight of 16. The molecular weights of oxygen (O_2) and ozone (O_3) are thus 32 and 48, respectively. A sample of an element contains 1 *mole* of particles (atoms or molecules) when its atomic or molecular weight is measured in grammes weight. A mole is now known to contain $6 \cdot 025 \times 10^{23}$ particles, this number being known as *Avogadro's number* (N_0). Thus, a single oxygen atom weighs $16 N_0^{-1}$, i.e. $2 \cdot 66 \times 10^{-23}$ g. The quantity $1 \cdot 66 \times 10^{-24}$ g (i.e. N_0^{-1} g) is the *atomic mass unit*. Atomic weights are given in Table 1.1. Their wide range should be noticed, particularly as the densities of liquids and solids depend mainly on the weights of their atoms. A sample of an element with density ρ and atomic weight A contains $N_0 \rho / A$ atoms/cm^3.

Atoms and molecules are of course very small. The volume of 1 mole of gas at NTP (i.e. 0°C and 760 mm of mercury pressure) is 22,400 cm^3. Divided into Avogadro's number, this gives *Loschmidt's number*, $2 \cdot 7 \times 10^{19}$ particles/cm^3, which corresponds to an average distance of about 33×10^{-8} cm from one particle to the next. The particles themselves are even smaller. When a vapour is condensed to a liquid or solid, it shrinks to about one-thousandth of its volume at NTP. The atoms or molecules are thus only about 3×10^{-8} cm across. Atomic sizes can now be measured accurately by X-ray diffraction analysis of crystals. Because of their smallness, it is convenient to use a special unit of length, the Ångstrom ($1 \text{ Å} = 10^{-8}$ cm) for them.

TABLE 1.1

Physical Properties of Elements

Atomic number	Element	Atomic weight	Density at 20°C, g/cm³	Melting point, °C	Boiling point, °C	Specific heat at 20°C, 0.1 cal g⁻¹ °C⁻¹	Coefficient of linear thermal expansion, 10⁻⁶ °C⁻¹	Thermal conductivity, 0-100°C, 0.1 cal cm⁻¹ sec⁻¹ °C⁻¹	Young's modulus, 10⁶ psi	Shear modulus, 10⁶ psi	Poisson's ratio	Viscosity at 20°C, 10⁻² poises
1	Hydrogen (H)	1·008	0·00009	−259·2	−252·7	34·5		4×10^{-3}				0·0085
2	Helium (He)	4·003	0·00018		−268·9	12·5		$3·4 \times 10^{-3}$				0·0194
3	Lithium (Li)	6·94	0·534	180·5	1330	7·9	56	1·7	1·66			
4	Beryllium (Be)	9·013	1·848	1277	2770	4·5	12·4	3·8	43	21	0·08	
5	Boron (B)	10·82	2·34	(2100)	(2550)	3·09	8·3					
6	Carbon (C) (Graphite)	12·011	2·25	3700ᵃ	4830	1·65	3	0·57	$\begin{cases} 144 \text{ (in basal plane only)} \\ 1 \text{ (bulk polycrystalline)} \end{cases}$			
	(Diamond)		3·52			1·22	1·2	38	120	50	0·25	
7	Nitrogen (N)	14·008	0·00125	−210	−196	2·47		$5·8 \times 10^{-4}$				0·0174
8	Oxygen (O)	16·00	0·00143	−219	−183	2·18		$5·8 \times 10^{-4}$				0·020
9	Fluorine (F)	19·00	0·0017	−219·6	−188	1·8						
10	Neon (Ne)	20·183	0·0009	−249	−246			$1·1 \times 10^{-3}$				0·031
11	Sodium (Na)	22·997	0·971	97·8	892	2·95	71	3·2	1·3			
12	Magnesium (Mg)	24·32	1·74	650	1105	2·45	26	3·8	6·5	2·5	0·33	

No.	Element											
13	Aluminium (Al)	26·98	2·699	660	2450	2·15	24	5·3	10	3·8	0·34	
14	Silicon (Si)	28·09	2·33	1410	2680	1·62	7·6	2·0	16			
15	Phosphorus (P) (white)	30·975	1·83	44·2	280	1·77	125					
16	Sulphur (S) (yellow)	32·066	2·07	119	445	1·75	64					
17	Chlorine (Cl)	35·457	0·0032	−101	−34·7	1·16		$1\cdot7\times10^{-4}$				0·0132
18	Argon (A)	39·994	0·0018	−189·4	−186	1·25		$3\cdot8\times10^{-4}$				0·0222
19	Potassium (K)	39·10	0·86	63·7	760	1·77	83	2·4	0·5			
20	Calcium (Ca)	40·08	1·55	838	1440	1·49	22	3·0	3	1	0·31	
21	Scandium (Sc)	44·96	2·99	1540	2730	1·34						
22	Titanium (Ti)	47·90	4·507	1670	3260	1·24	8·5	0·36	16·8	6	0·34	
23	Vanadium (V)	50·95	6·1	1860	3400	1·19	7·8	0·85	20	7·3	0·36	
24	Chromium (Cr)	52·01	7·19	1875	2665	1·1	6·2	1·6	36			
25	Manganese (Mn)	54·94	7·43	1245	2150	1·15	22		23			
26	Iron (Fe)	55·85	7·87	1536	3000	1·1	11·7	1·8	28·5	11·5	0·28	
27	Cobalt (Co)	58·94	8·85	1495	2900	0·99	12·3	1·65	30		0·31	
28	Nickel (Ni)	58·71	8·90	1453	2730	1·05	13·3	2·2	31	11·5	0·31	
29	Copper (Cu)	63·54	8·96	1083	2600	0·92	16·5	9·4	18	6·7	0·35	
30	Zinc (Zn)	65·38	7·13	419·5	906	0·915	40	2·7	14	5	0·35	
31	Gallium (Ga)	69·72	5·91	29·8	2240	0·79	18	1·5	1·4			
32	Germanium (Ge)	72·60	5·32	937	2830	0·73	6		11			
33	Arsenic (As)	74·91	5·72		613[a]	0·82	4·7		11			
34	Selenium (Se)	78·96	4·79	217	685	0·84	37					
35	Bromine (Br)	79·916	3·12	−7·2	58	0·70						
36	Krypton (Kr)	83·8	0·0037	−157	−152			$2\cdot1\times10^{-4}$				
37	Rubidium (Rb)	85·48	1·53	38·9	688	0·8	90		0·34			
38	Strontium (Sr)	87·63	2·60	768	1380				2·5	1	0·28	
39	Yttrium (Y)	88·92	4·47	1510	3030	0·71						

Table 1.1 (continued)

Atomic number	Element	Atomic weight	Density at 20°C, g/cm³	Melting point, °C	Boiling point, °C	Specific heat at 20°C, $0.1\ \mathrm{cal\ g^{-1}\ °C^{-1}}$	Coefficient of linear thermal expansion, $10^{-6}\ \mathrm{°C^{-1}}$	Thermal conductivity, 0–100°C, $0.1\ \mathrm{cal\ cm^{-1}\ sec^{-1}\ °C^{-1}}$	Young's modulus, 10^6 psi	Shear modulus, 10^6 psi	Poisson's ratio	Viscosity at 20°C, 10^{-2} poises
40	Zirconium (Zr)	91·22	6·49	1852	3580	0·67	5·8	0·4	14	5	0·34	
41	Niobium (Nb) / Columbium (Cb)	92·91	8·57	2470	4900	0·65	7·1	1·4	15	5·4	0·38	
42	Molybdenum (Mo)	95·95	10·22	2610	5550	0·66	4·9	3·5	50			
43	Technetium (Tc)	98		(2100)	(3900)				59			
44	Ruthenium (Ru)	101·1	12·2	(2500)	(4900)	0·57	9·1	2·1	60	27	0·25	
45	Rhodium (Rh)	102·91	12·44	1965	4500	0·59	8·3	1·7	42			
46	Palladium (Pd)	106·7	12·02	1552	4000	0·58	11·8		17	7	0·39	
47	Silver (Ag)	107·88	10·49	960·8	1761	0·56	19·7	10	11	4	0·38	
48	Cadmium (Cd)	112·41	8·65	320·9	765	0·55	30	2·2	8	4	0·29	
49	Indium (In)	114·82	7·31	156·2	2000	0·57	33	0·57	1·6			
50	Tin (Sn)	118·70	7·298	231·9	2270	0·54	23	1·6	6·8	2·5	0·36	
51	Antimony (Sb)	121·76	6·62	630·5	1380	0·49	9·5	0·45	11			
52	Tellurium (Te)	127·61	6·24	449·5	990	0·47	16·8	0·14	6			
53	Iodine (I)	126·91	4·94	113·7	183	0·52	93	0·01				
54	Xenon (Xe)	131·30	0·0059	−112	−108			$1·24 \times 10^{-3}$				

No.	Element										
55	Caesium (Cs)	132·91	1·903	28·7	690	0·48	97			0·25	
56	Barium (Ba)	137·36	3·5	714	1640	0·68			1·8		
57	Lanthanum (La)	138·92	6·19	920	3470	0·48			10		
	(Rare earth elements here)										
72	Hafnium (Hf)	178·58	13·09	2250	5400	0·35	6·5	6·0	20	5	0·37
73	Tantalum (Ta)	180·95	16·6	2980	5400	0·34	4·3	1·3	27	22	0·17
74	Tungsten (W)	183·86	19·3	3410	5900	0·33	6·7	4·5	50	30	0·26
75	Rhenium (Re)	186·22	21·04	3170	5900	0·33		1·7	70	34	0·25
76	Osmium (Os)	190·2	22·57	(3000)	5500	0·31	4·6		80	32	0·26
77	Iridium (Ir)	192·2	22·5	2455	5300	0·31	6·8	1·4	75	8	0·39
78	Platinum (Pt)	195·09	21·45	1769	4530	0·31	8·9	1·7	21	4	0·42
79	Gold (Au)	197·0	19·32	1063	2970	0·31	14·2	7·1	12		1·556
80	Mercury (Hg)	200·61	13·55	−38·36	357	0·33	61	0·2	1	0·4	0·45
81	Thallium (Tl)	204·39	11·85	303	1457	0·31	28	0·93	2·6	0·8	0·45
82	Lead (Pb)	207·21	11·36	327·4	1725	0·31	29·3	0·83	4·6		
83	Bismuth (Bi)	209·00	9·80	271·3	1560	0·31	13·3	0·2			
84	Polonium (Po)	210		250							
85	Astatine (At)	211		(300)							
86	Radon (Rn)	222	0·01	(−70)	−61·8						
87	Francium (Fr)	223		(27)							
88	Radium (Ra)	226·05	5·0	700							
89	Actinium (Ac)	227		(1000)							
90	Thorium (Th)	232·05	11·66	1750	(3850)	0·34	12		11	4	0·30
91	Protactinium (Pa)	231·1	15·4	(1200)							
92	Uranium (U)	238·07	19·07	1132	3820	0·28	17	0·63	25	11	0·24

(Transuranic elements here; e.g. 93 Np, 94 Pu, 95 Am, 96 Cm, 97 Bk, 98 Cf, 99 E, 100 Fm, 101 Mv, 102 No, etc.)

ᵃ Sublimes.

1.2   KINETIC THEORY OF GASES

The other outstanding success of the atomic theory in the 19th century was its quantitative explanation of the mechanical properties of gases. Gases are simple forms of matter. They have low densities, expand to fill their containers and are easily compressed. These properties suggested that the particles in gases are mostly too far apart to exert forces on one another. Moreover, because equal numbers of gas particles in equal volumes at equal temperatures produce the same pressure (at low densities and high temperatures), *whatever their chemical species*, it seemed that this pressure was produced by the particles acting in a purely mechanical way as point centres of mass.

A gas was thus pictured as a collection of small elastic spheres, each of mass m, velocity c, momentum mc and kinetic energy $\frac{1}{2}mc^2$, flying about ceaselessly as separate individuals, moving in straight lines and bouncing off the walls of their containers and also, when they happen to meet, off one another. When a gas is allowed to remain undisturbed in its container for a length of time, these collisions cause it to settle down into a *state of equilibrium* in which it is *uniform* (i.e. its particles are distributed impartially throughout the whole container—we ignore for the moment the effect of gravity), is *isotropic* (i.e. its particles move impartially in all directions) and is *constant* in its bulk properties such as pressure and temperature.

When the gas is in this state, the theory of its properties becomes much simpler. We can, for example, pretend that the gas particles bounce off the walls of their container like perfect elastic spheres rebounding from a perfectly smooth, reflecting surface, even though it is known experimentally that they do not do this but tend to stick to the surface for a little time before flying off in some unrelated direction. The reason is that, because the gas is uniform and isotropic, particles approach and leave the walls in random directions, so that the *overall* distribution of directions of motion is unaffected by the walls. Also, when the gas is constant in its properties, the particles on average approach and leave the walls in the same numbers and with the same kinetic energies. The walls thus behave to the gas as a whole, even though not to individual particles, as if they were perfect reflectors of its particles. By the same argument, the distribution of particle motions is unaffected by collisions among the particles themselves; on average, as many particles are knocked into a particular direction and speed as are knocked out of it. It also follows that the shape of the container does not affect the equilibrium properties of the gas; particles re-enter the gas from a wall in just the same numbers and with just the same velocities as they would if that wall were not there and they had come instead from another part of the gas behind it. The container can thus be altered in shape (at constant volume), by removing some walls and building others in other positions, without altering the properties of the gas.

The leading idea in the kinetic theory of gases is that gas pressure is caused by the impacts of gas particles on the walls of the container. The equilibrium properties mentioned above enable us to calculate the pressure in terms of the behaviour of a single gas particle moving in a cubical box with perfectly reflecting sides of length L, as shown two-dimensionally in Fig. 1.2. We resolve its velocity c into component velocities u, v and w,

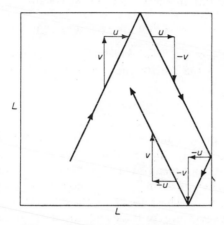

Fig. 1.2.   Path of a gas particle in a box.

parallel to the edges of the box. At each collision the velocity component perpendicular to the wall is exactly reversed and the other two components unchanged. Each collision which changes, say, $+u$ to $-u$, also changes the momentum of the particle by $2mu$ and hence, because momentum is conserved, gives an outward momentum $2mu$ to the wall in question. This wall is hit $u/2L$ times per second by the particle and so receives momentum from it at a rate mu^2/L per second. From Newton's second law (*force = rate of momentum change*) this gives a force mu^2/L on the wall. The *pressure* p, i.e. force per unit area, on this wall due to all particles in the gas is then given by

$$p = \frac{\sum mu^2}{L^3} = \frac{m}{V} \sum u^2, \tag{1.1}$$

where $V\,(=L^3)$ is the volume and the sum is over all particles. We see that each particle contributes independently to the total pressure. This remains true if the particles are of different chemical species and have different masses m_1, m_2, ... etc., the sum then being replaced by $m_1 u_1^2 + m_2 u_2^2 +$ ... etc. This gives *Dalton's law*: the total pressure exerted by a mixture of different gases is the sum of the *partial pressures* of each of these gases separately.

The pressures on the other faces are found similarly. Since the gas is uniform and isotropic, they are all equal. Hence,

$$\sum u^2 = \sum v^2 = \sum w^2 = \tfrac{1}{3} \sum (u^2 + v^2 + w^2) = \tfrac{1}{3} \sum c^2 = \tfrac{1}{3} N \overline{c^2}, \quad (1.2)$$

where $\overline{c^2}$ is the mean-square velocity and N is the number of particles in volume V. Substituting in Equation (1.1), we obtain

$$\tfrac{1}{3} N m \overline{c^2} = \tfrac{2}{3} E = pV, \quad (1.3)$$

where $E (= \tfrac{1}{2} N m \overline{c^2})$ is the total kinetic energy of the particles. Since $Nm = \rho V$, where ρ is the *density*, the mean-square speed is given by

$$\overline{c^2} = \frac{3p}{\rho}. \quad (1.4)$$

The densities of hydrogen and nitrogen at NTP are 9×10^{-5} and $1 \cdot 25 \times 10^{-3}$ g cm^{-3}, respectively. If $p \simeq 10^6$ dyn cm^{-2} (atmospheric pressure), this gives the speed (root-mean-square) of hydrogen molecules at NTP as about $1 \cdot 8 \times 10^5$ cm sec^{-1} ($\simeq 1$ mile sec^{-1}) and that of nitrogen molecules as about 5×10^4 cm sec^{-1}. These are about the muzzle velocities of rifle bullets. There is direct evidence for these high speeds from the fact that sound waves in air, which are transmitted by the motion of air molecules, travel at about $3 \cdot 3 \times 10^4$ cm sec^{-1} under ordinary conditions (cf. Section 12.4).

When a gas is heated at constant pressure, it expands; when heated at constant volume, its pressure rises. Thus, pV increases as heat energy is given to the gas. Since pV is proportional to the kinetic energy of the particles, we suppose that this kinetic energy is *heat energy*. The hotness of the gas, i.e. the *temperature*, is thus a measure of the average kinetic energy per particle. Of all the various ways in which we could define temperature in terms of this kinetic energy, the simplest and most useful is to make it directly proportional, i.e. $T \propto E$. In fact, the *absolute temperature* T on the *perfect gas scale* is defined from the relation

$$\tfrac{1}{2} m \overline{c^2} = \tfrac{3}{2} kT, \quad (1.5)$$

where k is a constant. Substituted into Equation (1.3), this becomes

$$pV = NkT, \quad (1.6)$$

the familiar *equation of state* of a *perfect* gas. It summarizes the two laws of a perfect gas: *Charles' law* (V is proportional to T at constant p) and *Boyle's law* (pV is constant at constant T). We also see that the *specific heat* (rate of increase of energy with temperature) belonging to the translational motion of the particles of a perfect gas is the same at all temperatures; this follows from the fact that we have *defined* temperature as proportional to the energy of such a gas.

For 1 mole of gas ($N = N_0 =$ Avogadro's number), the equation of state becomes

$$pV = RT, \tag{1.7}$$

where

$$R = N_0 k. \tag{1.8}$$

The constants k and R are *universal* constants, the same for all substances. This can be proved in a very general way from statistical mechanics. It also follows directly from Avogadro's hypothesis. Consider two gases with particles of masses m_1 and m_2 and mean-square velocities $\overline{c_1^2}$ and $\overline{c_2^2}$, which are at the same temperature T and which have the same volume V and pressure p. Then, by Avogadro's hypothesis, they have the same number of particles, $N_1 = N_2$. Since $pV = \frac{1}{3}N_1 m_1 \overline{c_1^2} = \frac{1}{3}N_2 m_2 \overline{c_2^2}$, it follows that

$$m_1 \overline{c_1^2} = m_2 \overline{c_2^2} \tag{1.9}$$

and, hence, that k in Equation (1.5) is the same for both gases.

We see that light particles move faster than heavy ones at the same temperature. This enables us to separate two gases from a mixture by allowing them to leak out of small holes in their container. The rate of *effusion* from such a hole is proportional to the thermal velocity of a gas particle, and so varies inversely with the square root of the mass (*Graham's law*). Thus, hydrogen escapes four times as fast as oxygen. The holes must be small to prevent the gas from streaming out in bulk and thereby sweeping heavy and light particles along together. A porous membrane (e.g. unglazed porcelain) is often used, in which case the effect is referred to as *transpiration*. The method is important for separating gases of similar properties, such as *isotopes* (atoms of different masses belonging to the same chemical element). When the isotopes differ only slightly in mass, as in the separation of uranium isotopes by the gaseous diffusion method, the degree of separation achieved by the passage through a single membrane is small. It is usual then to pass the gas (e.g. uranium hexafluoride) through a series or *cascade* of separations, each stage of which consists of a membrane with ingoing and outgoing chambers, together with pumps to send the enriched gas forward to later stages and the impoverished gas back to earlier stages.

To find the numerical value of R in Equation (1.7), we fit this equation to real gases; these always 'obey this equation at sufficiently low densities. When R and N_0 are known, then k in Equations (1.6) and (1.8) can be found. The temperature scale is chosen so that there are 100 degrees between the freezing and boiling points of water at 1 atm pressure; we then have $k = $ *Boltzmann's constant* $= 1 \cdot 38 \times 10^{-16}$ erg deg^{-1} and $R = $ *the gas constant* $= 8 \cdot 313 \times 10^7$ erg deg$^{-1} = 1 \cdot 986$ cal deg^{-1} mole^{-1}. At the freezing point of water a gas at normal pressure decreases in volume by about 1 part in 273 for each degree fall in temperature. Thus, a *perfect* gas (which obeys Equation (1.6) under all circumstances) should shrink to zero volume and lose all its kinetic energy at about $-273°$C. The temperature T in Equation (1.6) is the *absolute temperature*, measured in *degrees Kelvin* (°K). The *absolute zero* of temperature is 0°K $(= -273 \cdot 16°$C $= -459 \cdot 69°$F$)$.

1.3   THE MAXWELL–BOLTZMANN DISTRIBUTION LAW

Owing to gains and losses of energy in chance collisions, the actual energy of a particle at any instant fluctuates widely about its average value. Although these individual fluctuations are highly erratic, the average distribution of a large number of particles over various levels of energy is a steady and calculable property of a gas, provided there is thermal equilibrium among its particles.

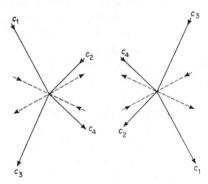

Fig. 1.3.   A collision and its inverse. The broken lines show each collision as seen by an observer moving with the centre of mass of the particles.

Let all particles have the same mass. Let f be the number of particles with velocity c; or, more precisely, let $f(c)\ dc$ be the number within a small range dc of velocities centred about the value c. We expect f to depend on c and so write it as $f(c)$. In Fig. 1.3 we show a collision in which two particles change their velocities from c_1 and c_2 to c_3 and c_4. The total kinetic energy of the particles is conserved; hence,

$$c_1{}^2 + c_2{}^2 = c_3{}^2 + c_4{}^2. \tag{1.10}$$

The chance of such a collision occurring in a given time is proportional to the numbers, $f(c_1)$ and $f(c_2)$, of both kinds of particles in the gas. The rate of such collisions is thus given by $\alpha f(c_1)f(c_2)$, where α is a proportionality factor. Consider next the inverse collision in which two particles change their velocities from c_3 and c_4 to c_1 and c_2 (Fig. 1.3). The rate of this may similarly be written $\alpha' f(c_3)f(c_4)$. Since the gas is in equilibrium,

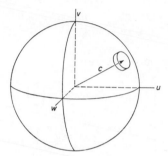

Fig. 1.4. Velocity diagram.

these two rates are equal. We must also have $\alpha = \alpha'$, since observers moving with the centres of mass of the particles would see the two collisions as *completely equivalent*, as shown by the broken lines in Fig. 1.3. Hence, $f(c_1)f(c_2) = f(c_3)f(c_4)$, i.e.

$$\ln f(c_1) + \ln f(c_2) = \ln f(c_3) + \ln f(c_4). \tag{1.11}$$

A solution of Equations (1.10) and (1.11) suggests itself immediately. It is simply $\ln f(c) \propto c^2$, or

$$f(c) = A\, e^{-\beta c^2}, \tag{1.12}$$

where A and β are constants. The minus sign is introduced since, for a finite kinetic energy, $f(c)$ must decrease as c^2 increases. It can be proved by further analysis that this equation is the only solution.

We sometimes need to know the numbers of particles moving at various speeds, *irrespective of direction*. In the velocity diagram of Fig. 1.4, c is represented by the vector shown and $f(c)\, dc$ is the number of particles whose velocities are represented by points lying in the element shown at the end of this vector. This element is of thickness dc and stands on unit area of a sphere of radius c. To find the total number of particles $N(c)\, dc$

with speeds ranging from c to $c + dc$, irrespective of direction, we must enlarge this element (keeping its thickness constant) until it covers the whole sphere; $N(c)\, dc$ is then the number of particles in this spherical shell. Since the gas is isotropic, $N(c)$ is simply $f(c)$ multiplied by the area of the sphere,

$$N(c) = 4\pi c^2 f(c) = 4\pi c^2 A\, e^{-\beta c^2}. \tag{1.13}$$

The constants A and β can be calculated from the total number N and the total energy E of particles in the gas. Thus,

$$N = \int_0^\infty N(c)\, dc, \tag{1.14}$$

$$E(= \tfrac{3}{2}NkT) = \int_0^\infty \tfrac{1}{2}mc^2 N(c)\, dc. \tag{1.15}$$

Making use of the standard integrals

$$\int_0^\infty x^2\, e^{-ax^2}\, dx = \frac{1}{4a}\left(\frac{\pi}{a}\right)^{1/2}, \tag{1.16}$$

$$\int_0^\infty x^4\, e^{-ax^2}\, dx = \frac{3}{8a^2}\left(\frac{\pi}{a}\right)^{1/2}, \tag{1.17}$$

after substituting Equation (1.13) into (1.14) and (1.15), we obtain

$$N(c) = 4\pi c^2 f(c) = 4\pi c^2 N\left(\frac{m}{2\pi kT}\right)^{3/2} e^{-mc^2/2kT}, \tag{1.18}$$

which is the celebrated *Maxwell–Boltzmann distribution law*.

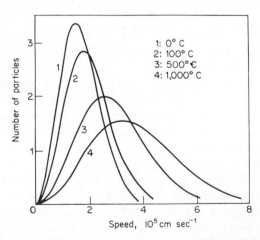

Fig. 1.5. The Maxwell–Boltzmann distribution of molecular speeds in hydrogen.

This distribution is shown graphically for hydrogen at various temperatures in Fig. 1.5. The area under each curve has to be made equal to N by a suitable choice of units for the $N(c)$ axis. We notice that the distribution becomes broader at higher temperatures and that each curve shows a maximum. The maximum occurs because $N(c)$ refers to speeds in all directions and the chance that a particle is moving slowly along all three co-ordinate axes is small. This contrasts with $f(c)$, which refers to speed in a particular direction and which has its greatest value at $c = 0$. The *mean speed* $\bar{c}$ of a particle is defined as the integral of $cN(c)/N$ and is equal to $(8kT/\pi m)^{1/2}$ for a Maxwellian distribution. It is thus slightly smaller than the *root-mean-square speed*, the value of which is given by Equation (1.5) as $(3kT/m)^{1/2}$.

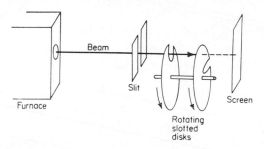

Fig. 1.6. Apparatus for measuring molecular speeds.

The Maxwell–Boltzmann law has been accurately confirmed in various ways. One experimental arrangement is sketched in Fig. 1.6. Hot gas particles stream out of a hole in a furnace and, after collimation, impinge on two rotating slotted disks. Those going at the right speed can pass through both slots and so reach the screen, where they can be counted. The experiment is made in vacuum to protect the beam from random molecular collisions.

The most important feature of the Maxwell–Boltzmann law is the exponential factor, $\exp(-\varepsilon/kT)$, where ε is the energy of the particle. It indicates the unlikelihood that a particle will, in chance exchanges of energy with other particles, gain an energy much higher than average. By a general theorem of statistical mechanics, it can be shown that this same exponential factor is not limited to gases and that ε can be regarded as the total energy (kinetic and potential) of a particle in thermal equilibrium. This theorem lies beyond the scope of the present treatment, but we can see the germ of it by the following simplified argument. Suppose that the energy of each particle in a substance can only be 0, ε or 2ε. Of the N particles in all, let N_0 have energy 0, N_1 have ε, and N_2 have 2ε.

According to the theory of permutations, the number of ways of arranging this distribution of particles among their energy levels is given by $W = N!/N_0! N_1! N_2!$. The central idea is that the equilibrium distribution is that which has the largest value of W consistent with the constant total energy $(N_1 + 2N_2)\varepsilon$ of the particles. Suppose that this equilibrium state exists. Because W is then a maximum, its value will not alter if we make a small change in the distribution. One such change would be to take two particles out of the level ε, put one in 2ε and the other in 0. This conserves the total energy. We then have $W = N!/(N_0 + 1)!(N_1 - 2)!(N_2 + 1)!$. This has the same value as before, provided that $N_1(N_1 - 1) = (N_0 + 1)(N_2 + 1)$. If N_0, N_1 and N_2 are very large numbers, this in effect becomes $N_2/N_1 = N_1/N_0$, i.e. the particles are distributed in a geometrical progression, which implies that the number N_i in the energy level ε_i must be given by a relation of the type $N_i \propto e^{-\beta \varepsilon_i}$, where β depends on the average energy available to each particle, i.e. on the temperature of the substance.

1.4 EARTH'S ATMOSPHERE

A simple example of an exponential distribution in potential energy is provided by the earth's atmosphere. Gravity pulls the air molecules down, but thermal energy prevents them settling. Let n and p, respectively, be the number of particles per unit volume and the pressure, both at the height h. The force exerted on unit area at this height is equal to the total weight of all the particles directly above this area, i.e.

$$\int_h^\infty nmg \, dh = p, \tag{1.19}$$

where g is the gravitational constant (980 cm sec^{-2}) and mg the weight of a particle of mass m. Since n refers to unit volume, we have, for a perfect gas, $p = nkT$. Hence, differentiating Equation (1.19) with respect to h,

$$nmg = -\frac{dp}{dh} = -kT \frac{dn}{dh}, \tag{1.20}$$

where the minus sign allows for the decrease in pressure with height. Rearranging and integrating, we obtain

$$\frac{n}{n_0} = \frac{p}{p_0} = e^{-mgh/kT}, \tag{1.21}$$

where n_0 and p_0 are the values at zero height and mgh is the gravitational potential energy of a particle of mass m at the height h. Taking $m = 5 \times 10^{-23}$ g as typical and assuming a constant temperature $T = 300°K$, we find that the density decreases by a factor e^{-1} (i.e. 0·368) for each 8 km increase in height. The actual variation in the earth's atmosphere is complicated by temperature gradients, but this is roughly correct.

According to the Maxwell–Boltzmann distribution, there are always a few particles with exceptionally high speeds. There are thus always some air molecules with thermal velocities greater than the escape velocity (11 km sec^{-1}) from the earth's gravitational field, so that we are (exceedingly slowly!) losing atmosphere by this effect. The corresponding escape velocity from the moon is only 2·4 km sec^{-1}, which is easily attainable by thermal motion; this is believed to be why the moon has no atmosphere. By far the most abundant elements in the universe are hydrogen and helium. Despite this, there is practically no helium or free hydrogen in the earth's atmosphere. Being so light, the particles of these gases achieve unusually high thermal velocities and are only weakly held by the force of gravity.

1.5 EQUIPARTITION OF ENERGY

Because it moves in three-dimensional space, a gas particle is said to have three *degrees of freedom* for its translational motion, i.e. three co-ordinates are needed to specify its position. Over a period of time its motion is divided impartially between these three degrees of freedom and so, on average, one-third of its translational thermal energy belongs to each. We say that there is an *equipartition of energy* among the degrees of freedom. In a simple monatomic gas this translational motion is usually the only kind we have to consider. However, if the particle consists of two atoms joined in the form of a dumb-bell-shaped molecule, or consists of some more complex polyatomic cluster, it possesses various internal modes of motion, vibration and rotations, as well as the three translational modes. How much energy exists in these various modes when the particle is in thermal equilibrium? The answer to this depends on three things: (1) the number of degrees of freedom possessed by the particle, (2) the ability of these degrees of freedom to absorb thermal energy and (3) the principle of equipartition of energy.

The number of degrees of freedom of a system of particles is the smallest number of co-ordinates needed to define the positions of all the particles. A single atom, regarded as a point in space, has three. A gas of n such atoms has $3n$. A diatomic (dumb-bell) molecule has six degrees of freedom, three for each atom. When atoms are joined together, as in molecules and solids, it is convenient to define the degrees of freedom in a way that recognizes the unity of the piece of matter to which they belong. Thus, for the dumb-bell molecule we define three translational degrees of freedom for the centre of gravity of the molecule, two degrees of freedom for the *rotation* of the atoms about one another (two rotational axes perpendicular to the dumb-bell axis), and one degree of freedom for the *vibration* of the atoms, against each other, along the dumb-bell axis. The total number of degrees of freedom remains, of course, at six.

. There is advantage in defining the degrees of freedom, wherever possible, in such a way that the total energy ε of the particle can be written as a sum of independent terms, $\varepsilon_1 + \varepsilon_2 + \varepsilon_3 + \ldots$, each of which belongs to a single degree of freedom. The exponential energy factor then breaks up into the product of separate factors, thus

$$e^{-\varepsilon/kT} = e^{\frac{-\varepsilon_1 - \varepsilon_2 - \ldots}{kT}} = e^{-\varepsilon_1/kT}\, e^{-\varepsilon_2/kT}\, e^{-\varepsilon_3/kT} \ldots . \tag{1.22}$$

The energy distribution within each degree of freedom can then be dealt with separately as an isolated problem. When this is possible, the degrees of freedom are said to be *separable*.

It often happens that these separate energy terms can be written as *squares of variables*. Thus, the total translational energy of a gas particle can be written as $\frac{1}{2}mu^2 + \frac{1}{2}mv^2 + \frac{1}{2}mw^2$, in which the squares of the velocity components u, v and w, appear. Similarly, rotational energy can be expressed by terms such as $\frac{1}{2}I\omega^2$, where ω is the angular velocity and I the moment of inertia about the rotational axis. A vibrating particle can often be regarded as making a simple harmonic motion, with a restoring force αx proportional to its displacement x from the centre of oscillation and to the strength of the "spring constant" α pulling it back to that centre. Its total vibrational energy can then be written in the form $\frac{1}{2}m\dot{x}^2 + \frac{1}{2}\alpha x^2$, where the first term is its kinetic energy and the second its potential energy. In all such examples the energy can be written in the form

$$\varepsilon = \sum_i \varepsilon_i = \sum_i \tfrac{1}{2}\gamma_i x_i^2, \tag{1.23}$$

where γ_i stands for m, I, α, etc., according to the mode of motion represented by ε_i, and x_i^2 stands for u^2, v^2, w^2, ω^2, $\dot{x}^2$, x^2, etc.

The ability of such an energy term to absorb thermal energy depends on the size of the smallest amount, or *quantum*, of energy which the particle can accept into (or release from) this particular mode of motion. We shall discuss the quantization of energy in later chapters. If the quantum is *large*, compared with the average thermal energy available per degree of freedom at the temperature considered, the particle is unable (except through a rare fluctuation) to accept thermal energy into this mode of motion. If the quantum is *small*, the fact that the energy can enter only in discrete units hardly restricts the flow of thermal energy freely into and out of this mode of motion; this is called *classical* behaviour.

Let us suppose that a mode represented by the energy term $\varepsilon_i = \frac{1}{2}\gamma_i x_i^2$ is classical at the temperature T, and so can absorb thermal energy freely. The probability P that the particle will have an energy ε_i in this mode is then, from Equation (1.22), given by

$$P = C \exp\left[-\frac{\gamma_i x_i^2}{2kT}\right], \qquad (1.24)$$

where C must be such as to satisfy the condition that

$$\int_{-\infty}^{+\infty} P \, dx_i = 1. \qquad (1.25)$$

The mean value of ε_i is then

$$\bar{\varepsilon}_i = \int_{-\infty}^{+\infty} \varepsilon_i P \, dx_i = \frac{\dfrac{\gamma_i}{2}\displaystyle\int_{-\infty}^{+\infty} x_i^2 \exp\left[-\dfrac{\gamma_i x_i^2}{2kT}\right] dx_i}{\displaystyle\int_{-\infty}^{+\infty} \exp\left[-\dfrac{\gamma_i x_i^2}{2kT}\right] dx_i} = \frac{1}{2}kT, \qquad (1.26)$$

where we have used Equation (1.16) and also the standard integral

$$\int_{-\infty}^{+\infty} e^{-ax^2} \, dx = \left(\frac{\pi}{a}\right)^{1/2}. \qquad (1.27)$$

We have thus reached the important result that, in equilibrium, an average energy $\frac{1}{2}kT$ belongs to each classical squared term in the expression for the total energy of a particle. This is the *principle of equipartition*. It should be noticed that in a vibrational mode of motion there are two squared terms, $\frac{1}{2}m\dot{x}^2$ and $\frac{1}{2}\alpha x^2$. Thus, the average energy in such a mode (if classical) is kT; i.e. $\frac{1}{2}kT$ of kinetic energy and $\frac{1}{2}kT$ of potential energy.

1.6 BROWNIAN MOTION, FLUCTUATIONS AND THERMAL NOISE

A good example of translational thermal energy in large particles is provided by *Brownian motion*. When still air is viewed through a microscope, small specks of dust and smoke can be seen in ceaseless irregular motion. This effect, discovered by the botanist Brown in 1828, provided the first direct evidence for the reality of molecular motion. Such a speck behaves as a gas particle. Although much heavier than a gas molecule, it is continually knocked to and fro by the impacts of these molecules. In this way its various modes of motion acquire thermal energy, including an average of $\frac{1}{2}kT$ in each of its three translational modes. Brownian motion has been used to determine Boltzmann's constant, Avogadro's number and molecular masses. The simplest way of using it is through the *sedimentation* of specks in a vertical column. Gravity pulls them down but thermal motion keeps them up, and an equilibrium distribution is reached in which n, the number of specks per unit volume, varies with height h as in Equation (1.21).

Brownian motion is a typical *fluctuation* effect of thermal motion. Another example is the colour of the sky, caused by fine-scale fluctuations in the density of the atmosphere. The number of air molecules in a given volume fluctuates irregularly as molecules pass in and out of the region considered. The smaller the volume, the stronger are these fluctuations. Light rays are scattered when they pass through regions of varying density. The short waves at the blue end of the solar spectrum are more strongly scattered by the fine-scale atmospheric fluctuations than are the red waves, the wavelength of which is too large to respond sensitively to such fine-scale effects. Hence, the sky at day-time appears uniformly diffused with blue light.

Brownian motion sets a limit to the accuracy with which we can measure physical quantities. A galvanometer mirror or instrument pointer, for example, trembles continually as a result of the molecular bombardments it receives. These oscillations are very small, but they can be shown by optical levers and are important in sensitive instruments. Closely related is the *thermal noise*, which sets a limit to the smallness of signals which can be detected or transmitted through electrical apparatus. This noise, so named because it can be heard in an amplifier and speaker unit, is caused by random fluctuations of voltage due to Brownian motion in the apparatus. The intensity of Brownian motion and thermal noise is proportional to kT, and so, when making measurements of great delicacy, it is sometimes necessary to work at low temperatures.

1.7 SPECIFIC HEATS OF GASES

The equipartition principle is useful for discussing specific heats. We have seen that when a system is in thermal equilibrium, an average energy $\frac{1}{2}kT$ belongs to each independent, classical squared term in the formula for its energy. The specific heat of n such terms is thus $\frac{1}{2}nk$. In particular, the specific heat of a monatomic perfect gas (three translational degrees of freedom only) is $3k/2$ per particle, or

$$C_V = \tfrac{3}{2}R \qquad (1.28)$$

per mole of particles.

This is written as C_V because it refers to heating at *constant volume*. If the gas is heated at *constant pressure*, it expands and does mechanical work against its surroundings. From the perfect gas law, we have

$p\delta V = R\delta T$, where δV is the expansion brought about by the temperature rise δT. Taking $\delta T = 1°$, the work $p\delta V$ done by the expanding gas is then equal to R. The energy for this work has to be supplied by the heater. The specific heat of a perfect monatomic gas at constant pressure is thus

$$C_p = C_V + R \qquad (1.29)$$

per mole. Defining

$$\gamma = \frac{C_p}{C_V}, \qquad (1.30)$$

we then arrive at Clausius' relation, $\gamma = \frac{5}{3}$, for gas which satisfies Equation (1.28).

This relation is obeyed well by the inert gases, He, Ne, A, etc. Gases of polyatomic particles usually have different values of γ, however. This is because various internal modes of motion, rotations and vibrations exist in such particles. These internal modes do not affect the pressure exerted by the gas, so that Equation (1.29) remains valid, but they do absorb thermal energy, so that Equation (1.28) is no longer applicable. For example, in a dumb-bell molecule such as O_2 each atom may rotate round the other. These rotational modes can take place about two perpendicular axes and $\frac{1}{2}kT$ can go to each. We then have $C_V = \frac{5}{2}R$, $C_p = \frac{7}{2}R$ and $\gamma = \frac{7}{5}$. Air and many diatomic gases have these values. It may be asked why there is not a further contribution R to C_V and C_p from the *vibration* of the atoms along the axis of the molecule. The reason is that the energy quantum for vibration is usually larger than kT at room temperature. The vibrational modes do not usually become classical until higher temperatures are reached. We shall discuss this question in Section 6.6. An even more striking quantum effect is that, whereas the rotation of one atom round another contributes to the specific heat, the rotation of a single atom or dumb-bell molecule round its own axis does not. The quantum is fairly small in the first case, very large in the second; this is connected with the fact that practically all the mass of an atom is concentrated at its centre, so that the moment of inertia is much larger for the rotation of one atom about another than for the rotation of an atom about its own axis.

1.8   ADIABATIC EXPANSION

When a perfect gas is compressed or expanded *isothermally*, its pressure changes inversely with volume. During this change some energy enters or leaves the gas as mechanical work $(p\delta V)$. However, the internal energy of the gas is constant ($\frac{1}{2}kT$ per classical squared term) at constant temperature. Hence, in an isothermal volume change some heat must leave or enter the gas to balance this mechanical work. To produce an isothermal change, the gas must therefore be in thermal contact with its surroundings and the volume must be changed slowly to allow time for the necessary flow of heat. We know from using air pumps that when a gas is altered quickly in volume (or slowly, if in a thermally insulated container), its temperature changes. In the limit of a completely *adiabatic* change in which no heat enters or leaves, the mechanical work $p\delta V$ is entirely provided from or given to the internal energy of the gas. Consider 1 mole of a perfect gas which makes a small change $V \to V + dV$, $p \to p + dp$, $T \to T + dT$. From the perfect gas law,

$$(p + dp)(V + dV) = R(T + dT) \tag{1.31}$$

and hence, subtracting $pV = RT$ and neglecting products of the small quantities dp and dV,

$$p\,dV + V\,dp = R\,dT, \tag{1.32}$$

which is a general relation valid for any small change of p, V and T in a perfect gas. Since the internal energy of this gas depends only on T, this small general change contributes $C_V\,dT$ to the internal energy of the gas. Thus, if the change is completely adiabatic,

$$C_V\,dT = -p\,dV, \tag{1.33}$$

the minus sign indicating that when the gas does work *on* its surroundings, by expanding, its internal energy and temperature decrease. Substituting Equation (1.33) into Equation (1.32) and making use of Equations (1.29) and (1.30), we obtain

$$\frac{dp}{p} + \gamma\,\frac{dV}{V} = 0. \tag{1.34}$$

When γ is independent of temperature, this integrates to give

$$pV^{\gamma} = \text{constant}, \tag{1.35}$$

which differs from Boyle's law. Since $\gamma > 1$, the pressure changes more sharply with volume under adiabatic conditions.

Some refrigerators and liquefiers make use of this difference between isothermal and adiabatic changes. A gas is compressed and then passed through a heat exchanger to remove its heat of compression and so make this part of the cycle nearly isothermal. This compressed unheated gas then enters an expansion chamber, where it is allowed to push back a

piston adiabatically and so cool down. It then passes through the re-
frigerator back to the compressor, where it repeats the cycle. To reach
really low temperatures, the incoming compressed gas is usually cooled
through a heat exchanger which brings it into thermal contact with the
outgoing gas. This cycle provides a form of *heat pump*. The mechanical
work done by the pistons sucks heat into the gas circuit from the re-
frigerator chamber and expels it from the gas circuit into the compressor
and heat exchanger.

1.9 MEAN FREE PATH

An old argument against the kinetic theory of gases was: if the
particles do really move quickly, why do different gases mix together so
slowly? For example, if a trace of chlorine is released at one end of a still
room, it is several minutes before the gas can be smelt at the other end of
the room. Similarly, a woollen blanket owes its value to the poor thermal
conductivity of still air, despite the great speeds of air molecules.

The problem was solved by Clausius, who pointed out that gas particles
do not travel in uninterrupted lines but are continually knocked to and
fro into random zigzag paths by collisions with one another. A particle
may thus be knocked backwards as often as it is knocked forwards. The
mean distance travelled between successive collisions, called the *mean free
path l*, is an important property of a gas. Each particle presents a "target
area" $\pi\sigma^2$ to the others, where σ is its *gas-kinetic collision diameter*. The
mean free path thus depends on the size of the particles as well as on the
number n per unit volume.

Simultaneous collisions between three or more particles are usually rare
in a gas and we can neglect them. A binary collision occurs each time the
centre of one particle approaches within a distance σ of the centre of
another. To estimate the mean free path, we imagine the diameter of one
particle increased to 2σ and all the other particles reduced to points. As
the enlarged particle travels a distance l, it sweeps out a volume $\pi\sigma^2 l$. On
average, there is one other particle in this volume and, hence, a collision
when $\pi\sigma^2 ln = 1$, i.e.

$$l = \frac{1}{\pi\sigma^2 n}. \tag{1.36}$$

This is an approximate formula, because we have neglected various additional effects such as the compressibility of particles, the fact that all the
particles are in motion, and the tendency of a particle to retain some of its
original motion after a collision. More exact treatments produce only
slight numerical changes, however. To appreciate the scale of collisions,

we take $\sigma \simeq 5$ Å. At room temperature this gives $l \simeq 500$ Å at 1 atm pressure and $l \simeq 40$ cm at 10^{-4} mmHg pressure. In air under ordinary conditions a molecule makes about 10^{10} collisions per second.

1.10 THERMAL CONDUCTIVITY, VISCOSITY AND DIFFUSION

We shall now discuss some properties of gases which depend on the mean free path. In particular, we shall discuss *transport properties*, i.e. the ability of the gas particles, through their random migrations, to carry physical quantities such as energy and momentum from one part of the gas to another. We shall deal only with effects due to individual particles moving through the gas. Processes of mass flow such as convection and the propagation of pressure waves will be discussed later.

The properties to be considered are thermal conductivity, viscosity and diffusion. *Thermal conduction* in a *still* gas (which is difficult to arrange experimentally because the temperature gradient through the gas tends to set up convection currents) is due to the random migration of particles from hot to cold regions of the gas and vice versa. *Viscosity* appears when one layer of gas streams as a whole past another layer. Some molecules in the slow stream, by chance thermal movements, jump into the fast stream and slow it down. Similarly, the fast stream loses particles to the slow one, which is thereby speeded up. The overall mechanical effect is equivalent to a frictional drag along the interface between the layers. *Diffusion* is simply the mixing of the particles among themselves. To recognize it experimentally, we have to distinguish one type of particle from another. Thus, in practice, diffusion usually refers to the mixing together of two chemically different gases which exist initially in different concentrations (though at the same total pressure) in different parts of their container. Different gases usually have different molecular speeds and collision diameters, which complicates the analysis of diffusion, but the process becomes particularly simple when these differences can be ignored. This can be realized experimentally by using radioactively "labelled" atoms; these enable *self-diffusion*, i.e. the diffusion of a pure substance into itself, to be measured.

The theory of the transport properties of a gas is very simple, provided we content ourselves with approximate results. We regard each gas particle, of which there are n per unit volume, as a carrier of some physical quantity j ($j =$ thermal energy, momentum or concentration). Suppose that j increases along the z co-ordinate axis through the gas and is constant over any plane perpendicular to z. Thus, dj/dz is the *gradient* of j in the gas. Consider a plane of unit area perpendicular to z. We take the number of particles with mean speed $\bar{c}$ which cross this plane from a given side in unit time to be $\frac{1}{6}n\bar{c}$, the $\frac{1}{6}$ appearing because we can regard the total random motion of the gas as roughly equivalent to six independent streams moving up and down the three co-ordinate axes. These $\frac{1}{6}n\bar{c}$ particles bring with themselves values of j which are typical of the regions where they last made collisions, i.e. regions at a distance of order l from the plane. Thus, the particles coming in from the high-j side of the plane bring the value $j + (dj/dz)l$, and those from the low-j side bring the value $j - (dj/dz)l$, where j is the value of the physical quantity at the plane itself. The *net* amount J of j flowing through the unit plane in unit time is thus approximately

$$\tfrac{1}{6}n\bar{c}\left[\left(j + \frac{dj}{dz}\,l\right) - \left(j - \frac{dz}{dj}\,l\right)\right],$$

i.e.
$$J \simeq -\tfrac{1}{3}n\bar{c}l\frac{dj}{dz}, \tag{1.37}$$

where the minus sign indicates that the net flow is *down* the gradient of j.

This is the *transport equation*. We see that the rate of flow is proportional to the gradient and that the *coefficient of proportionality*, $-\frac{1}{3}n\bar{c}l$, is independent of the gas pressure p at constant temperature, since $n \propto p$ and $l \propto n^{-1}$. At high pressure there are more particles but each moves a shorter distance between collisions.

Applied to thermal conductivity, J becomes the quantity of heat carried through the unit plane in unit time and j is the average heat energy per particle, which we take to be $\frac{1}{2}skT$ for a particle with s classical, squared, energy terms at a temperature T. Substituting this for j in Equation (1.37) and defining

$$c_v = \frac{sk}{2m} = \frac{C_V}{M} \tag{1.38}$$

and
$$\rho = nm \tag{1.39}$$

as the specific heat of unit mass (at constant volume) and the density, respectively, where m is the particle mass and M the atomic or molecular weight, we obtain

$$J = -\tfrac{1}{3}\rho c_v \bar{c}l\frac{dT}{dz}. \tag{1.40}$$

The *coefficient of thermal conductivity* κ is the amount of heat carried in unit time through a unit plane perpendicular to a unit temperature gradient. Hence,

$$\kappa \simeq \tfrac{1}{3}\rho c_v \bar{c}l. \tag{1.41}$$

We shall discuss this result after we have considered viscosity.

The *coefficient of viscosity* is defined in terms of a state of flow as in Fig. 1.7. A flat plate is moved at velocity u_0 parallel and relative to a similar plate spaced a small distance z_0 from it. The fluid between these plates, through its viscosity, then flows in a *laminar* manner in which the velocity u at a distance z from the stationary plane is given by

$$\frac{u}{u_0} = \frac{z}{z_0}. \tag{1.42}$$

The viscous drag between neighbouring layers of fluid and between the fluid and the plates causes the relative motion u_0 of the plates to distribute itself uniformly across the fluid, so that the *rate of shear, du/dz*, is

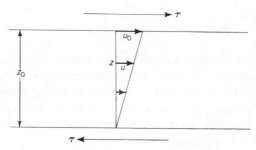

Fig. 1.7.   Viscous flow of a fluid.

constant throughout. The viscous drag is felt as a force opposing the motion of the moving plate. To maintain this motion, a tangential force, or *shear stress*, τ per unit area, must be applied to the plate. At low rates of shear *Newton's law of viscosity*,

$$\tau = \eta \frac{du}{dz}, \tag{1.43}$$

is obeyed by all fluids. Here η is the *dynamic viscosity* of the fluid, usually measured in *poises* ($\eta = 1$ poise when a stress 1 dyn cm^{-2} produces a shear rate 1 sec^{-1}). Typical values at or near room temperature are: air, $1 \cdot 8 \times 10^{-4}$; water, 10^{-2}; treacle, 10^3; pitch, 10^{10} poises. In some problems the *kinematic viscosity*,

$$v = \eta/\rho, \tag{1.44}$$

where ρ is the density, is used. For air under ordinary conditions $v \simeq 0 \cdot 15$ cm^2 sec^{-1}.

To derive Equation (1.43) from Equation (1.37) we note first that u is not the individual thermal velocity of a particle but the overall *stream* velocity of a whole layer of particles. The random thermal motion causes neighbouring layers of the gas to exchange particles, but the quantity carried by these particles which produces viscosity is the *stream momentum* *mu*. Hence, $j = mu$ and J is the total momentum which passes through a layer of unit area in unit time. Since force is equal to rate of momentum change, $J = \tau$ and, hence, Equation (1.43) follows, with

$$\eta \simeq \tfrac{1}{3}\rho \bar{c} l. \tag{1.45}$$

Since ρl is constant at constant temperature, the coefficient of viscosity is independent of gas pressure at constant temperature (*Maxwell's law*) (as are the other coefficients which derive from Equation (1.37)). This surprising result and its experimental confirmation (except at very low densities; see Section 1.12) was a triumph for the kinetic theory of gases. It means, for example, that the motion of a swinging pendulum dies away owing to viscous drag of the air equally quickly in both dense and rarefied air, an observation which Boyle made as early as 1660.

We can use Equation (1.45) to estimate mean free paths. For example, $\eta = 8\cdot5 \times 10^{-5}$ poises in hydrogen at NTP. With $\rho = 9 \times 10^{-5}\,\mathrm{g\,cm^{-3}}$ and $\bar{c} = 1\cdot7 \times 10^{5}\,\mathrm{cm\,sec^{-1}}$, this gives $l \simeq 10^{-5}$ cm. Substituting this in Equation (1.36), we obtain $\sigma \simeq 3\,\text{Å}$ for the collision diameter. The mean free path in this example is thus about 400 times the collision diameter.

Combining Equations (1.41) and (1.45), we obtain a relation between the coefficients of conductivity and viscosity,

$$\kappa = \alpha \eta c_v, \tag{1.46}$$

where α is a numerical factor. According to our simple theory, $\alpha = 1$, but this is not accurate. More elaborate calculations enable accurate estimates to be made. In practice α varies from about 2 for the common diatomic gases to about 2·5 for monatomic gases such as helium. For hydrogen we take $\alpha = 2$ and, with $c_v = 2\cdot42\,\mathrm{cal\,g^{-1}\,deg^{-1}}$ and the above value of η, obtain $\kappa = 4\cdot1 \times 10^{-4}\,\mathrm{cal\,cm^{-1}\,sec^{-1}\,deg^{-1}}$; the observed value is $3\cdot97 \times 10^{-4}$.

Turning now to self-diffusion, we write $j = n_1/n$, where n_1/n is the concentration of one type of particle (n_1 per unit volume) in the gas. We then have $J \simeq -\tfrac{1}{3}\bar{c}l(dn_1/dz)$. The *coefficient of diffusion D*, i.e. the net number of particles which flow in unit time through a unit plane perpendicular to a unit concentration gradient, is then given by

$$D \simeq \tfrac{1}{3}\bar{c}l. \tag{1.47}$$

This has the same dimensions as kinematic viscosity (e.g. $\mathrm{cm^2\,sec^{-1}}$) and, to the accuracy of the above estimates, is numerically equal to it. More accurate calculations show small differences. In fact,

$$D = \alpha\eta/\rho,\qquad(1.48)$$

where this α usually varies between 1·2 and 1·4. In air under ordinary conditions $D \simeq 0\cdot18$ cm^2 sec^{-1}.

For the interdiffusion of two gases which have different particle velocities, $\bar{c}_1$ and $\bar{c}_2$, and different mean free paths, l_1 and l_2, a more general analysis is necessary. For instance, if one gas diffuses faster than the other, a *mass flow* of gas occurs into the region from which the fast gas is diffusing, to equalize the total distribution of gas. The diffusion coefficient is found in this case to be

$$D = \frac{n_2\alpha_1 l_1\bar{c}_1 + n_1\alpha_2 l_2\bar{c}_2}{3(n_1 + n_2)},\qquad(1.49)$$

where the subscripts 1 and 2 refer to the two gases and where $\alpha_1 \simeq \alpha_2 \simeq 1$.

1.11 RANDOM WALK PROCESSES

Conduction and diffusion in a still gas are slow because the collisions cause each particle to make a *random walk* which carries it only uncertainly from its starting point. Let us consider how far, on average, such a particle moves from its starting point in a time t. The total length of its zigzag path is $\bar{c}t$, but this consists of $\bar{c}t/l$ separate straight segments, each of length l. These segments point in all directions, but we may say very roughly that one-third of them point either away from or towards the starting point. Thus, roughly, the particle has taken

$$N \simeq \frac{\bar{c}t}{3l} \simeq \frac{Dt}{l^2}\qquad(1.50)$$

steps of length l away from or towards its starting point. There is a general theorem of statistics that the most probable sum of N equal positive or negative quantities is $\sqrt{N}$ times one such quantity. Thus, a particle migrates by random diffusion during a time t an average distance

$$x \simeq l\sqrt{N} \simeq \sqrt{Dt}\qquad(1.51)$$

from its starting point.

This gives the order of magnitude of the distance over which diffusion takes place in a given time. For example, in the problem of the spread of a gas across a still room, mentioned at the beginning of Section 1.9, we may take $D = 0 \cdot 1 \text{ cm}^2 \text{ sec}^{-1}$ and $x = 1000 \text{ cm}$. Then $t \simeq 10^7 \text{ sec} \simeq 4$ months! In practice, of course, *convection* causes the gas to spread much more quickly than this.

A similar argument can be applied to thermal conduction, which is the diffusion of heat through the gas. Although a unit of heat does not retain its identity through the collisions, we can pretend that it does and so regard such units as carried along in a random walk with the gas particles.

1.12 GASES AT LOW PRESSURES

It is not difficult to reduce the pressure of a gas to the level at which the mean free path for collisions between particles is greater than the width of the container. The transport properties are then drastically altered, since a particle rarely meets another during its flight from wall to wall. Under these so-called *Knudsen conditions* the mean free path l_0 is effectively equal to the width of the container. Since this width may remain constant as the gas pressure is altered, it is no longer true that the viscosity is independent of pressure. Knudsen conditions are important in the design of high-vacuum equipment. Connecting tubes from the vacuum pump need to be wide and short, without constrictions, otherwise the pressure drop necessary to force the gas along them may be greater than the pressure required in the vacuum chamber and the pumping speed will then be limited by the characteristics of the connecting tubes rather than those of the pump.

In the simplest analysis of Knudsen flow it is assumed that, during its flight from wall to wall, a molecule is on average accelerated down the tube by an occasional molecular collision, the effect of which can be expressed through the gradient of gas pressure along the tube, and that the molecule loses this extra velocity at each collision with a wall. In a tube of radius r the time of flight from wall to wall is about $2r/\bar{c}$, and, hence, under a pressure gradient along the tube, the extra velocity acquired during one such flight is on average about $(2r/\bar{c}\rho)(dp/dx)$. Since this is lost at each collision, the mean drift velocity is one-half of it. The volume V of gas transported per second down the tube is then given by

$$V = \frac{\pi r^3}{\bar{c}\rho} \frac{p_1 - p_2}{L}, \qquad (1.52)$$

where $p_1 - p_2$ is the pressure drop per length L of the tube. It may happen that the extra velocity is not entirely lost at a collision with a wall; this can be allowed for by introducing a correction factor called the *slip coefficient* into the formula.

We see that, for a given pressure drop, the volume of flow increases as r^3. This contrasts with the ordinary viscous flow of a fluid, which goes, according to Poiseuille's law (cf. Section 7.8), as r^4. For intermediate pressures a mixed formula of the type

$$V = \frac{r^3}{\bar{c}\rho} \frac{p_1 - p_2}{L}\left(A + B\frac{r}{l}\right),\qquad (1.53)$$

where l is the mean free path in the gas, may be used. The adjustable constants A and B provide a range of behaviour, from Poiseuille flow, proportional to r^4/η, at high pressures, to Knudsen flow, proportional to $r^3/\bar{c}$, at low pressures. This formula is important in the design of high-vacuum apparatus.

FURTHER READING

PAULING, L., *General Chemistry*, Freeman, San Francisco, 1956.

JEANS, SIR JAMES, *An Introduction to the Kinetic Theory of Gases*, Cambridge University Press, 1946.

KITTEL, C., *Elementary Statistical Physics*, Wiley, New York, 1958.

GOMER, R., *Field Emission and Field Ionization*, Oxford University Press, 1961.

PROBLEMS

1.1   One g of iodine gas at 1 atm pressure and at 184°C has a volume of 148 cm³. Deduce from this the number of atoms in the iodine molecule.
                                                                                    [2]

1.2   The mass of an ordinary (light) hydrogen atom is 1·0078 atomic mass units; that of the heavy hydrogen (deuterium) atom is 2·0143. The atomic weight of hydrogen gas is 1·0080. Deduce approximately the number of light hydrogen atoms for each deuterium atom in hydrogen gas.   [5000]

1.3   Assuming that air is 80% N_2 and 20% O_2, calculate its density in the atmosphere (sea level) at 25°C.   [0·00118 g cm⁻³]

1.4   Show that the *most probable speed* of a particle in a Maxwellian gas, as defined by the position of the maximum in the Maxwell–Boltzmann distribution law, is equal to $(2kT/m)^{1/2}$.

1.5   A vessel of gas is divided into two chambers A and B by a porous membrane. These chambers are maintained at two different temperatures T_A and T_B. Prove that the gas pressures p_A and p_B in these chambers become steady when $p_A/p_B = (T_A/T_B)^{1/2}$.

1.6   Show that the ratio of the specific heats of a gas at constant pressure and constant volume can be written in the form

$$\gamma = 1 + \frac{2}{s},$$

where s is the number of classical, squared energy terms per particle. Discuss the significance of the following observed approximate values of γ: 1·67 (mercury vapour), 1·4 (hydrogen at room temperature), 1·6 (hydrogen at $-180°C$), 1·3 (carbon dioxide at room temperature), 1·2 (carbon dioxide at $600°C$), 1·024 (ethyl ether).

1.7 Particles of colloidal platinum have a mass of $2·5 \times 10^{-15}$ g. Show that their root-mean-square thermal velocity at room temperature is about 7 cm sec^{-1}. If these particles were suspended in water, would you expect to be able to measure this velocity by observing them with a microscope?

1.8 Show that the work required to compress 1 mole of particles of a perfect gas from volume V_1 to V_2 is $RT \ln (V_1/V_2)$ when the compression is done isothermally at temperature T; and is

$$C_V T \left[\left(\frac{V_1}{V_2} \right)^{\gamma - 1} - 1 \right]$$

when the compression is done adiabatically, starting at the temperature T.

1.9 Neglecting temperature variations in the atmosphere, air molecules at high altitudes have the same kinetic energy as those at low altitudes, despite the fact that their gravitational potential energy is higher. Explain this by considering the molecules as projectiles shot upwards from low altitudes with a Maxwellian distribution of velocities. Hence show that the Maxwellian distribution (for vertical components of velocity) can be deduced from the barometer formula (Equation (1.21)) and from the observation that the distribution of velocities is the same at every height.

1.10 A parallel beam of electrons passes through a gas at room temperature and at a pressure of 10^{-4} mmHg. Assuming that an electron is removed from the beam whenever it passes within 2 Å of the centre of a gas particle, show that the beam is reduced by about 40% in intensity in passing through 1 m of the gas.

chapter **2**

Condensed States of Matter

We shall now discuss the effects of forces between atoms and molecules. As well as producing deviations from perfect gas behaviour, these forces cause gases to condense into liquids and crystals when cooled. Crystals are stable at low temperatures where thermal motion is relatively weak and the forces are able to pull the particles together into orderly arrangements. The equation of state of an imperfect gas shows, however, that a condensed state of matter, whose structure is continuous with that of a gas, must also exist at certain temperatures and pressures. A crystal, with orderly structure, cannot meet this requirement. Hence, there are two condensed states: liquid and crystalline. After discussing these questions we shall briefly describe how the various main kinds of forces (van der Waals, ionic, covalent, metallic) originate from the electronic structure of matter.

2.1 CONDENSATION

The perfect gas law $pV = RT$ obviously fails for highly compressed gases in which most of the space is taken up by the gas particles themselves. We would expect an equation such as $p(V - \beta) = RT$, where β is proportional to the molecular volume, to describe things better. Experiments do, in fact, show that the pressure rises steeply, above the perfect gas value, when V is very small. They also show the opposite effect, however,—a pressure below the perfect gas value—at slightly larger values of V. This is due to attractive interatomic or intermolecular forces which pull nearby gas particles together, forces which condense gases into liquids and solids at low temperatures.

32

Figure 2.1 shows the p,V curves of a gas at several temperatures. The effect of the attractive forces is best seen at temperatures below the *critical temperature* T_c. Thus, when the gas is compressed at the temperature T_2 to the point A, it then, on further compression, shrinks *finitely at constant pressure* to the much denser state at point B. It has *condensed*. At temperatures where it can be condensed by pressure it is described as *vapour*.

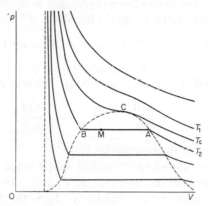

Fig. 2.1. Compression of an imperfect gas at various temperatures, $T_1 > T_c > T_2$.

Points on the branch BC of the curve ACB represent a condensed state of matter, in equilibrium with vapour represented by points along AC. This condensed state must have a gas-like structure, since in the limit, as the critical point C is approached, its structure changes *continuously* into that of a gas. By a careful sequence of changes we could start with a gas at A and take it round the curve ACB to B without at any stage altering its structure more than infinitesimally. The state at B is nevertheless finitely different from the gas at the same temperature. It is much denser and, unlike gas, does not expand to fill its container. Though gas-like in structure, it is a distinct state of matter. It is of course the *liquid* state. The shape of the p,V curves near the critical point, particularly the narrowing of the discontinuity AB to vanishing point at C, thus ensures the existence of the liquid state of matter.

Near C a liquid and its vapour are so similar that the *meniscus* or interface between them becomes practically invisible. At lower temperatures they differ more strongly. For example, at 100°C the specific volume of steam (1 atm) is 1600 times that of water. States with intermediate specific volumes can of course be produced. Thus, we could put some water in a cylinder and withdraw a piston by an arbitrary amount. These states are, however, *heterogeneous*, at least for a system in equilibrium. Some of the particles exist as vapour, others are clustered together as liquid.

These two *phases* of the system coexist in dynamical equilibrium with each other. Some particles evaporate from the liquid into the vapour, others condense from the vapour into the liquid, and in equilibrium these two counter-current processes occur at the same rate. Referring to Fig. 2.1, let V be the volume of the system in the heterogeneous state M and let V_A and V_B be its volumes at A (all vapour) and B (all liquid) respectively. Let c and $(1-c)$ be the proportions (by weight or by numbers of particles) of vapour and liquid, respectively, in the state M. Then

$$V = cV_A + (1-c)V_B, \quad \text{i.e.} \quad c = (V-V_B)/(V_A-V_B).$$

The differences of volume, $V-V_B$ and V_A-V_B, are proportional to the lengths MB and AB in Fig. 2.1, respectively. We thus obtain the *lever rule*,

$$c = \frac{MB}{AB}, \qquad (1-c) = \frac{AM}{AB}, \tag{2.1}$$

for the proportions of the two phases at M.

The pressure corresponding to the line AMB is the lowest at which the stable liquid can exist at the temperature T_2 and is the *vapour pressure* of the liquid at this temperature. Similarly, T_2 is the *boiling point* at this pressure.

Table 2.1 gives some critical points; the 1 atm boiling point is usually about $\frac{2}{3}T_c$ (°K). All the substances shown have roughly similar critical volumes, about 60 cm³ mole^{-1}, i.e. 10^{-22} cm³ particle^{-1}, which corresponds to a sphere of radius about 3 Å for each particle. This is the order of particle spacing below which the attractive forces strongly distort the

TABLE 2.1
Liquid–Vapour–Gas Critical Points

Substance	Temperature, T_c (°K)	Pressure, P_c(atm)	Volume V_c (cm³ mole⁻¹)
Carbon (graphite)	4170	100	
Mercury	1172	180	
Water	647·4	218·3	56
Carbon dioxide	304·2	72·9	94
Ammonia	405·5	111·3	72·5
Oxygen	154·8	50·1	78
Nitrogen	126·2	33·5	90·1
Hydrogen	33·3	12·8	65
Helium 4	5·3	2·26	57·8
Helium 3	3·34	1·15	

p, V curves. These are thus *short-range* forces, strong only when the particles are within about one atomic spacing of one another.

2.2 COHESIVE FORCES AND BINDING ENERGIES

The short-range attractive force between two gas atoms or molecules can be represented by curve 1 in Fig. 2.2. The strong resistance to compression which sets in when the particles come "into contact" can be represented by the even shorter-range repulsive force of curve 2. These curves add to give a resultant force, curve 3, which gives a position of stability at a spacing a_0; any small deviation from this spacing is opposed

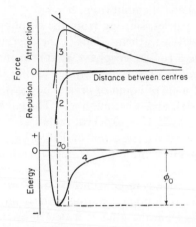

Fig. 2.2.   Forces and energy of interaction between particles.

by a restoring force. A similar argument can be applied to the condensation of a number of particles to a liquid or crystal, and in this case an equilibrium point similar to a_0 determines the average spacing of particles in the condensed phase.

Interatomic and intermolecular forces between particles with centres a distance a apart are often described mathematically by formulae of the type

$$f(a) = Aa^{-n} - Ba^{-m}, \qquad (2.2)$$

where Aa^{-n} is the attraction, Ba^{-m} the repulsion, and $m > n$. For example, for two xenon atoms $n = 7$, $m = 12$, $A = 3 \cdot 16 \times 10^{-57}$ c.g.s. units and $B = 5 \cdot 58 \times 10^{-94}$ c.g.s. units. It is convenient to describe the interactions of atoms and molecules in terms of energy. Since mechanical work $f(a)\, da$ is done when the force $f(a)$ moves a distance da along its direction of application and since this is equivalent to change of energy, we deduce the *energy of interaction* by supposing the particles to start with zero energy at infinite spacing and then subtract from this zero energy the work done by them as they move together to various spacings. This gives curve 4, or energy of interaction $\phi(a)$ obtained by integrating Equation (2.2),

$$\phi(a) = -\frac{A}{n-1} a^{-(n-1)} + \frac{B}{m-1} a^{-(m-1)}, \qquad (2.3)$$

with a minimum value,

$$\phi_0 = \frac{A(m-n)}{(n-1)(m-1)} a^{-(n-1)}, \qquad (2.4)$$

at the spacing of mechanical equilibrium, $a = a_0$.

To *dissociate* the particles, starting from the equilibrium point, a total energy ϕ_0 has to be given to them. This is the *energy of dissociation* or *binding energy* and, for a condensed phase, is approximately* equal to the *latent heat of vaporization* or *sublimation*, i.e. to the thermal energy required to dissociate 1 mole of liquid or crystal into gas. Values have been measured by calorimetric and other methods. The latent heat of evaporation of water to steam at 25°C is 10·5 kcal mole^{-1}, i.e. 586 cal g^{-1}. Some values for chemical elements are given in Table 2.2. These values vary roughly as the boiling and melting points, although there are exceptions, e.g. hydrogen.

This energy of interaction is responsible for the *Joule–Thomson* effect in gases. When a gas seeps through a porous plug from a full vessel to an empty one, it emerges from the plug at a detectably lower temperature. Air, for example, loses 0·05 cal g^{-1} of kinetic energy when the pressure drop is 1 atm. Since it is expanding into a closed vessel, it does no work on its surroundings. Its loss of kinetic energy must therefore be due to the work done against the intermolecular forces when the molecules move apart into the empty vessel. A complicating factor, for which a correction has to be made, is the work done by the dense gas in pushing the more rarefied gas ahead of it through the plug.

The familiar cooling of a liquid by evaporation also depends on the energy of interaction. A particle approaching the surface from inside the liquid cannot pass through to join the vapour unless its total energy is large enough to overcome the cohesive forces holding it to the liquid. Only those particles high in the exponential energy distribution have enough energy. The liquid thus cools through the loss of these high-energy

---

* Approximate unless corrected for various additional effects such as the energy of motion of the particles in the condensed phase and the work done in expanding the vapour against the external pressure.

particles. This argument enables us to understand the well-known exponential dependence of vapour pressure on temperature. Let the heat of vaporization be l_v per particle or L_v per mole and let n_l and n_v be the number of particles in unit volumes of liquid and vapour, respectively.

TABLE 2.2
Binding Energies of Elements, in kcal mole^{-1}

Metals (to monatomic vapour at room temperature):

K	Na	Li	Zn	Mg	Pb	Ca	Al	Ag	Cu	Au	Ni	Fe	Mo	W
19·8	25·9	39	27·4	36·3	47·5	47·8	55	68	81	92	85	94	160	210

Covalent solids (to monatomic vapour at room temperature):

C (graphite or diamond)	B	Si	Ge	Sn (grey)
170	115	85	85	78·6

van der Waals solids (to monatomic vapour at the melting point):

He	Ne	A	Kr	Xe
0·052	0·52	1·77	2·67	3·76

van der Waals solids (to molecular vapour at the melting point):

N_2	O_2	H_2	Cl_2	CO_2	I_2
1·50	1·74	2·44	6·0	8·24	18·9

From the exponential law, only a fraction s, proportional to $\exp(-l_v/kT)$, of the particles in the liquid have enough energy to pass into the vapour. For equal rates of flow of particles in both directions between liquid and vapour, the number of particles per unit volume capable of making the transition must be the same in both phases, i.e. $n_v = n_l s$. Thus,

$$n_v = \alpha n_l \, e^{-l_v/kT} = \alpha n_l \, e^{-L_v/RT}, \tag{2.5}$$

where α is a constant. The pressure p of a dilute vapour is given by the perfect gas law, $p = n_v kT$, and so

$$p = A \, e^{-L_v/RT}, \tag{2.6}$$

where the parameter A contains kT. This equation enables binding energies to be found from variations of vapour pressure with temperature. Similar equations describe many kinds of physico-chemical equilibria (e.g. the distribution of a solute between two solvents), the common feature being that when particles attempt to pass from one state to another, the energy difference, e.g. l_v, between those states provides a "filter" which lets only a fraction, e.g. $\exp(-l_v/kT)$, of particles through from the low-energy to the high-energy state, but lets all through in the opposite direction. For dynamical equilibrium the density of particles in the high-energy state must therefore be reduced by this same fraction.

2.3 THERMODYNAMIC EQUILIBRIUM BETWEEN PHASES

We have discussed the equilibrium between a liquid and its vapour as a dynamical equilibrium between the counter-currents of particles flowing from one to the other. Equilibrium between other phases, e.g. crystal and liquid, could be discussed in the same way, but we now ask whether there is any more general property of a system from which we could determine phase equilibrium. Total energy is clearly not this property, since, for example, the energy of a vaporized system is higher than that of the liquid, partly through the breaking of the cohesive bonds and partly through the mechanical work done by the system in expanding to vapour and pushing back the surrounding atmosphere. Thus, if we vaporize 1 mole of liquid at a constant pressure and temperature, p and T, and thereby increase its *internal* energy (i.e. total energy of its particles) by ΔE and increase its volume by ΔV, we have to provide a total energy of $\Delta E + p\Delta V$, where $p\Delta V$ is the work done by the system on its surroundings and may be regarded as *external* energy of the system. Since $\Delta p = 0$, the vaporization leads to an increase $\Delta(E + pV)$ in what is called the *enthalpy* $(E + pV)$ of the system.

The vapour is stable despite its higher energy, because of the greater freedom of space allowed to its particles; compared with those in the liquid, they have a much greater specific volume. A similar principle applies generally to all systems. A substance possesses a heat capacity because it allows heat energy to distribute itself randomly among its particles. When a crystal melts, its particles escape from the regimentation of the orderly crystal structure into the disorderly jumble of the liquid structure. When a substance dissolves, its particles distribute themselves randomly throughout the entire volume of the solvent, like the particles of a gas in a container; their tendency to do this, in fact, gives rise to the force of *osmotic pressure* which is directly analogous to gas pressure.

This randomness of distributions is measured numerically by a quantity W, called the *thermodynamical probability* of a system. In energy calculations W is usually used in the form

$$S = k \ln W, \tag{2.7}$$

where k is Boltzmann's constant ($1 \cdot 38 \times 10^{-16}$ erg deg^{-1}) and S is called the *entropy* of the system. This has various advantages, among which is the fact that entropy can be measured experimentally from the heat absorbed by the system, through the relation

$$dS = \frac{dQ}{T},$$ (2.8)

where dS is the increase in entropy of a system in equilibrium at temperature $T(^\circ K)$ when this system *reversibly* absorbs a small amount of heat dQ from its surroundings. A reversible change is one in which the system is in equilibrium at every stage during the change. The creation of heat from mechanical work by friction is an irreversible change. Equation (2.8) is not valid for irreversible changes, since in these the system may increase its entropy by frictional heating.

To obtain an energy criterion for stability, we have to subtract from the enthalpy that part of the total energy of the system that is "bound up" with the entropy. It can be shown that this is equal to TS. Thus, the remainder, i.e.

$$G = E + pV - TS$$ (2.9)

is a measure of the thermodynamic instability of the system and must be as small as possible if the system is to be stable at constant temperature and pressure. It is called the *free energy* of the system, i.e. it is that part of the total energy of the system that is *free* to do mechanical work. The general connexion between equilibrium and mechanical work lies in the following fact. If there is not equilibrium, then particles can flow preferentially from one part of the system to another, as, for example, when particles evaporate from a liquid into a vacuum. This one-way flow of particles can be opposed by placing obstacles such as paddles or pistons in the way of the particles, so obtaining mechanical work from the displacement of these obstacles by the particles. When there is equilibrium, however, the obstacles are pushed equally forwards and backwards by the two streams moving in opposite directions, and no mechanical work can be extracted from them.

The formal derivation of the above relations is given in books on thermodynamics.

2.4 VAN DER WAALS' EQUATION

Many formulae have been suggested to replace the perfect gas law as the *equation of state* of a gas. The familiar *van der Waals equation*

$$\left(p + \frac{\alpha}{V^2}\right)(V - \beta) = RT$$ (2.10)

is simple and enlightening. The term α/V^2 represents the attractive forces between the gas particles, although not accurately except at low gas densities. Figure 2.3 shows a family of van der Waals p, V curves. They are like the curves of Fig. 2.1, but the horizontal line AB has become a

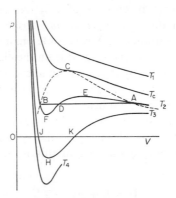

Fig. 2.3.   van der Waals isothermals $(T_1 > T_c > T_2 > T_3 > T_4)$.

continuous curve AEDFB with a maximum and minimum. Each point on this curve represents an *unstable homogeneous* state of the substance, intermediate between liquid and vapour. If the substance could be prepared in the homogeneous state D, for example, its pressure would *increase* with increasing volume, since dp/dV is positive at D. Thus, if part of the system chanced to expand slightly, at the expense of the rest, it could then continue to expand, since its pressure would then be higher than elsewhere, until it reached the state represented by point A. Similarly, the remaining part, compressed by this expansion, would continue to shrink, since its pressure would become smaller until it reached the point B. The homogeneous system at D would thus break down into vapour at A and liquid at B.

It can be shown in various ways that the regions AEDA and BFDB must have equal areas. One argument is that, if this were not so, it would be possible in principle to produce a perpetual motion by taking the system from A to B along the homogeneous path (AEDFB), returning to A along the heterogeneous path (horizontal line AB), or vice versa, obtaining more mechanical work from the system along the one path than is put into it along the other.

In the region FE, where dp/dV is positive, even the smallest density fluctuation in the unstable homogeneous system triggers off the change to the stable heterogeneous state. In AE and FB, however, dp/dV is negative and the homogeneous state is *metastable*, i.e. stable against small fluctuations but unstable against ones large enough to create *seeds* or *nuclei* of the phase at the other end of the AB line. With care it is, in fact, possible to prepare metastable *superheated* liquids and *supersaturated* vapours. The erratic, almost explosive, boiling of pure water in a clean smooth glass vessel ("bumping") is a familiar example of this. The true boiling point is slightly exceeded, because there are no seeding points on which vapour bubbles can nucleate easily. Regular, smooth boiling can be produced by adding a nucleating agent, such as a piece of unglazed earthenware, the pores of which provide nuclei for vapour bubbles. Almost as familiar are the rain-making techniques in which fine crystals of nucleating substances such as silver iodide are dropped into clouds to stimulate the condensation of supersaturated water vapour. Electrically charged atomic particles also act as nucleating agents, and this is exploited in the *cloud chamber* and *bubble chamber*, used for detecting the tracks of such particles by the lines of liquid or vapour droplets nucleated along them.

Returning to Fig. 2.3, at the position of the critical point C the p, V curve is horizontal and also has zero curvature, since it is the point where the maximum, minimum and point of inflexion on the curve AEDFB all coalesce when the temperature is raised to T_c. Thus,

$$\frac{dp}{dV} = \frac{d^2 p}{dV^2} = 0 \tag{2.11}$$

on the isothermal at C. Applying these conditions to van der Waals' equation, we obtain

$$V_c = 3\beta, \qquad p_c = \frac{\alpha}{27\beta^2}, \qquad T_c = \frac{8\alpha}{27R\beta}, \tag{2.12}$$

and hence,

$$p_c V_c = \tfrac{3}{8} R T_c \tag{2.13}$$

at the critical point. Thus, pV is well below its perfect gas value at this point. Using the critical quantities V_c, p_c, T_c as units of volume, pressure and temperature, we can rewrite van der Waals' equation as

$$\left(p + \frac{3}{V^2}\right)(V - \tfrac{1}{3}) = \tfrac{8}{3} T, \tag{2.14}$$

in which form it is called a *reduced equation of state*.

Equation (2.13) can be used to test the accuracy of van der Waals' equation. From Table 2.1 we find, e.g. for carbon dioxide, that the predicted value of V_c, i.e. $3RT_c/8p_c$, is 130 cm^3 mole^{-1}, whereas the observed value is 94. For most substances the predicted value is, in fact, about 1·4 times the observed value. This shows the quantitative limitations of van der Waals' equation, which spring from the simplicity of the terms, αV^{-2} and β, used to represent the interactions of the gas particles. Several better equations have been developed, but van der Waals' equation remains useful for the simple picture it gives of the behaviour of dense gases and condensation effects.

2.5 TENSILE STRENGTH OF A LIQUID

Figure 2.3 shows that at temperatures such as T_3 the pressure is negative over part of the range, e.g. in JHK. Thus, we could expand a homogeneous liquid (with no vapour nuclei) from J to H by pulling on it with a piston in a cylinder, so applying a *hydrostatic tension* numerically equal to the negative pressure at H. Ideally, this liquid should not "break", i.e. form vapour, until H is reached. The hydrostatic tension at this point is thus the *ideal tensile strength* of the liquid.

To estimate this strength, we rewrite Equation (2.14) as

$$p = \frac{8T}{3V - 1} - \frac{3}{V^2} \qquad (2.15)$$

and then apply the condition that $dp/dV = 0$ at H, which gives

$$(3V_m - 1)^2 = 4V_m{}^3 T \qquad (2.16)$$

for V_m, the value of V at H. Substituting this into Equation (2.15), we then obtain p_m, the limiting value of p, as

$$p_m = \frac{3V_m - 2}{V_m{}^3}. \qquad (2.17)$$

Since $p_m = 0$ when $V_m = \frac{2}{3}$, a van der Waals liquid has tensile strength only at temperatures below that for which $V_m = \frac{2}{3}$. As the temperature is lowered, the tensile strength, $-p_m$, becomes increasingly large, until at low temperatures where $V_m \simeq \frac{1}{3}$ (cf. Equation (2.16)) it approaches the value $27p_c$. Thus, from Table 2.1 the limiting tensile strength of water should be of order 6000 atm ($\simeq 90,000$ pounds per square inch), which compares with the (uniaxial) tensile strength of strong steel. This estimate is very rough, but it shows the great mechanical strength of molecular forces.

Experiments on the tensile strength of liquids are difficult, not only because vapour-nucleating agents must be totally excluded, but also because in a cylinder and piston the liquid usually tears away from the wall while the tension is still quite low. Nevertheless, liquids have been stressed to over 500 atm ($\simeq 7500$ psi) before failure. Dynamical experiments in which sharp pulses of tensile stress are produced inside the liquid by intense high-frequency sound waves have demonstrated still higher strengths. The liquid fails in this case by *cavitation*, i.e. by the formation of small vapour bubbles inside it. These bubbles often then collapse violently, emitting strong pressure waves. Cavitation is an important practical problem of ships' propellers. At high propeller speeds strings of cavities form on the trailing edges of the blades, and the repeated mechanical blows applied to the blades by the collapse of these cavities gradually wear away the blade metal by tearing small fragments out of its surface.

2.6 LIQUID AND CRYSTALLINE STATES OF MATTER

It is fairly easy to see that the gaseous and crystalline states of matter should exist. At high temperatures the thermal forces overcome the cohesive forces and the particles fly apart. At low temperatures, where the cohesive forces are almost unopposed by thermal forces, each new particle which condenses on to the material takes up the best position in the force field of its neighbours and then, in this position, helps them provide best positions for other condensing particles. In this way a regular periodic arrangement of particles builds up, governed by the laws discussed in Chapter 3.

It may seem strange that *two* condensed states, liquid and crystal, should exist. However, we have seen that there must be a liquid state for reasons of continuity near the critical point. This behaviour at the critical point is due ultimately to the short-range character of the interatomic and intermolecular forces, for it is this that produces p,V curves such as those of Fig. 2.1, which give a discontinuity AB that merges into a single critical point at C. The liquid state thus appears as an interpolated state, between the gas and crystal, which is necessary because the crystal, with its *long-range ordered* structure, cannot provide that continuity of structure with the gas near the critical point that the short-range forces demand.

If we ignore the long-range order of the crystal and those properties that depend critically on it (e.g. X-ray diffraction, anisotropy), then the structure and properties of a liquid, just above the melting point, are usually not grossly different from those of its crystal. Metals, for example, expand only 3-5% on melting (bismuth, like ice, contracts), so that the number and packing of nearest neighbours cannot alter much. Similarly, their latent heat of melting is only about 3-5% of their latent heat of vaporization. It is as if 3-5% of the crystal sites became empty and their free volume were then broken up and distributed between the particles in such a way as to destroy the long-range order of the structure. (Metals are exceptional in the smallness of their changes at the melting point; most other simple substances expand about 10–30% on melting.)

Despite this similarity, the melting point is always a sharp transition across which most properties of the substance change finitely. The liquid–crystal system could be represented on a p, V curve by a line similar to AB in Fig. 2.1, with a range of heterogeneous (liquid + crystal) states between, but there is no evidence that this range ever narrows down to a single critical point, and it seems very unlikely, since long-range order is so profoundly different from long-range disorder, that such a point could exist. The liquid and crystalline states differ finitely at all temperatures and pressures.

The concept of *free volume* is useful when discussing the liquid–crystal transition. The properties of a dense collection of (spherical) particles appear to depend mainly on the difference

$$v_f = v - v_0 \qquad (2.18)$$

between the average volume allowed per particle, v, and the actual volume of a particle, v_0. Temperature is thought to exert its effect indirectly, by determining how much free volume v_f the system possesses at a given pressure. By applying great pressure to a liquid, to reduce its free volume at high temperature, it is thus possible experimentally to reproduce values of properties, such as viscosity and diffusion, normally associated with much lower temperatures.

This idea leads naturally to a simplified picture of dense matter as a collection of rigid spheres, like marbles, which exert no forces on one another except when they touch and repel, and which are spread out to a certain density in a box. This model does not entirely ignore the attractive, repulsive and thermal forces on the particles. The average balance of these forces at a given free volume is allowed for in the choice of the size of box in relation to the number and size of particles in it. The predominance of the repulsive forces at close distances is also allowed for, roughly. What is not allowed for (except in refined versions of the model) is the precise variation of the forces with spacing between particles; since, however, the liquid state is common to all substances, irrespective of their detailed laws of force, it is reasonable to suppose that this does not invalidate the model for discussing general features of the liquid state.

Models of this kind have been studied mathematically, using computers and random numbers to describe the positions of particles; also experimentally, in two dimensions by spreading out ball bearings or marbles on flat horizontal trays and in three dimensions by supporting spheres in a liquid of the same specific gravity. At low densities the spheres spread themselves randomly, like a gas. At the highest density they crystallize

into a close-packed array. At slightly lower densities than this (e.g. 80 %
of maximum) they fluctuate between two states when the box is shaken:
a crystalline state which, being close-packed, does not fill the box and so
does not exert pressure on it; and a disordered state which, being in-
efficiently packed, fills the box and so exerts pressure. This second state
represents liquid or dense gas. The drop in pressure obtained in going from
this state to the crystalline one persists until the box is contracted finitely
to take up the free space. This is reminiscent of the pressure drop from
E to F in Fig. 2.3, and shows that the change from disordered to crystalline
state is finite and discontinuous.

A vapour can of course condense directly to crystal, as when hoar frost
forms in cold dry weather, and a crystal can sublimate directly to vapour,
as when solid carbon dioxide ("dry ice") evaporates in air. The liquid
state does not usually exist at low vapour pressures (helium is exceptional).
Figure 2.4 shows the *phase diagram* of water at low pressures. Points on

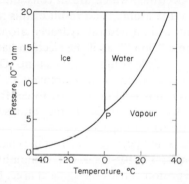

Fig. 2.4. Phase diagram of water at low pressures.

a phase line give pressures and temperatures at which the phases on the
two sides of the line are in equilibrium. We see that liquid water does not
exist as a stable phase at any vapour pressure below 0·006 atm (4·58 mmHg).
The point P, at this pressure and at $+0·0098°C$, is the *triple point*,
at which water, ice and vapour all coexist in equilibrium. The triple
point lies near the ordinary melting point, because the liquid–crystal

TABLE 2.3
Vapour–Liquid–Crystal Triple Points

Substance:	H_2	Ne	O_2	N_2	CO_2	H_2O	I_2	As	C
Temperature (°K):	13·9	24·6	54·8	63·1	216·6	237·17	387·3	1091	3900
Pressure (atm):	0·068	0·425	0·002	0·124	5·11	0·006	0·118	35·8	100

equilibrium is only slightly sensitive to small pressures. Values of triple points are given in Table 2.3. We notice the high vapour pressures at the triple point of those substances that sublimate readily.

2.7 ATOMIC STRUCTURE AND THE ORIGIN OF COHESIVE FORCES

Although we are concerned in this book with the effects, rather than the origins, of atomic forces, it will be useful to consider briefly how these forces originate from the structures and properties of atoms.

The *protons* and *neutrons* which cluster together in the atomic nucleus and the *electrons* which swarm about that nucleus are all exceedingly small particles ($\simeq 10^{-13}$ cm) compared with the radius of the atom as a whole ($\simeq 10^{-8}$ cm). We can picture the atom as a mainly spherical "cloud" or "gas", centred about a small heavy nucleus, very dense near the nucleus and tenuous beyond a radius of about 10^{-8} cm from the nucleus. This cloud represents the electrons, which are in constant motion throughout the space occupied by the atom. As a result of this motion, even if only one electron is present, as in a neutral hydrogen atom, it is best to think in terms of an electronic "cloud", as if the electron had somehow become "smeared out" over the whole volume of the atom. Despite its seemingly nebulous construction, an atom is an extremely stable unit of matter. The electronic cloud is firmly held in place by forces from the nucleus and it strongly repels the electronic clouds of other atoms when these attempt to penetrate deeply into its own space.

Each electron carries a unit negative electrostatic charge, $-e$ ($e = 4 \cdot 80 \times 10^{-10}$ electrostatic units; 3×10^9 e.s.u. = 1 coulomb), and is very light ($9 \cdot 1 \times 10^{-28}$ g). The proton carries the same charge, but of opposite sign ($+e$), and is 1836 times heavier. The neutron is uncharged and slightly heavier than the proton. The weight of an atom depends mainly on the *mass number A*, i.e. the total number of protons and neutrons in its nucleus, but the chemical and physical properties depend mainly on the *atomic number Z*, i.e. the number of protons, because this is also the number of electrons which the atom possesses in its normal, electrically neutral, state. Each *chemical element* has its own atomic number (Table 1.1), and the *isotopes* of an element are atoms with the same numbers of protons but different numbers of neutrons. In atoms of the heavier elements there are more neutrons than protons.

The only important force exerted on the electrons of an atom, or on one atom by another, is the *electrostatic force* due to the charges on the particles. There are also *magnetic* forces, induced by the motions of the particles (including spinning motion), but these are of secondary importance. The electrostatic force f (dyn) between two charged particles

z_1e and z_2e (e.s.u.) separated in space by a distance r (cm) is given by Coulomb's law as

$$f = \frac{z_1 z_2 e^2}{r^2},$$ (2.19)

and the electrostatic potential energy V (erg), i.e. the work done by this force in bringing the particles to this distance, is given by

$$V = \int f\, dr = -\frac{z_1 z_2 e^2}{r} + \text{constant.}$$ (2.20)

The integration constant is chosen so that $V = 0$ when $r = \infty$, which gives negative potential energies for charges of opposite signs. Thus, the potential energy of an electron and proton is given by

$$V = -\frac{e^2}{r}$$ (2.21)

and has the form shown in Fig. 2.5. It is usual to measure V, and all energies in atomic problems, in terms of the *electron-volt* (eV) unit, the work done when an electron falls through a potential difference of 1 V ($1\text{ eV} = 1\cdot6 \times 10^{-12}\text{ erg} = 1\cdot6 \times 10^{-19}\text{ J}$).

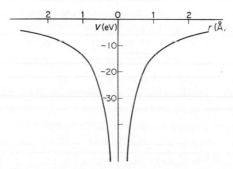

Fig. 2.5. Potential energy of an electron in the field of a proton.

A hydrogen atom is a single electron held in the electrostatic field of a single proton. The movement of this electron towards the proton, so "falling down" the potential energy "well" of Fig. 2.5, is analogous to a stone falling down a hillside under gravity. There is no friction of course, but, because the electron is a charged particle, it can release, by emitting electromagnetic radiation, the kinetic energy it acquires as it spirals in to the nucleus. We might thus expect the atom to shrink until the electron has joined the proton. The same argument applies to all other atoms.

The fact that atoms are stable with electrons at distances of order 10^{-8} cm from the nucleus shows that something prevents this shrinking process from going too far. We have to accept as an observed fact of nature that the electrons are in some way prevented from doing all that the classical laws of electricity and mechanics would allow them to do. These *"quantum"* effects are the dominating features of modern physics, and their elucidation and quantitative description has been one of the decisive advances of all science. It is now possible to explain a vast range of atomic and molecular behaviour in terms of the quantum-restricted motions of electrons in electric fields.

The electron of a hydrogen atom has a potential energy $-e^2/r$ and a kinetic energy $p^2/2m$, where $p(=mv)$ is its momentum, m its mass, and v its velocity. Its *total energy* is

$$E = -\frac{e^2}{r} + \frac{p^2}{2m}.$$ (2.22)

A low-energy electron cannot escape far from the nucleus, since, even if it were to convert all its kinetic energy to potential energy, this would not be enough to allow it to move completely away. The same is true of any field of electrostatic force which "boxes in" the electron within a certain region of space. The quantum effect observed in all such cases is that, if an electron is forced to exist within a limited volume of space (being boxed in by an electrostatic field), then it can *never be at rest*. It must always be in motion and always have kinetic energy, and the smaller the volume allowed to it the larger is this kinetic energy. This is why the atom maintains a certain minimum size. The electrostatic force tries to shrink the atom into a smaller volume but is resisted by the increase in kinetic energy associated with this localization of the electron. This effect is expressed quantitatively by the *Heisenberg principle*; if $\bar{r}$ and $\bar{p}$ are the average distance and momentum of the electron, then $\bar{r}\bar{p}$ cannot be smaller than $h/2\pi$, where h is *Planck's constant* ($6\cdot626 \times 10^{-27}$ erg sec). Substituting $h/2\pi\bar{r}$ for $\bar{p}$ in Equation (2.22), we have

$$E = -\frac{e^2}{\bar{r}} + \frac{h^2}{8\pi^2 m \bar{r}^2}.$$ (2.23)

We can then find the stable minimum size of the atom by minimizing this with respect to $\bar{r}$, i.e. by taking $dE/d\bar{r} = 0$. Thus,

$$\bar{r} = \frac{h^2}{4\pi^2 m e^2} = 0\cdot53 \times 10^{-8} \text{ cm},$$

$$E = -\frac{2\pi^2 m e^4}{h^2} = -13\cdot5 \text{ eV}.$$ (2.24)

These correctly give the observed *ionization energy* (i.e. the energy needed to dissociate the electron and proton) and the order of size of the hydrogen atom in its most stable state, or *ground state*, as it is called.

By solving *Schrödinger's equation*, which is the general equation governing the quantum-restricted motion of an electron in a force field, we could describe this atom more completely. It can be shown, for example, that in this ground state the electron explores equally all directions round the nucleus and is thus equivalent over a period of time to a spherical ball of electrical charge, dense at the centre and tenuous at the edges, centred on the nucleus. There are also other states of motion for the electron, in which the electron has a higher total energy than in the ground state, and these "excited" states occur only at certain definite *energy levels*. In these

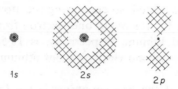

1s   2s   2p

Fig. 2.6. Some electronic quantum states in a hydrogen atom. The shaded areas represent those regions of space about the nucleus in which the electron spends most of its time. The ground state, denoted as 1s, and the excited state 2s are both spherically symmetrical: the excited state 2p has rotational symmetry about the vertical axis.

quantum states the electron on average lies further out from the nucleus than in the ground state; it is tending to dissociate from the nucleus. In some of these states the electronic cloud remains spherically symmetrical and centred on the nucleus but is made up of concentric shells of higher and lower density (Fig. 2.6). In other states the electron confines itself to the neighbourhood of one particular axis through the atom, and the electron cloud then has a dumb-bell shape (Fig. 2.6). More complex shapes are developed at still higher energies. An electron can, under certain conditions, transfer from one quantum state to another, and when it does so there is a corresponding absorption or emission of a "quantum" of electromagnetic radiation which brings in or takes away the energy due to the difference in the two energy levels. This is the basis of the optical and X-ray spectra of atoms.

The quantum states of the heavier atoms are qualitatively similar to those of hydrogen. The electronic clouds are pulled in closer to the nucleus, of course, because the higher nuclear charge, ze, can combat the kinetic energy of localization more strongly. The two electrons in the ground state of the helium atom ($z = 2$), for example, are strongly held in a small 1s state (Fig. 2.6.).

We might expect from this that all atoms would become smaller and more stable as we move to the heavier elements. This does not happen, however, because another quantum restriction, expressed by *Pauli's principle*, is observed when several electrons share the same field of force; *there is room for only two electrons* (*which must spin in opposite directions*) *within a single quantum state.* Because of this, many electrons cannot crowd into the same quantum state. The third electron of the lithium atom ($z = 3$), for example, cannot go into the $1s$ state but must go into a higher state, preferably the $2s$ state (Fig. 2.6). This makes the atom large again, despite the higher nuclear charge, and the $2s$ electron is only weakly held, because the nuclear field $+3e$ acting on it is largely *shielded* by the opposing field $-2e$ of the two electrons lying close to the nucleus in their highly shrunken $1s$ "shell". This is reflected by the ease with which this $2s$ electron can be dissociated, so converting the atom into a *positive ion*. The ionization energy for removing one electron from a neutral atom of hydrogen, helium and lithium, respectively, is 13·5, 24·5 and 5·37 eV. However, to remove a second electron from lithium, (from the $1s$ state) requires 75 eV.

A striking feature of the heavier atoms is the formation of "shells" holding eight electrons. This number comes from the fact that each such shell contains one state of s type and three of p type (along the three coordinate axes), each of which can hold two electrons of opposite spins. When filled, each such shell is spherically symmetrical; the electrons in it are strongly held by the nucleus, because each of them is about as near to the nucleus as any other and so can participate in the nuclear field with the minimum of shielding from its companions in that shell. Atoms whose outermost shells are of this type are particularly stable. Together with helium they form the *inert gas* series of elements.

The atoms of most elements in their neutral state have the wrong numbers of electrons for completely filled shells. Either they have too many, in which case the surplus ones have to go into quantum states outside the filled shells and become similar to the third electron of lithium in their remoteness and semi-dissociation from the nucleus; or they have too few, in which case there are "electron holes" in their outermost shells, i.e. places capable of accepting further electrons which could enjoy quite strongly the field of their nucleus. The first type are the *electropositive* elements, or the *electron-donors*, which can easily give up electrons and become positive ions. *Metals* belong to this class. The second are the *electronegative* elements, or the *electron-acceptors*, which readily accept extra electrons to become negative ions. Elements such as the halogens (F, Cl, Br, I), oxygen and sulphur belong to this class.

2.8 TYPES OF COHESIVE FORCES

We have seen that the electrons in an atom try to participate as much as possible in the electrostatic field of the nucleus but are prevented from going right in by the kinetic energy of localization and by the limited room in the quantum states near the nucleus. These same factors dominate the behaviour when one atom approaches another. For example, when the electronic clouds of these atoms interpenetrate deeply, Pauli's principle requires some of these electrons to enter higher quantum states, which is equivalent to a repulsion (*the filled-shell repulsion*) between the atoms. This is the basis of the incompressibility of matter.

The attraction which all atoms show towards one another, when near neighbours, is due to electrostatic forces. Even the inert gas atoms, the electrons of which all participate well in their own nuclear fields and which are thereby chemically inactive, are weakly attracted together by *van der Waals forces*. These forces, which cause inert gases to liquefy and solidify at low temperatures, are due to the fact that, although the average electrical field of a neutral spherical atom is zero, the instantaneous field is not zero but fluctuates with the movements of the electrons in the atom. When two atoms approach, they are able to correlate their electronic motions so that the electrical charge in one surges towards the other when the fluctuations in this second atom happen to leave its nuclear field rather more exposed in this particular direction than usual.

Far better opportunities for the outer electrons of neighbouring atoms to participate in one another's nuclear fields exist when these electrons are not in completely filled shells. Atoms with "electron holes" in their outer shells electrostatically attract the electrons of other atoms into these holes and so tend to adhere to those atoms. Atoms with weakly attached electrons outside filled shells offer these electrons to other atoms to allow them to participate in their nuclear fields. These effects lead to three broad classes of interatomic bonds, *ionic*, *homopolar* and *metallic*.

The ionic bond is simplest. When, for example, a sodium atom meets a chlorine atom, it may give up its single outer electron to fill the electron hole in the chlorine. Both atoms have then achieved a filled-shell structure, but the sodium has become a positive ion and the chlorine a negative ion. These ions attract electrostatically and cohere as an ionic molecule. It is easy to estimate roughly the energy of this bond. If $r = 3 \times 10^{-8}$ cm (i.e. distance between nuclei), Equation (2.21) gives

$$\frac{e^2}{r} = \frac{(4 \cdot 8)^2 \times 10^{-20}}{3 \times 10^{-8} \times 1 \cdot 6 \times 10^{-12}} = 4 \cdot 8 \text{ eV}. \qquad (2.25)$$

Although only rough, this indicates the order of magnitude.

This bond is moderately strong, as is shown by the fact that alkali halide crystals have some hardness and their melting points approach 1000°C. Equation (2.21) shows that small ions (i.e. small r) should be more strongly bonded. This is so, as the following melting points show:

	NaF	NaCl	NaBr	NaI
Interionic distance (Å):	2·31	2·79	2·94	3·18
Melting point (°C):	988	801	740	660

These compounds are all *univalent*, composed of atoms with *single* valencies, i.e. that donate or accept only one electron to achieve a filled shell. In *divalent* compounds, composed of atoms that donate or accept two electrons, the electrostatic binding forces are much stronger, as is shown by the high melting points of the following oxides:

	MgO	CaO	SrO	BaO
Interionic distance (Å):	2·10	2·40	2·57	2·77
Melting point (°C):	2640	2570	2430	1933

Strong multivalent ionic bonds are important for *refractories* and also for hard materials, such as sapphire (Al_2O_3; m.p. = 2020°C), used in jewelled bearings and cutting tools.

A third factor which increases the cohesion of ionic solids is *polarizability*. The electric field of one ion distorts the electronic cloud of the other so that there is a preponderance of opposite charge next to it. The positive ion, having lost electrons and become small and "hard", is generally the polarizing ion; and the negative ion, having gained electrons and become large and "soft", is the polarized ion. Because the electrons are to some extent pulled back towards the donor ion, this polarization represents a partial transition from ionic towards covalent binding. In solids of trivalent and tetravalent elements the transition is almost complete. Owing to the large number of electrons participating in their bonds, very high melting points are achieved, e.g.:

	BN	NbC	SiC	TaN	TaC	TiN	TiC	VC	WC	ZrN	ZrC
Melting point (°C):	3000	3500	2700	3090	3800	2940	3150	2800	2750	2950	3500

The pure ionic bond is *non-saturating*, i.e. the attraction of one ion for another is not diminished by the attachment of other ions to it, and is *non-directional*, i.e. is equally exerted in all directions round the ion. The crystal structure depends partly on *valency* (e.g. in CaF_2 and TiO_2 two negative ions must go with each positive ion) and partly on the *relative sizes* of the ions. In, for example, NaCl the sodium ion is too small to allow more than six chlorine ions to be packed round it, but in CsCl the caesium ion is large enough to accept eight chlorine ions and a different structure is then obtained (cf. Fig. 3.7). Much more complicated ionic crystal structures are formed in substances such as the silicates, carbonates, nitrates and sulphates, in which one of the partners is not a single atom but a *molecular ion*; e.g. the negative SiO_4 ion which is the basis of the silicates. *Layer structures*, formed by the parallel stacking of layers of material, occur commonly in some types of ionic crystals. For example, in $CdCl_2$ and CdI_2 each layer consists of a sheet of positive ions sandwiched between two sheets of polarized negative ions. The binding between these layers is not ionic and is relatively weak, involving mainly van der Waals forces. Such materials are mechanically weak, being easily cleaved between the layers. Because of its weakly bonded layer structure, molybdenum sulphide (MoS_2) is a good lubricant. The flaky cleavage of mica and the softness of talc are very familiar. Both of these substances have layer structures consisting of silicon–oxygen sheets held together weakly by forces from metallic ions in the material, e.g. muscovite mica, $(K,Na)(OH)_2Al_2(Si_3AlO_{10})$; talc, $Mg_3Si_4O_{10}(OH)_2$. Many hydroxides, such as $Ni(OH)_2$ and $Fe(OH)_2$, have layer structures and precipitate in the form of flakes which have a "soapy" feel. Boron nitride, with a partly covalent bond, crystallizes in the form of thin slippery flakes and is a possible solid lubricant. Its structure is similar to that of graphite, another material that slides easily on layer planes. When heated to 2500°C under a pressure of 50,000 atm, however, boron nitride transforms to a cubic structure similar to that of diamond (Fig. 3.7).

The covalent bond appears in its purest form in molecules such as H_2, Cl_2, CH_4, C_2H_6 and CCl_4, and in diamond. Its essential feature is the *sharing* of electrons between atoms rather than transfer from one to the other. Consider the simplest possible example, two protons and one electron. The electron is equally attracted to both protons, of course, although, when these are far apart, it settles on one of them, giving a neutral hydrogen atom and a separate bare proton (Fig. 2.7). As the proton draws near, the electronic cloud of the atom is polarized and pulled towards it. Eventually a stable state is achieved in which the electron distributes its motion equally between both protons and spends most of its time in the region between them, where it enjoys the electrostatic attraction of both. The electron is now in a *bonding* quantum state between the two protons. This covalent bond is thus simply the electrostatic attraction of two close protons to the electron they share in common. As in a single atom, quantum restrictions prevent the electronic cloud from shrinking too much, and it is this that maintains the spacing of the nuclei in the molecule. By repeating the kind of analysis made for the single hydrogen atom, one can find the ground state for the electron in this molecule. This ground state is analogous to the 1s state of the single atom (Fig. 2.6) and can hold two electrons of opposite spins.

Two protons and one electron form a singly charged positive ion capable of accepting one more electron to become a neutral H_2 molecule. This second electron can join the first, in the ground state, so forming an

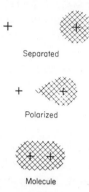

Fig. 2.7.   Approach of a proton to a hydrogen atom to form a hydrogen molecular ion.

electron-pair bond between the nuclei. This two-electron bond is not quite twice as strong as a one-electron bond, because of the electrostatic repulsion between the two electrons.

If a third hydrogen atom is brought up to this molecule, we could imagine the ground state spreading itself over all three nuclei. But there is room for only two electrons in it, and the third electron must therefore enter some higher quantum state, as in the lithium atom. The attraction of the third hydrogen atom is greatly weakened by this effect, and in fact it does not bond itself chemically to the other two. In recognition of this we say that H_2 is a chemically *saturated* molecule. It is the molecular equivalent of the helium atom.

Because it depends on quantum states which extend from one atom to another, the covalent bond reflects, far more than the ionic bond, the quantum restrictions which govern the electronic structures of atoms. It is predominantly an *electron-pair* bond, because each bonding quantum state between two atoms has room for two electrons with opposite spins. It is often a *directed* bond, because it is formed from quantum states which often extend preferentially in certain directions out of the atom (cf. Fig. 2.6). It is also a bond which can be *saturated*, because only a limited number of bonding quantum states can be formed between an atom and its neighbours.

These characteristics lead to many features of molecular chemistry. Because of the limited number of places available in a partly filled outer electron shell, an electron with N electrons in that shell can bond with only $8 - N$ neighbours by sharing electrons with them. It follows that in the halogen elements ($N = 7$) the atoms join up in pairs, as in the diatomic gases fluorine and chlorine, and as in bromine and iodine, which form at room temperature *molecular liquids* and *solids*, materials in which chemically saturated molecules (Br_2 and I_2) are held together weakly by van der Waals forces. Many organic substances, e.g. waxes, oils and polymers, similarly consist of chemically saturated covalent molecules held together mainly by van der Waals forces.

When $N = 6$, each atom can form electron-pair bonds with two neighbours. This is the condition for long chains of atoms to form, as observed in some forms of selenium and sulphur. When $N = 5$, the ability to bond with three neighbours leads in substances such as arsenic to *layer structures*, in which the atoms are covalently bonded into sheets, these sheets being held together mainly by van der Waals forces. Graphite is a similar layer structure, although in this case there is a fourth electron ($N = 4$ in carbon) which is *free* (see below) and provides the material with partly metallic properties. The ideal example of a covalent solid is diamond. In this, as also in silicon, germanium and grey tin, each atom is bonded equally to four neighbours by tetrahedrally directed bonds, and the entire crystal consists of a single covalent molecule, in the sense that every atom is linked to every other atom by a network of electron-pair bonds. The high strength possible in the covalent bond is shown by the great hardness and high melting point ($> 3500°C$) of diamond. Silicon carbide and many of the metallic carbides are also predominantly covalent substances of great hardness and high melting point.

Not all the possible covalent bond states are always used to form electron-pair bonds. In graphite only three of the four electrons are so used. Many substances, particularly those with $N < 4$, prefer to form *metallic* rather than electron-pair bonds. The metallic bond, easily recognizable from the electrical conductivity and optical properties of metallic substances, is very common among the heavier elements. It is a form of covalent bond in which the positively charged ionic cores of the metal atoms are attracted electrostatically to the electrons moving between them. Its particular feature is that these bonding electrons are *free*. They are no longer localized, in pairs, in particular regions between the atoms but move freely through the entire set of positive ions as a kind of free-electron "gas". The flow of this gas through the metal under an applied electrical field is responsible for the high electrical conductivity of metals.

The metallic bond is similar in strength to the electron-pair bond. For example, the measured binding energies of metallic lithium and sodium are 1·7 and 1·13 eV per atom, respectively. The corresponding energies of diamond and silicon are 7·4 and 3·7 eV per atom, but these covalent crystals have four bonding electrons per atom; expressed per bonding electron, these values become 1·85 and 0·925 eV, respectively, i.e. comparable with the monovalent metals. Some of the multivalent metals, particularly the *transitional* metals such as iron, chromium, nickel, niobium, molybdenum, tantalum and tungsten, have several bonding electrons per atom and form crystals of great strength and high melting points.

The factors which favour the metallic form of covalent bonding are the presence of far-ranging, weakly bonded electrons in the free atom and the existence of many alternative quantum states of similar energies for the outer electrons of the atom. The first of these is responsible for the metallic character of light elements such as lithium and sodium, in contrast, for example, to hydrogen, which prefers the electron-pair bond of the H_2 molecule. The second enables most of the heavier elements to be metals. The reason in the first case is that a far-ranging, weakly bonded electron is so weakly held by the electric field of its own ion, or of any single neighbouring ion of the same type, that it passes in its movements through the fields of several ions; in this way it loses its allegiance to any particular ion and wanders away. In the second case the reason is that there are more bonding quantum states than electrons to fill them. The metallic bond is thus an *unsaturated* covalent bond. An electron may jump from one bonding quantum state to another at the same energy, and so wander through the material, without ever finding the bonding states along its path already saturated with electrons.

These features of the metallic bond lead to several characteristic properties of metals and alloys. Since there is no saturation of the bonding states, the $8 - N$ rule no longer limits the number or type of neighbours which an atom may have. Close-packed structures are thus common. The bond that an atom makes with a single neighbour is weak because of the small number of bonding electrons per atom (with the exception of the multivalent transition metals), but each atom has such a large number of nearest neighbours that the overall cohesion is strong. Thus, metals compare well with ionic and other covalent solids as regards strength and melting point. Metals (e.g. many of the transitional metals) which can accommodate small atoms such as carbon and nitrogen in the spaces between their own atoms, without losing the metallic bond or the close packing of their own atoms, can be further strengthened by this method for further increasing the number of bonding electrons in them. This is the basis of the hard metallic carbides, nitrides, silicides, etc.

Since there is no saturation of bonding states, the normal valency restrictions which govern the chemical bonding of atoms no longer apply when different metals are brought together. This is the basis of the exceptional ability of metals to form *alloys*. Their atoms, especially if they are of similar sizes, are able to intermingle freely on a common crystal lattice over wide ranges of composition, so forming *solid solutions* (cf. Section 3.3). By continuously varying the composition, while preserving the same crystal structure, it is possible in such alloys to achieve a continuous range of mechanical, physical and chemical properties to suit a variety of purposes.

The tendency to close-packing causes many metals and alloys to crystallize in simple structures. We shall see later that it is the simplicity of these structures, together with the free electron bond, that is responsible for the unique plastic properties of metals and alloys. In certain alloys, however, advantage is taken of the differences in sizes of the participating atoms to form much more complex crystal structures in which an even closer packing of the atoms is achieved. These special packings can of course occur only at certain particular ratios of the participating atoms, e.g. $MgCu_2$, $CuAl_2$, KNa_2. Because they are formed at simple compositions and are made up of precisely arranged groupings of atoms, these materials are generally regarded as "intermetallic compounds", even though their formation is governed by the sizes rather than the valencies of their atoms. The *sigma phases* (e.g. FeCr), which make some stainless steels of high chromium content brittle, also have complex crystal structures based on atomic sizes.

FURTHER READING

SLATER, J. C., *Introduction to Chemical Physics*, McGraw-Hill, New York, 1939.

TEMPERLEY, H. N. V., *Properties of Matter*, University Tutorial Press, London, 1953.

FRENKEL, J., *Kinetic Theory of Liquids*, Oxford University Press, 1946.

MOELWYN-HUGHES, E. A., *States of Matter*, Oliver and Boyd, Edinburgh, 1961.

PAULING, L., *Nature of the Chemical Bond*, Cornell University Press, 1945.

RICE, F. O. and TELLER, E., *The Structure of Matter*, Wiley, New York, 1949.

TURNBULL, D., *Trans. AIME*, **221**, 422 (1961).

PROBLEMS

2.1   By considering the size of region round each particle into which the centre of no other particle may enter, and by noting that such a region round one particle can enter that round another, show that β in van der Waals' equation is approximately four times the volume of the molecules.

2.2 In a container of total volume V_1 is a partition which separates a region of volume V_2 from the remainder, $V_1 - V_2$. In this region V_2 is 1 mole of a perfect gas at a temperature T. The partition is removed and the gas particles allowed to spread themselves randomly throughout the whole volume V_1. By considering the probability that any given particle will then be found in the smaller volume V_2, and by supposing that each particle moves independently of the others, show that when a gas is expanded isothermally and reversibly from V_2 to V_1, its entropy increases by $R \ln (V_1/V_2)$. Discuss the significance of this in relation to the work of isothermal compression, given in Problem 1.8.

2.3 One mole of gas, between the particles of which there is negligible attraction, has a volume of 2486 cm³ at a pressure of 10 atm and of 246,000 cm³ at 0·1 atm and at the same temperature. Making use of the result given in Problem 2.1, calculate the approximate diameter of a gas particle.   [2·7 Å]

2.4 The *pressure coefficient* k_p of a gas heated at constant volume is defined in terms of the gas pressures p_1 and p_0 at absolute temperatures T_1 and T_0, respectively, by the formula

$$p_1 = p_0[1 + k_p(T_1 - T_0)].$$

Derive an expression for k_p for a van der Waals gas and show that, for a given density of such a gas, k_p is independent of temperature.

$$\left[k_p = \left(1 + \frac{\alpha}{p_0 V^2} \right) \frac{1}{T_0} \right]$$

For hydrogen, when p_0 and T_0 refer to NTP, it is found that

$$k_p = 0 \cdot 00366254.$$

Discuss the physical significance of this value.

2.5 The critical point of a gas occurs at a temperature of 10°C, a pressure of 51 atm and a density of 0·22 g cm⁻³. Assuming that van der Waals' equation is valid, calculate the molecular weight of the gas.   [37·4]

2.6 Using the argument of Section 1.2, show that the number of gas particles which strike unit area of a surface per second is approximately

$$z = p/(4mkT)^{1/2}.$$

Then show that the rate of evaporation from a condensed phase into a vacuum is approximately $G = p(M/4RT)^{1/2}$ g cm⁻² sec⁻¹, where p is the vapour pressure and M the molecular weight. Note that more refined calculations, which allow for the angular distribution of velocities, give $z = p/(2\pi mkT)^{1/2}$ and $G = p(M/2\pi RT)^{1/2}$.

2.7 Calculate the second ionization energy (i.e. the energy to remove an electron from a singly charged positive ion) for helium.   [54 eV] Why is the first ionization energy for helium smaller than this, but larger than that for hydrogen?

2.8 Calculate the characteristic orbit radius of the $1s$ state in a uranium atom. [0·0058 Å]

2.9 Calculate the kinetic energy, potential energy and speed of an electron at the characteristic distance $\bar{r}$ (Equation (2.24)) from the nucleus in the ground state of a hydrogen atom.

[13·5 eV, -27 eV, $2·2 \times 10^8$ cm sec^{-1}]

chapter 3

Structure of Crystals

We now consider the geometrical laws which govern the arrangements of atoms and molecules in crystals. First we discuss the framework, or space lattice, of a crystal and the various lattices permitted by the requirements of symmetry. Then we enquire into how the lattice framework is filled up with atoms to produce a crystal. Most real crystals are flawed, but the regularity of the crystal structure also imposes itself on the structures of the various imperfections that can exist in crystals, which can thus be defined and described in a precise, systematic way. In the final sections of this chapter we shall survey some of the imperfections which are of most importance in crystals, i.e. foreign atoms, vacant atomic sites, interstitial atoms, stacking faults, twin interfaces, edge and screw dislocations, and crystal boundaries in polycrystals.

3.1 SPACE LATTICES

Crystals are the outstanding examples of *orderliness* in nature. In a crystal the site of any particle is fixed precisely by the positions of its neighbours. Many geometric figures have this property. For example, the position of any corner in a regular pentagon is fixed precisely by the positions of the other corners. Each individual obeys the rule of pattern set by its neighbours. The regularity of the crystal goes beyond this, however. In a perfect crystal the pattern goes on *indefinitely* in all directions, repeating itself identically in every respect at every site, just as the pattern repeats itself periodically on a wallpaper. Having a repeatable pattern, the crystal can grow indefinitely by simply adding more particles to continue this pattern. Only certain geometric figures can be repeated in this way. A square obviously can, but a regular pentagon cannot, as Fig. 3.1 shows. Each site is correctly placed in one pentagon, but the

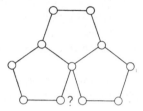

Fig. 3.1. A pattern which gives dissimilar sites when repeated.

pentagons cannot be fitted together without producing non-pentagonal groups of sites between them.

What is the general condition for a pattern to be infinitely repcatable? It is that it must form a *space lattice*. To make a space lattice, we take three *basic* or *unit* vectors, **a**, **b** and **c**, as in Fig. 3.2, and apply each in turn to a fixed point chosen as the origin of the lattice. The ends of these vectors then mark the positions of three other *lattice points*. We apply the same vectors again to each of these points and so locate more lattice points. We repeat this continually at each new lattice point until we have made an infinite three-dimensional array of points. This is a space lattice. We notice that the view of lattice points is always the same, seen from any

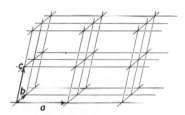

Fig. 3.2. A space lattice.

lattice point; also that each lattice point is a *centre of symmetry*, i.e. for each lattice point on one side of it there is always another identically placed on the opposite side of it. A real crystal of course, being finite in size, occupies only a limited region of this infinite framework.

A space lattice can equally well be constructed from the points of intersection of three intersecting families of parallel planes, all planes in one family being equidistant from one another. The *unit cells* into which space is divided by these planes are called *primitive cells*; they have lattice points at their eight corners and so possess the equivalent of one lattice point each. Some crystals are more conveniently described, however, by *non-primitive* cells, with more than one lattice point in each cell. An example is the *cubic* cell of the *face-centred cubic* lattice, shown in Fig. 3.7(a). This cell, which displays the cubic symmetry of the lattice, possesses four lattice points (i.e. a one-eighth share in each of the eight points at its corners and a one-half share in each of the six points at the centres of the cube faces).

Lattices are classified according to their symmetry. Bravais (1848) showed that fourteen different classes of these, now called *Bravais lattices*,

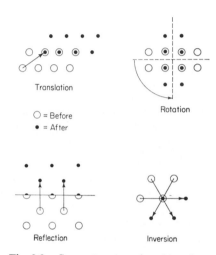

Fig. 3.3. Symmetry operations in a plane.

are possible. To study the symmetry of a two-dimensional plane lattice, or *Bravais net*, we can, for example, make copies of the net on paper, place one on top of the other with their lattice points in coincidence, and then find how many different movements (e.g. translations, rotations and turning over into mirror-image orientations) we can give the top sheet to carry its points into new positions of coincidence. Each movement which brings a lattice into self-coincidence is called a *symmetry operation*. Figure 3.3 shows examples. They consist of *translations* by whole numbers of lattice vectors; of *rotations* about axes which may or may not pass through lattice points; of *reflections* into a mirror-image orientation about a *reflection line* (two-dimensional nets) or *reflection plane* (three-dimensional lattices); and of *inversion*, in which each lattice point is taken through a point which is called a *centre of symmetry* to an equidistant position on the opposite side of this centre.

Every Bravais lattice possesses both a centre of symmetry and translation symmetry, the latter from the fact that the same vector

$$\mathbf{r} = l\mathbf{a} + m\mathbf{b} + n\mathbf{c}, \tag{3.1}$$

where l, m, and n are integers, describes both the positions of lattice points and translation symmetry operations. One lattice, the *triclinic*, has no more symmetry than this, but all others possess some rotation and reflection symmetry. There are, for example, nine orientations for a reflection plane through the centre of a cubic unit cell. The thirteen rotation

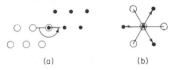

(a)   (b)

Fig. 3.4. Symmetry in Bravais nets: (a) necessary twofold symmetry; (b) increase of symmetry by inversion.

axes which also pass through the cube centre, about which the cube can be rotated into self-coincidence, consist of three *fourfold* axes through centres of opposite cube faces, four *threefold* axes through opposite corners and six *twofold* axes through centres of opposite edges. An *n-fold* axis is one about which a rotation through an angle $2\pi/n$ brings self-coincidence. Since n such rotations total 2π, which is always a self-coincidence rotation, n must be an integer.

We now examine the symmetry of a Bravais net. Because each lattice point is a centre of symmetry, there is a twofold rotation axis (perpendicular to the net plane) through each lattice point (Fig. 3.4) and there is also one through the midpoint between two lattice points (not always true in three dimensions). It follows that if there is, for example, a threefold axis, then this is also a sixfold axis, since each three points symmetrically placed round it can be inverted to give three more, as shown in Fig. 3.4.

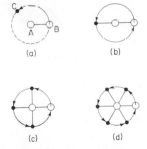

(a)   (b)

(c)   (d)

Fig. 3.5. (a) Generation of lattice points by rotation: (b) twofold axis; (c) fourfold axis; (d) sixfold axis.

Generalizing, if there is an n-fold axis through a lattice point in a Bravais net, and n is odd, then there is also a $2n$-fold axis through that point.

To find the permissible rotation axes, we select two nearest neighbour lattice points, A and B, in the net and rotate B about the axis through A to generate points such as C in Fig. 3.5. The direct distance from B to C cannot be less than that from A to B, since the latter are nearest neighbours. Hence, $n > 6$ is impossible. Similarly, $n = 5$ is impossible because it implies $n = 10$. Thus, since $n = 3$ implies $n = 6$, the only axes through a lattice point in a Bravais net are two-, four- and sixfold.

By stacking replicas of each type of net on top of one another in regular stacking sequences, the fourteen Bravais space lattices can be constructed. These are shown in Fig. 3.6 and their main features summarized in Table 3.1.

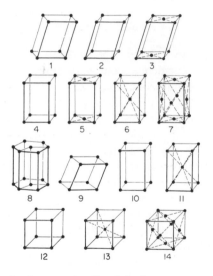

Fig. 3.6. Conventionally chosen unit cells of the fourteen space lattices. 1, Triclinic; 2, monoclinic, simple; 3, monoclinic, base-centred; 4, orthorhombic, simple; 5, orthorhombic, base-centred; 6, orthorhombic, body-centred; 7, orthorhombic, face-centred; 8, hexagonal; 9, rhombohedral; 10, tetragonal, simple; 11, tetragonal, body-centred; 12, cubic, simple; 13, cubic, body-centred; 14, cubic, face-centred.

3.2   CRYSTAL STRUCTURES

The change from space lattice to crystal structure is made by placing identical particles (single atoms, groups of atoms, or molecules) identically round each lattice point. These particles, particularly if complex molecules, may have unsymmetrical shapes, in which case some of the

TABLE 3.1
The Seven Crystal Systems and Fourteen Space Lattices

System	Features	Lattices
Triclinic	Three axes not at right angles, of any lengths	(1) Simple
Monoclinic	Three axes, one pair not at right angles, of any lengths	(1) Simple (2) Base-centred
Orthorhombic	Three axes at right angles, of any lengths	(1) Simple (2) Base-centred (3) Body-centred (4) Face-centred
Tetragonal	Three axes at right angles, two equal	(1) Simple (2) Body-centred
Cubic	Three axes at right angles, all equal	(1) Simple (2) Body-centred (3) Face-centred
Hexagonal	Three axes in a plane at 120° to each other, and equal; fourth axis at right angles to them, of any length	(1) Simple
Rhombohedral	Three axes equally inclined, not at right angles, all equal	(1) Simple

symmetry (rotation, reflection, inversion) of the space lattice may no longer exist in the crystal.

The simplest crystals are obtained by placing single atoms at lattice points; or additionally at symmetrical positions between these points. Figure 3.7 shows some simple crystal structures with cubic lattices. The atomic spheres, which in reality are large enough to touch one another, have been reduced almost to points in these diagrams, for clarity.

The f.c.c. is a common structure for elements whose atoms bond without valency restrictions, such as metals and inert gases; the *co-ordination number*, i.e. number of nearest neighbours to a particle, is high (12), in agreement with the lack of valency restrictions on the number of neighbours a particle may have. The lower co-ordination number (8) of the b.c.c. structure is to some extent compensated by the next nearest neighbours (6), which lie only slightly further away. The diamond structure has a low co-ordination number (4), in accord with the tetravalent bonding between the atoms of elements that form this structure. The characteristic tetrahedral pattern of neighbours round an atom emphasizes this aspect of the structure. The simplest crystals of chemical compounds have structures of the sodium chloride or caesium chloride types. In these each atom has atoms of its chemical partner as nearest neighbours, in accord with their chemical affinity. The pattern of atomic *sites* is simple cubic in

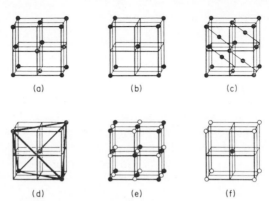

Fig. 3.7. Some cubic crystal structures (open and full circles represent chemically different atoms).

(a) Face-centred cubic (f.c.c.) (Cu, Ni, Al, Pb, Au, Ag, Pt, Fe at high temperatures, solid inert gases);	(b) Body-centred cubic (b.c.c.) (Li, K, Na, W, Cr, Mo, V, Nb, Fe at low temperatures);	(c) Diamond (C, Si, Ge, grey Sn);
(d) Tetrahedron round atom in diamond structure;	(e) Sodium chloride (NaCl, KCl, MgO, many alkali halides);	(f) Caesium chloride (CsCl, CsBr, NiAl, ordered CuZn).

sodium chloride and b.c.c. in caesium chloride. However, the Bravais lattice is f.c.c. in sodium chloride and simple cubic in caesium chloride, in both cases with two atoms per lattice point, one from each partner of the compound.

Figure 3.8 shows the close-packed hexagonal (c.p. hex.) structure. Equal hard spheres can be packed together equally closely in the c.p. hex. and f.c.c. structures, and there is no other crystal structure which packs together a large number of such spheres so closely. The c.p. hex. structure consists of hexagonal close-packed layers (cf. Fig. 3.4(b)) stacked on top of one another in an alternating staggered sequence. The atomic sites do not form a space lattice, since they are not centres of symmetry. To construct the lattice, we must take two atoms per lattice point, from neighbouring hexagonal layers. Ideally, the c.p. hex. structure has a co-ordination number of 12 and the hexagonal cell has a height 1·633 times the distance between neighbours in a hexagonal layer. In metals a distorted form is often observed in which the axial ratios differ from 1·633 (e.g. Zn, Mg).

Fig. 3.8. Close-packed hexagonal structure (Mg, Be, Zn, Cd, Co, Zr, Ti, He).

The f.c.c. and c.p. hex. structures are both made up of close-packed hexagonal layers stacked on top of one another. In the f.c.c. these layers are called *octahedral planes*, because an octahedron can be formed from them as shown in Fig. 3.9, and in the c.p. hex. they are called *basal planes*. The stacking sequence in the two structures can be understood from Fig.

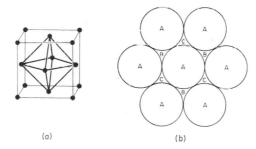

(a)   (b)

Fig. 3.9. Close-packing: (a) an octahedron of close-packed planes inscribed in the f.c.c. cell; (b) plan view of close-packed layers.

3.9(b). The close-packed layer of spheres in A positions provides, above the hollows between three touching spheres, two distinct sets of positions, B and C, for the spheres of the next layer. Any sequence of stacked layers such as ABACBCABAC ... , which avoids configurations such as ... AA ... , is fully close-packed. The two simplest close-packed crystalline sequences are ABCABCABC ... (f.c.c.) and ABABAB ... (c.p. hex.).

3.3 GEOMETRICAL IMPERFECTIONS IN CRYSTALS

A real crystal is rarely, if ever, perfect. Typically, however, less than one atom in about ten thousand is out of place. Such minor imperfections may appear trivial, but in fact some of the most important properties of crystalline materials, e.g. strength of steel, optical sensitivity of photographic emulsions, colour of gemstones and semi-conductivity of non-metals, depend vitally upon what happens at the rare places where the crystal structure goes wrong.

These places are called *crystal imperfections* or *lattice defects*. We exclude elastic deformation of the lattice from this category, because it is not an inherent feature of a crystal but is the response of the crystal to an applied force and disappears when the source of that force (which may be an externally applied load or an imperfection, such as a foreign particle too large for its hole in the crystal) is removed. An imperfection is an irregularity that persists through any elastic distortion which the crystal may suffer, just as a hole in a piece of rubber persists when the rubber is pulled into various shapes. There are imperfections in the *occupation* of crystal sites and imperfections in the *pattern* of sites. In the first the sites may be perfectly positioned but not all identically occupied (cf. a missing brick in an otherwise perfectly regular brick wall). In the second the sites may be identically occupied but not all perfectly positioned (cf. irregularities of pattern at the junctions of badly aligned strips of wallpaper).

The main imperfections of occupation (other than the surface of the crystal) are *foreign atoms* and *point defects*. No substance is completely pure. Even the purest germanium and silicon crystals for transistors and other electrical uses contain one part in 10^{10} of important impurities. Ordinary "pure" substances contain usually $0 \cdot 01$–1% of impurities. In *alloys* foreign atoms are added to a metal deliberately, often in large amounts (e.g. 1–50%), to obtain special properties. Foreign atoms can exist in the host crystal in various forms. They may be grouped into clusters or even exist as small distinct crystals, such as *inclusions* or *precipitates* inside the main crystal. Often they are dispersed as individual atoms (or molecules) throughout the crystal, and they are then described as *solute* in *solid solution*. If sufficiently small compared with the size of the host or *solvent* atoms, these foreign atoms usually form an *interstitial* solid solution, i.e. occupy sites between the host atoms (Fig. 3.10). For example, carbon atoms in steel fit themselves into the spaces between the array of iron atoms. When fairly similar in size to the solvent, the solute atoms usually occupy some of the sites of the main crystal lattice itself, i.e. form a *substitutional* solid solution (Fig. 3.10). For example, the zinc and copper atoms in α-brass (e.g. 70%Cu, 30%Zn) coexist substitutionally on an f.c.c. array of sites.

The simplest and most important point defects are the *vacancy* and the *interstitial*. The first is an empty lattice site, the second an atom of the parent crystal substance lodged in an interstitial site. A pure crystal which contains vacancies and interstitials is, in effect, a solid solution with space as substitutional "solute" and with some of the crystal substance itself as interstitial "solute".

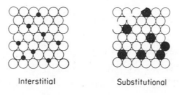

Interstitial   Substitutional

Fig. 3.10. Solid solutions.

One of the most direct ways of producing vacancies and interstitials is by *radiation damage*. The crystal is bombarded by fast particles, e.g. neutrons in a nuclear reactor, which pass through it and occasionally collide with one of its atoms. Such collisions are usually so violent that the struck atom is knocked out of its crystal site, which becomes a vacancy, and injected into the nearby lattice, where it eventually comes to rest as an interstitial (Fig. 3.11), often after having itself knocked some other

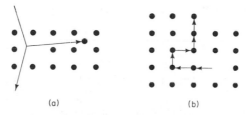

(a)   (b)

Fig. 3.11. (a) Creation of a vacancy–interstitial pair by a knock-on collision. (b) Creation of vacancies at a surface, which subsequently migrate into the interior.

atoms off their sites. In graphite, for example, about 3 % of the atoms are displaced per year by neutron bombardment in a medium-power reactor, with important consequences for the use of graphite as a reactor material (change of dimensions, diminution of thermal conductivity, spontaneous energy releases and temperature increases, increased rate of oxidation, hardening and embrittlement).

Vacancies exist naturally in all crystals at high temperatures. Just as matter "dissolves" in space at high temperatures, i.e. forms vapour, so also does space "dissolve" in matter by processes such as that shown in Fig. 3.11(b). The same exponential law that governs the equilibrium density of vapour particles in space also governs (though with a different energy of formation) the equilibrium density of vacancies formed thermally in a crystal. If the energy of vacancy formation is E_F, per mole of vacancies in dilute solution, the number n of vacancies among N crystal sites is given by

$$n = \alpha N \, e^{-E_F/RT}, \qquad (3.2)$$

where $\alpha(\simeq 10)$ is a numerical factor.

Values of E_F are given in Table 3.2. For most metals E_F is about one-third of the binding energy. At the melting point about one site in 10^4 or

TABLE 3.2
Average Experimental Values of Energies to Form (E_F) and Move (E_M)
Vacancies, the Activation Energy of Self-diffusion (E_D), and the Binding Energy
(ϕ), in kcal mole^{-1}

	Au	Cu	Ag	Al	Pt	Ni	Graphite
E_F :	23	(28)	25	18	30	35	70–85
E_M :	18	(19)	(21)	13	28	(23)	60–90
$E_F + E_M$:	41			31	58		130–175
E_D :	42	47	46	32	58	64	160
ϕ :	92	81	68	55	127	85	171

10^5 is usually vacant. By cooling a sample very rapidly from a high temperature, e.g. by quenching a hot metal filament in water, it is often possible to retain these vacancies at low temperatures in a state of super-saturated solution. If the sample is then *aged* or *annealed* at a moderate temperature, to allow the vacancies to diffuse about inside it, it is possible to detect the changes of physical properties, such as length and electrical resistivity, brought about by the escape of these vacancies or by their grouping into clusters inside the material. In this way the properties of vacancies can be studied. Individual vacancies have been seen in the field-ion microscope and vacancy clusters in the electron microscope.

Equation 3.2 with other values of E_F can be applied to the formation of various types of point defects in crystals. We see that defects with a large E_F cannot exist in thermal equilibrium in a crystal of finite size. For example, at the melting point T_m of a crystal of 10^{22} sites, $n < 1$ when $E_F > 53RT_m$. For copper, $53RT_m = 144$ kcal mole^{-1}. Thus, only small defects, such as vacancies, with small values of E_F can exist in large numbers in thermal equilibrium. Even thermally formed interstitials cannot exist in significant numbers in metals such as copper, silver and gold, because too much strain energy (about 100 kcal mole $^{-1}$) is required to push the surrounding atoms aside to make room for them.

The most obvious place for vacancies to form thermally is at the free surface; they can then move into the crystal by the process shown in Fig. 3.11(b). In substances which cannot form thermal interstitials, thermal vacancies cannot be created inside the crystal, because they would there have to be created as vacancy–interstitial pairs. They can be created only at certain vacancy *sources*, e.g. free surfaces and grain boundaries (see below), where atoms can be moved aside to make vacancies without their becoming interstitials.

3.4 STACKING FAULTS, TWINS AND DISLOCATIONS

Imperfections of pattern are usually large faults, affecting thousands or millions of atoms. As a simple example, we consider the stacking of close-packed layers. We saw in Section 3.2 that random sequences such as

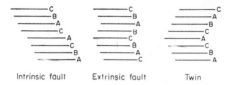

Intrinsic fault   Extrinsic fault   Twin

Fig. 3.12. Side view of close-packed f.c.c. layers, showing stacking faults and a twin.

ABACBCABA ... are fully close-packed. If the atomic forces insist only that the atoms be closely packed, such random sequences should be preferred, since they have higher entropy than ordered ones. We thus take the fact that nature prefers the ordered f.c.c. and c.p. hex. sequences to mean that atoms are to some extent sensitive to the positions of their more distant neighbours. These long-range atomic interactions are, however, fairly weak, with the result that in at least some close-packed crystals (e.g. Co; f.c.c. stainless steel; plastically deformed Au and Cu; and f.c.c. alloys based on these metals) the ordered sequence of close-packed planes contains *stacking faults*. Figure 3.12 shows faults in the f.c.c. sequence. The *intrinsic* fault is equivalent to the omission of one layer from the sequence, the *extrinsic* to the insertion of an extra layer. Both have a few layers of c.p. hex. packing in the fault itself; CACA in the first; BAB and BCB in the second. Closely related to them is the *twinning fault*, an interface across which the stacking sequence inverts into mirror-image orientation. Twinning faults are very obvious in crystals, because a single such fault forms the boundary between two *twin* crystals, easily recognizable by their large difference of lattice orientation.

The occurrence of stacking faults and annealing twins in the metals mentioned above is a consequence of free-electron bonding (cf. Section 2.8). The positively charged ions are all drawn together by the free-electron gas in which they are embedded. To make the total volume as small as possible, with the least possible overlapping of their closed electronic shells, the ions crystallize in close-packed arrays. The various close-packed stacking sequences are almost equally acceptable as regards the free-electron energy and the filled-shell overlap energy.

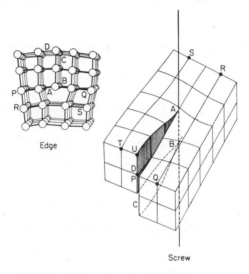

Edge

Screw

Fig. 3.13. Crystal dislocations.

The most important imperfection of pattern is the *dislocation*. Dislocations vitally affect the mechanical strength and plasticity of crystals, particularly metals and alloys. They are also important in crystal growth, chemical reactivity, and the electrical and magnetic properties of crystals. To understand dislocations, we recall a well-known problem of architecture: the building of a curved structure such as an arch from ordinary bricks. This is solved by matching $n + 1$ bricks along one curved layer against n along the next. A similar principle applies in crystals and allows them to exist in various bent and twisted shapes, such as may be produced by plastically deforming a soft metal wire. Dislocations are the imperfections of crystal architecture which produce these curvatures.

Figure 3.13 shows two basic types of dislocations. For ease of illustration we consider only a simple lattice. Each dislocation is a *line defect* which runs through the crystal from A to B. The *edge* dislocation, named from the edge AB of the half-plane ABCD, matches $n + 1$ crystal sites along PQ against n along RS and is directly analogous to the curved brick arch. The *screw* dislocation is analogous to a spiral staircase. We can understand its structure by imagining an initially perfect crystal, in the second diagram of Fig. 3.13, to be cut by a vertical knife along the surface ABCD, as far as the line AB of the intended dislocation, and this dislocation then made by sliding one side of the cut face one crystal spacing vertically downwards, past the other, before rejoining it to the other side. For example, the atom at Q, once bonded to U, is now bonded to P. The spiral structure round the dislocation line can be seen by moving round a path such as PQRSTU. The crystal layers pierced by the dislocation are no longer separate sheets but successive leaves of a single spiral surface which composes the entire crystal. This spiral surface ends at two *ledges* AD and BC, which run across the top and bottom faces of the crystal as far as the *points of emergence*, A and B, of the dislocation line.

Many ways of detecting and observing dislocations in crystals are now known. Dislocation lines have been seen in practically all types of crystals under the electron microscope. The field-ion microscope, and in special cases the electron microscope also, has enabled the atomic structure at the centre of a dislocation to be seen.

The dislocations of Fig. 3.13 are *perfect, unit* dislocations of the crystal lattice. Only at the very centre of such a dislocation is the crystal "bad" in the sense that the original lattice there has become distorted beyond recognition; outside this centre the original lattice is clearly recognizable from the crystal structure, which remains "good" despite its elastic strains. This is the basis of *Frank's method* of defining a perfect dislocation. Referring to Fig. 3.14, we arbitrarily choose A → B as the positive direction

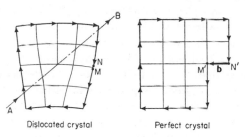

Dislocated crystal   Perfect crystal

Fig. 3.14. Definition of a dislocation.

along the dislocation line AB and then make a clockwise closed circuit round this line, M → N, moving always through good crystal along local lattice vectors. We then make a second circuit, M' → N', following *exactly* the same sequence of lattice vectors in a similarly oriented piece of perfect crystal. If this second circuit *fails to close*, then AB is a dislocation line.

The *extra vector* N′ → M′ needed to complete the second circuit is called the *Burgers vector* **b** of the dislocation and is its most important feature. The Burgers vector is independent of the circuit chosen to define it and of the point on the dislocation line round which the circuit is made. A single dislocation line has the same Burgers vector at all points along its length, even if the line itself follows an arbitrary curved path through the crystal. The dislocation line is an edge when perpendicular to its Burgers vector; a screw when parallel; and a *mixed* type at intermediate angles. We notice that a dislocation line cannot end inside a crystal, because the Burgers vector, being constant, cannot become zero at some point along the line. The line must run out to the free surface, or be a closed loop inside the crystal, or be joined to other dislocation lines forming a network through the crystal.

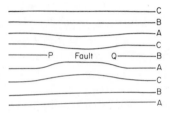

Fig. 3.15. Stacking fault and imperfect dislocations, P and Q, produced by removal of part of one layer from an f.c.c. sequence.

A unit dislocation has a Burgers vector equal to a unit lattice vector. Other Burgers vectors are possible, provided they leave the atoms of the crystal in mechanically stable positions. Any dislocation with a Burgers vector equal to a whole number of lattice vectors (cf. Equation (3.1)) is a *perfect* dislocation. In some crystals *imperfect* dislocations also occur, in which **b** is not a lattice vector. Mostly these are *partial* dislocations, in which **b** is smaller than a unit lattice vector. An imperfect dislocation cannot be entirely surrounded by good crystal, because at some point in any circuit round it we must pass from one crystal site to another along a vector which is not a lattice vector. Imperfect dislocations are in fact found along the edges of stacking faults which do not run completely across a crystal. When taking a circuit round an imperfect dislocation, we must start and end at the stacking fault, in order not to pass through bad crystal.

Figure 3.15 shows a simple example of imperfect dislocations and associated stacking fault in an f.c.c. crystal. By removing part of the central B layer and collapsing the neighbouring layers together, we obtain a stacking fault of CACA type with imperfect dislocations at its ends,

P and Q. Dislocations of this type are sometimes found in f.c.c. metals which have been quenched from high temperatures. The supersaturated vacancies caught in the crystal by the rapid cooling cluster together to form disk-shaped holes on close-packed planes. When these disks grow large, their sides collapse and they become closed loops (typically about 10^{-5} cm across) of imperfect dislocation line. The observation of such loops in the electron microscope provides one method of studying effects of vacancies in crystals.

Edge dislocations are geometrically capable of acting as *sources* and *sinks* of vacancies. Atoms may, for example, migrate from lattice sites near the dislocation into positions on the edge of the half-plane of the dislocation, so leaving their initial sites vacant, as shown in Fig. 3.16.

Fig. 3.16.   Creation of vacancies by the migration of atoms to an edge dislocation.

Conversely, vacancies may migrate to the edge of the half-plane and anni-hilate themselves there. One result of these processes is that the half-plane grows longer or shorter as atoms are added to or removed from its edge. The position of the dislocation in the crystal thus alters. This so-called *climb* motion of edge dislocations, which plays an important part in the annealing of badly distorted crystals, must not be confused with the *glide* motion of dislocations, which is responsible for the plastic properties of crystals and which we shall discuss in Chapter 9. Screw dislocations can also absorb vacancies; the observed effect in this case is that the dis-location line twists up into a helix to become partly of edge character, in which form it can absorb the vacancies.

3.5 POLYCRYSTALS

Many crystalline materials exist normally as conglomerates of small crystals. Such conglomerates are called *polycrystals*. Most ordinary pieces of metal consist of large numbers of small irregular-shaped crystals, called *grains*, joined together at *grain boundaries*. Occasionally this grain structure can be seen quite easily by eye, e.g. on an old, well-worn, brass door-knob, but usually special methods are needed to display it. In the *metallographic* method, which is widely used for examining metals and alloys, a flat surface is ground on the specimen, then made smooth by rubbing with successively finer grades of emery paper, then polished with a fine abrasive powder (e.g. alumina or magnesia) on a damp, soft cloth, and finally lightly etched with a dilute acid or other chemical reagent. The grains can then be seen, by eye or by means of a reflection microscope, from the way they reflect a beam of light in various directions according to their crystal orientations. In the *petrological* method, used to study transparent minerals, thin slices are cut and viewed in the microscope by transmitted light.

A polycrystal must not be thought of as a loose conglomerate of un-connected grains, like a heap of sand. In a properly prepared metal, for example, there is no free space between the grains; each internal grain is joined at every point of its boundary to its neighbours. It is known, e.g.

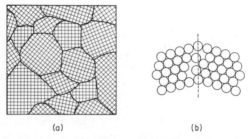

(a)   (b)

Fig. 3.17. Polycrystals: (a) polyhedral grains joined by boundaries; (b) atomic structure in a boundary.

from direct observation in the field-ion microscope, as well as by indirect methods, that the grain boundary in a pure metal is a thin transition region, one or two atomic diameters thick, across which the orientation of atomic packing changes sharply from that of the one crystal to that of the other (Fig. 3.17). The crystal structure of the grains continues almost perfectly, right up to the boundary, and the atoms in the boundary are packed about as closely as the irregularity of the structure there will allow, in positions which are a compromise between the rival crystal sites on the surfaces of the adjoining grains.

When a liquid such as a metal casting is allowed to freeze in a mould, crystallization usually begins from many points, particularly points near the mould wall, which are the first to cool. These crystal *nuclei* grow at first with long branching arms called *dendrites* spreading out through the liquid from them. Figure 3.18 illustrates this. These dendrites continue to grow (along certain crystal axes) until prevented by running into den-drites from other nuclei. The remaining liquid then crystallizes on to the sides of the dendrites, which thicken up and become grains. The grain boundaries are then the regions between the grains where the last traces of freezing take place. The grains themselves have irregular polyhedral shapes which bear little or no relation to the symmetry of their crystal structure but are determined by the almost accidental distribution of the nuclei and dendrites and by the competition between one grain and another for the last traces of freezing liquid. The popular concept of a crystal as a solid bounded by beautifully regular plane surfaces parallel to its main crystallographic planes, does not apply to the grains of a polycrystal. Crystals with crystallographic faces can be grown even in metals, however, provided that growth is allowed to occur very slowly (e.g. from a vapour or dilute solution) on isolated seed crystals.

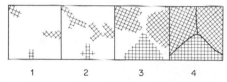

1   2   3   4

Fig. 3.18. Schematic view of stages in the freezing of a liquid to a polycrystal by the nucleation and growth of dendrites.

We expect the preferred structure of a grain boundary to be one which is as economical of energy as possible. If the change of crystal orientation across the boundary is small (less than about 10°) a *semi-coherent* boundary structure is possible. Through most parts of the boundary the crystalline structure continues coherently, although with change of direction, from the one crystal to the other. Alternating with these coherent regions are incoherent regions of the boundary across which there is a distinct break of crystalline continuity. In many cases these incoherent regions are simply dislocation lines. The entire boundary can then be regarded simply as a surface containing a regular array of dislocation lines. Figure 3.19 shows a boundary of this type, which consists of a "wall" of edge dislocations regularly arranged one above another. Dislocation boundaries of this and other types are commonly seen in imperfect crystals. There is an obvious connexion between the spacing h of the dislocations, the length b of the Burgers vector and the difference in orientation θ between the crystals. When θ (in radians) is small, this relation is

$$\theta = b/h. \tag{3.3}$$

Measurements of small-angle boundaries have accurately confirmed this.

In polycrystals the boundaries are mostly of the large-angle type (e.g. 10–40°). The structure of these can still be described by means of dislocations, but there is not much advantage in this, because the dislocations now lie so close together that it is difficult to distinguish one from another. It is better to picture such a boundary, like that of Fig. 3.17, as a disordered monolayer of material between the grains, with as much crystallographic coherence from one grain to the other as the change of orientation will

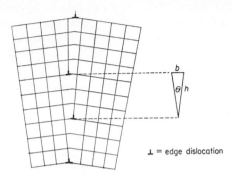

Fig. 3.19. A simple small-angle boundary formed from edge dislocations.

allow. Despite its irregular structure, such a boundary is usually mechanically coherent. In fact, grain boundaries are often sources of mechanical strength in solids, and rather special conditions are necessary before a polycrystal will fracture along them.

3.6 SOLID SOLUTIONS AND MULTIPHASE POLYCRYSTALS

When a liquid containing some dissolved foreign substance is allowed to freeze, the first crystals to form are usually much purer than the liquid from which they have grown. A simple reason for this is that the liquid, having a disorganized structure, can accommodate foreign atoms or molecules of odd sizes with less strain than can the crystal, with its strictly controlled lattice dimensions. As a result, the foreign particles tend to avoid the crystal, since they would have higher energy there. This tendency is of course opposed by the entropy, which tries to spread the solute randomly throughout the whole system, liquid and solid. At a given temperature an equilibrium is reached between dilute solid solution and more concentrated liquid solution. The solute rejected by the crystals becomes concentrated in the remaining liquid, the composition of which thus becomes even more unfavourable for crystallization. As a result, the freezing point of the liquid decreases. The sharp freezing point of the pure substance is thus broadened out, through the presence of the dissolved substance, into a range of freezing temperatures. On further cooling, more layers of material crystallize out, slightly richer in solute than before but still purer than the liquid, on to the crystals. This process continues until, at a sufficiently low temperature, the last traces of liquid, now highly concentrated with solute, freeze in the regions near the grain boundaries. There is thus a gradual increase of solute concentration from the centres to the boundaries of the grains.

These effects have many practical consequences. Salt or calcium chloride is thrown on icy roads to lower the freezing point of water and so thaw them. Arctic explorers observed that fairly pure water could be obtained by digging out the centres of sea-ice crystals. Extremely pure crystals are manufactured by the process of *zone-refining*, in which the impurities are swept along to the far end of a solid bar by running a succession of heated, molten zones along it. Plumbers exploit the wide freezing range in lead–tin alloys to maintain their solder in a "pasty" condition, part liquid and part solid, in which form they can "wipe" it on joints. To set against these advantages, there is the fact that when a polycrystal is heated, it may fall apart, by local melting of fusible impure regions at the grain boundaries, at a temperature well below the average melting point of the material. In metals and alloys this effect is known as *hot-shortness* and often causes difficulties when they are forged or heated at high temperatures.

If, during freezing, the remanent liquid becomes sufficiently rich in solute, it often freezes as a distinctly different crystalline phase from that of the solvent. The mechanical strength of copper, for example, is destroyed by the presence of a trace of bismuth in the metal, because a film of brittle, low-melting-point bismuth metal then forms along the copper grain boundaries. In many cases the remanent liquid freezes as a compound; e.g. the intermetallic compound $CuAl_2$ in aluminium–copper alloys (duralumin). In alloys containing several substances and in the complex, impure minerals of geology the freezing process is often very complicated, and many different types of crystals separate out in the solid.

A common result of multiphase crystallization is the *eutectic* structure. Suppose that we have a liquid solution of a substance B in A and that, on freezing, fairly pure A separates out. As the temperature falls, the remaining liquid becomes richer in B, so that eventually we might think of this liquid as a solution of A in B. This suggests a paradox, since we might then expect B to crystallize out, not A. What is observed in such cases is that, for all liquids up to a certain critical content of B in A, the A-rich phase separates out first on freezing and, for all liquids containing more B than this critical amount (called the *eutectic composition*), the B-rich phase separates first. At the eutectic composition itself the A and B phases separate out simultaneously, at a single temperature (the *eutectic temperature*), as a fine mixture of the two types of crystals. This mixture is called the *eutectic structure*. It usually consists of fine plates, needles or spheres of the one crystalline phase dispersed among the other phase.

One of the phases in a liquid not of eutectic composition thus freezes in two distinct stages. First, *primary* crystals of it form, before the eutectic temperature is reached. Then, at the eutectic temperature, the remanent liquid, now of eutectic composition, freezes to form a eutectic mixture of both this and the other phase.

.Many crystallized materials form eutectics, often much more complex than the simple type we have described here. Tinman's solder (60% Sn, 40% Pb) is an alloy of eutectic composition, chosen because of its low freezing point. Lime is added to silica-rich metallic ores to form a fusible CaO–SiO_2 eutectic which can be run out of metal extraction furnaces as a liquid slag. An important constituent of carbon steels, called *pearlite*, consists of fine parallel lamellae of b.c.c. iron (*ferrite*) and the iron–carbon compound Fe_3C (*cementite*); the structure in this case is formed by a solid-state reaction analogous to eutectic formation.

Because of the difference in crystal structures, the boundaries between grains of different crystalline phases are generally more complicated than those described in Section 3.5. The same principle of an energetically economical transition region still applies, however. In fact, the adjoining phases often choose to crystallize in a particular orientation relative to each other, so as to obtain the best possible "matching" of the two lattices across their interface.

FURTHER READING

BRAGG, W. L., *The Crystalline State*, Volume 1, Bell, London, 1933.

WANNIER, G. H., *Elements of Solid State Theory*, Cambridge University Press, 1959.

PHILLIPS, F. C., *An Introduction to Crystallography*, Longmans, London, 1946.

GUY, A. G., *Elements of Physical Metallurgy*, Addison-Wesley, New York, 1959.

BARRETT, C. S., *Structure of Metals*, McGraw-Hill, New York, 1952.

READ, W. T., *Dislocations in Crystals*, McGraw-Hill, New York, 1953.

COTTRELL, A. H., *Dislocations and Plastic Flow in Crystals*, Oxford University Press, 1953.

MCLEAN, D., *Grain Boundaries in Metals*, Oxford University Press, 1957.

VAN BUEREN, H. G., *Imperfections in Crystals*, North-Holland, Amsterdam, 1960.

PROBLEMS

3.1 Construct a rhombohedral primitive cell for the f.c.c. lattice, with an angle between the rhombohedral axes of 60° and a lattice constant $1/\sqrt{2}$ that of the cubic cell. Do likewise for the b.c.c. lattice, with an angle between axes of 109·5° and a lattice constant $\sqrt{3}/2$ that of the cubic cell. What is the shortest Burgers vector which a perfect dislocation may have in each of these two lattices?

3.2 Determine the symmetry elements of an ordinary building brick.

3.3 Show that a face-centred tetragonal cell can be regarded as a body-centred tetragonal cell and that this equivalence does not apply to the corresponding orthorhombic cells.

3.4 Starting from the primitive orthorhombic cell, explain why a Bravais lattice is still formed if lattice points are added to the centres of either one or three pairs of opposite faces, but not to two pairs.

3.5 Show that the axial ratio for an ideal c.p. hex. structure is 1·633.

3.6 Show that the maximum proportion of total volume which may be filled by equal hard spheres packed in various structures is as follows: f.c.c. (0·74); c.p. hex. (0·74); b.c.c. (0·68); simple cubic (0·52); diamond (0·34).

3.7 Sodium and chlorine weigh 22·997 and 35·457 g mole^{-1}, respectively. The density of a NaCl crystal is 2·165 g cm^{-3}. Show that the lattice constant of its cubic cell is 5·63 Å.

3.8 The atomic radius of an atom in a crystal of silver is 1·441 Å. Calculate the lattice constant of the f.c.c. silver cell and calculate the density of solid silver. [4·077 Å; 10·58 g cm^{-3}]

3.9 By taking various lattice circuits, prove that the Burgers vector of a dislocation is independent of the circuit taken and is the same at all points along the dislocation line.

3.10 A cutting and sliding process was used to describe the screw dislocation in Fig. 3.13. Explain how the edge dislocation could be made in a similar way. On what planes would the cut have to be taken if this dislocation were to be made by (a) pure sliding, (b) no sliding, but insertion of material, (c) no sliding, but removal of material? By means of lattice circuits, prove that these operations all produce the same dislocation.

3.11 What pair of dislocations is equivalent to (a) a row of vacancies, (b) a row of interstitials?

3.12 The twin interface in Fig. 3.12 coincides with the crystal plane of the twinning fault. Show that, when the interface is slightly inclined to the fault plane, an imperfect dislocation lies along each line where the interface jumps from one close-packed layer to the next.

3.13 The screw dislocation in Fig. 3.13 is right-handed. A dislocation line runs as a right-handed screw along the $+X$ axis of co-ordinates as far as the origin and then changes direction to run along the $+Z$ axis as an edge dislocation. On what side of the origin does the half-plane of this edge dislocation lie?

3.14 A crystal can form edge dislocations along its X axis, with a Burgers vector of length b. This crystal acquires a permanent bend about this axis, to a radius of curvature r, through the formation of edge-dislocation walls of the type shown in Fig. 3.19. Prove that $\rho = 1/rb$, where ρ is the number of such dislocations crossing a plane of unit area which is perpendicular to the X axis.

chapter **4**

Elasticity

The obvious property of a solid is its mechanical stability,
i.e. resistance to changes of size and shape. We consider
this in an elementary way for crystalline solids, and see how
Hooke's law of elastic deformation can be understood in terms
of interatomic forces; also the factors that determine the
magnitudes of elastic constants in materials. In Section 4.6
we go on to a more general analysis of the concept of strain
in a solid, and then, after similarly considering the concept
of stress, we derive general elastic stress–strain relations
and discuss the elastic constants of solids.

4.1 MECHANICAL STABILITY OF SOLIDS

Matter is often classified in terms of mechanical properties. Thus, *solids* and *liquids* have stable volumes, but *gases* expand to the limits of their containers. Solids have stable shapes, but unsupported *fluids* (gases and liquids) do not.

For a simple discussion of stability, we consider a large group of equal spherical atoms (or molecules) and suppose that between any pair of these there exist short-range forces which act along the line of centres and vary only with the distance a between centres. We suppose the forces to be as in Fig. 2.2, the atoms to be crystallized and the temperature to be $0°K$. If the forces are sufficiently short-ranged to make the interactions of all but nearest neighbours negligible, the binding energy ϕ_z of an atom is given by

$$\phi_z = \tfrac{1}{2}z\phi_0,\qquad(4.1)$$

since each atom has a half share, $\tfrac{1}{2}\phi_0$, in each of the z bonds (energy ϕ_0 at the equilibrium bond length a_0; see Fig. 2.2) it makes with its z nearest neighbours.

82

We expect the crystal structure to be f.c.c. or c.p. hex., since these have large z and, hence, large ϕ_z. Consider the f.c.c. structure. Each particle is individually in mechanical equilibrium, since it is at a centre of symmetry and so experiences equal and opposite forces from its neighbours on opposite sides. All the nearest-neighbour bonds have the same length, and the crystal is free to adjust this to the value a_0. The crystal is then stable against overall changes of volume, i.e. dilatations. Figure 4.1(a) shows that each atom is bonded tetrahedrally to neighbours. Each tetrahedron is stable against *shear distortions*, since it cannot shear without altering some of its bond lengths. The tetrahedra cannot move independently of one another, since their faces also form octahedra (cf. Fig. 3.9). The crystal is thus *fully constrained kinematically*. We notice that a simple

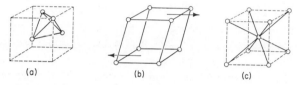

(a)   (b)   (c)

Fig. 4.1. Mechanical stability: (a) stable tetrahedron of bonds in f.c.c.; (b) collapse of simple cubic structure; (c) nearest-neighbour bonds in b.c.c.

cubic structure would be unstable with such bonds, since it could collapse by a mode of deformation such as that shown in Fig. 4.1(b), without altering any nearest-neighbour distances. This deformation mode is a mixture of shear and volume change at large strains. The b.c.c. structure, shown in Fig. 4.1(c), can also collapse without altering bond lengths if we remove the "struts" along the cube edges in recognition of the fact that these do not represent nearest-neighbour bonds. The mechanical stability of b.c.c. in practice must be due to other factors, such as next-nearest-neighbour bonds, directed bonds and thermodynamic effects.

Arguments of the above type do not establish the *thermodynamical* stability of a crystal structure. A structure may be mechanically stable yet may not exist, because other structures have lower *free energies*. This is particularly important at high temperatures, where entropy contributes strongly to free energy and a simple criterion based on bond energies is no longer applicable. Many substances, in fact, exist in different crystal structures over different ranges of temperature. On heating or cooling through the *critical point* which separates two such temperature ranges, a *solid-state transformation* then occurs, provided the atoms are mobile and there is no nucleation difficulty (cf. Section 8.8). Many of these transformations are important. *Quartz* (SiO_2) undergoes several when heated or cooled, and the volume changes which result from them can crack the material, which is troublesome in silica firebricks. *Tin* is stable as *white tin* (tetragonal, metallic, ductile) above 13°C and as *grey tin* (diamond structure, almost non-metallic, brittle) below 13°C. Grey tin forms as a loose powder, owing to the large volume increase which accompanies the transformation. This is sometimes a trouble when tin is used in cold climates, causing *tin pest*, but fortunately the change is difficult to nucleate. White tin usually remains metastable at low temperatures, provided it is not brought into contact with grey tin. *Iron* is stable as b.c.c. below 910°C and above 1400°C, and as f.c.c. between these temperatures. These changes are exploited in the important *quench-hardening* process for strengthening steel.

4.2   ELASTIC DEFORMATION

We now consider the mechanical stability of a solid when forces are applied to it. Let us pull two atoms apart or push them together by the balanced forces f, as in Fig. 4.2. These atoms

Fig. 4.2.   Forces on an atomic bond.

may be an isolated molecule or two neighbours in a solid, e.g. a wire in tension. By *balanced forces* on a body we mean a set of applied forces which exert no resultant and no couple and so do not set the body in motion. When $f = 0$, the equilibrium spacing is given by $a = a_0$ (Fig. 2.2). When f is small, the atoms can find a new equilibrium spacing a in which the atomic and applied forces are balanced. Referring to Fig. 2.2 and defining the *displacement* u as

$$u = a - a_0, \tag{4.2}$$

the equilibrium condition is that

$$f = \frac{d\phi(u)}{du}, \tag{4.3}$$

where $\phi(u)$ is the bond energy at the displacement u. If f is tensile, then $a > a_0$; if compressive, $a < a_0$. This equilibrium condition can also be expressed in terms of the *principle of virtual work*. Suppose that, in equilibrium under the force f, one of the atoms moves a small additional distance δu in the direction of this force. The work done by f through this *virtual displacement* is $f\delta u$; but the corresponding increase in bond energy

$\phi(u + \delta u) - \phi(u)$ is, from Equation (4.3), also equal to $f\delta u$ when the system is in equilibrium.

The force in the bond is a function of the displacement. To each displacement u there corresponds a characteristic value of the force $f(u)$, and the deformation of the bond is reversible in the sense that, when the displacement returns to some initial value u after an excursion, so also does the force return simultaneously to its corresponding value $f(u)$. These are the identifying features of *perfect elastic deformation*. The bulk elastic behaviour of large solid bodies is simply the aggregate effect of the individual deformations of the bonds in them, the applied force being transmitted from one loading point to another along the network of bonds running through the material.

4.3 HOOKE'S LAW

When the applied forces are sufficiently small, the elastic displacement is always proportional to force. This is *Hooke's law*. It follows from three things: (1) the continuous variation of $\phi(u)$ with u; (2) the minimum, $d\phi/du = 0$ at $u = 0$; (3) the smallness of the displacement, $u \ll a_0$. The first allows us to express the bond energy as a Taylor series,

$$\phi(u) = \phi_0 + \left(\frac{d\phi}{du}\right)_0 u + \frac{1}{2}\left(\frac{d^2\phi}{du^2}\right)_0 u^2 + \text{higher terms,} \qquad (4.4)$$

where ϕ_0 is the energy at $u = 0$ and where all differential coefficients are measured at $u = 0$. The second allows us to delete the second term from the series; and the third allows us to delete the terms beyond the third, because they contain higher powers of the small quantity u. Hence,

$$\phi(u) = \phi_0 + \frac{1}{2}\left(\frac{d^2\phi}{du^2}\right)_0 u^2 \qquad (4.5)$$

and

$$f = \frac{d\phi(u)}{du} = \left(\frac{d^2\phi}{du^2}\right)_0 u. \qquad (4.6)$$

Since $(d^2\phi/du^2)_0$ is the curvature of the ϕ,u curve at the minimum point and is therefore independent of u, it is a constant. We thus have $f \propto u$, i.e. Hooke's law. When expressed in terms of stress and strain, this curvature becomes the *elastic constant* of the substance. We see that it has the same value for both tension and compression, since it is independent of the sign of u.

The elastic constant is thus a measure of the sharpness of the minimum of the energy curve. It bears only an indirect relation to the binding energy, as may be seen from Fig. 4.3. When the repulsive interaction is very short-ranged, the binding energy ϕ_0 is almost equal to the *attractive* interaction energy ϕ_1, but the elastic constant is contributed almost entirely by the curvature of the *repulsive* interaction curve. Such correlation as does exist between binding energies and elastic constants is due to the fact that, when the attractive force is strong, the atoms are pulled together to the point where the repulsive curve rises sharply.

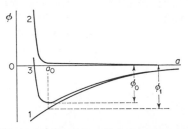

Fig. 4.3. Contributions of attractive (curve 1) and repulsive (curve 2) interactions to the total bond energy (curve 3).

Deformations that obey Hooke's law are called *linear* or *Hookeian*. They have a mathematical property important in the analysis of elastic problems. Suppose that an elastic body suffers the displacements u_1 under applied forces f_1, and u_2 under f_2, where f_2 need not be parallel to or even act at the same points as f_1 (Fig. 4.4). Then, if u_1 and u_2 are sufficiently

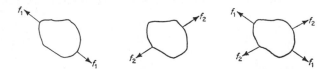

Fig. 4.4. Forces applied separately and jointly to a body.

small and the deformation is linear, the combined forces $f_1 + f_2$ produce the combined displacements $u_1 + u_2$. This is the *principle of superposition*. To deduce it we note: (1) that if f_1 is applied first and f_2 second, the elastic constants by which the body resists f_2 are unchanged by the presence of f_1, since f_1 does not strain the atomic bonds beyond the range of constant curvature at the minimum of the energy curve; (2) even though the displacements u_2 may cause the forces f_1 to move and do some work, this work is, by the principle of virtual work, compensated by an equal energy change in the atomic bonds and so does not interfere with the response of the body to f_2. Hence, the body responds to f_2 as if f_1 were not present.

4.4 FAILURE OF HOOKE'S LAW

When the deformations of the bonds become large and the higher terms in Taylor's series are no longer negligible, Hooke's law fails. The non-linearity of large elastic deformations can be detected in various ways, some of which involve physical properties, such as thermal expansion and conductivity (cf. Chapter 6). The obvious mechanical experiment is to pull a straight bar or wire strongly in tension. Figure 4.5 shows examples.

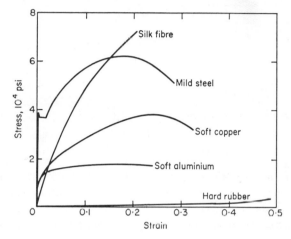

Fig. 4.5. Tensile stress–strain curves of various materials.

Here we have plotted *nominal tensile stress* (tensile force divided by initial cross-sectional area) against *tensile elongation* or *nominal tensile strain* (increase in length divided by initial length). The proportionality of stress to strain extends only a little way from the origin. However, what usually fails in such experiments is not the linearity of elastic deformation but the resistance of the material to other, quite different, *non-elastic* processes of deformation. For example, when a ductile metal is pulled beyond a certain stress (the *elastic limit* or *yield stress*) it deforms *plastically*; atoms slide in groups past their neighbours to new equilibrium positions with other neighbours (cf. Chapter 9). The elastic strains in the atomic bonds are still usually small and obey Hooke's law, but they are overshadowed by the much larger plastic strains. It is easy to separate the two types of strain by unloading the specimen, as in Fig. 4.6. The total strain is no longer in one-to-one correspondence with the stress. The elastic strain disappears on unloading, following Hooke's law with the same elastic constant, but the plastic strain mostly remains as a *permanent set*.

Sometimes a small part of the plastic strain also disappears and a *hysteresis loop* is then formed on reloading.

A genuine failure of Hooke's law can be seen easily by stretching materials such as rubber to large deformations. The failure in this case is

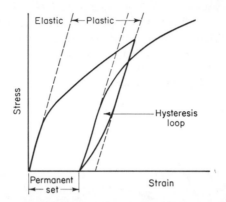

Fig. 4.6. Effect of unloading and reloading, showing reversibility of elastic strain, irreversibility of plastic strain (which leaves a permanent set), and a hysteresis loop.

due not to large tensile strains in the atomic bonds but to the special structure of such materials. They are not held together by dense three-dimensional networks of bonds but are *polymers* and consist of separate molecules in the form of long chains. Rubber is a chain of *isoprene* (C_5H_8) molecules. The chain itself is a line of carbon atoms; these are

Fig. 4.7.   Schematic diagram of a slightly cross-linked polymer.

bonded directly together but are also *side-linked* to the hydrogen atoms and CH_3 groups that make up each isoprene molecule of the chain. The chains in such a polymer lie loosely together in irregular tangles, one being bonded or *cross-linked* to another only at occasional intervals along its length, so that the molecular structure is somewhat as shown in Fig. 4.7.

The great elastic deformability of such a structure is due to the flexibility of these long chains. They can easily be crumpled up or straightened out. Only when they have become nearly straight, after large extensions, does the innate elastic strength of the bonds show itself in the stress–strain diagram (Fig. 4.5). At smaller strains the elastic resistance of the material is due to the thermal motion of the molecules, which tries to wriggle the chains up into random lengths in opposition to the applied forces which try to straighten the chains out or to crumple them up completely.

The clearest examples of non-linear elastic strains in crystalline solids are obtained by subjecting such materials to great hydrostatic pressures

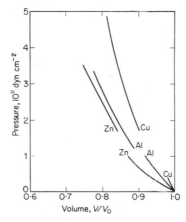

Fig. 4.8. Compression of materials at great pressures (after P. W. Bridgman, *Proc. Am. Acad. Sci.*, **72**, 207 (1938) and J. M. Walsh and R. M. Christian, *Phys. Rev.*, **97**, 1544 (1955)).

and measuring their volume changes. Bridgman succeeded in reaching pressures of order 10^5 atm ($\simeq 1 \cdot 5 \times 10^6$ psi) with special presses and leak-proof pistons. Even higher pressures have been obtained in shock waves produced by detonating explosive charges near the specimen. Figure 4.8 shows some results. The increased "hardness" of the atoms, above that expected from Hooke's law, at these high pressures is clear.

It is much more difficult to make the corresponding tensile (or uniaxial compression) experiments, because the material usually yields plastically or breaks while the stress is still relatively low. However, materials in the form of fine fibres or "whiskers", with diameters about 10^{-4} cm, can often withstand exceptionally high tensile stresses without yielding or breaking. Elastic strains of order $0 \cdot 1$ have been obtained on such fibres and, as Fig. 4.9 shows, these are large enough to show deviations from Hooke's law.

Elastic bodies sometimes fail as a whole to obey Hooke's law, even when their atomic bonds still obey the law. As anyone who has ever handled a thin piece of celluloid will know, *buckling* or *elastic instability* occurs when

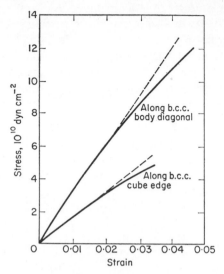

Fig. 4.9. Elasticity of an iron "whisker" (after S. Brenner, *J. Appl. Phys.*, **27**, 1484 (1956)).

a slender body is bent by compressive thrusts at its ends. In Fig. 4.10 we show a slightly bent strut ABC compressed by a force F at its ends. This force is of course transmitted along the strut as a compression, but in addition the arms BA and BC can be regarded as levers by which the force exerts a *couple* of magnitude $y_0 F$ on the material at B. If the strut begins to buckle under these forces the couple increases (yF), which encourages further bending and leads to a complete *elastic collapse*. The slenderness

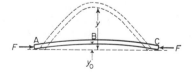

Fig. 4.10. Elastic instability.

of a body enables small Hookeian deformations of the atomic bonds to produce large changes of overall shape and, hence, change the leverage of applied forces. The principle of superposition obviously does not hold in such cases.

4.5 ELASTIC PROPERTIES AND ATOMIC FORCES

As an example of the quantitative dependence of elastic constants on atomic forces, we shall consider ionic crystals of the NaCl type (cf. Fig. 3.7). The interaction energy of a pair of univalent ions at a distance r can be written as

$$\phi(r) = \pm \frac{e^2}{r} + \frac{B}{r^s}, \tag{4.7}$$

where + and − refer to like and unlike ions, respectively, and where $s \simeq 9$. A given ion interacts with all others in the crystal. The repulsive interaction need be summed only over the six nearest neighbours because of its short range, but the long-range (i.e. r^{-1}) electrostatic interaction has to be summed over alternate shells of like and unlike ions out to large distances. Such sums have been made, and the total interaction energy of an ion can be written as

$$\phi_z = -\frac{A\,e^2}{r} + \frac{6B}{r^s}, \tag{4.8}$$

where A is called the *Madelung constant*, equal to 1·7476 for crystals of the NaCl type, and r is now the spacing of neighbouring unlike ions. We can eliminate B by applying the condition $d\phi_z/dr = 0$ at the equilibrium spacing, $r = r_0$. Thus,

$$\left(\frac{d\phi_z}{dr}\right)_{r=r_0} = \frac{A\,e^2}{r_0^2} - \frac{6sB}{r_0^{s+1}} = 0, \tag{4.9}$$

which gives

$$B = \frac{A\,e^2 r_0^{s-1}}{6s} \tag{4.10}$$

and

$$\phi_z = -\frac{A\,e^2}{r}\left[1 - \frac{1}{s}\left(\frac{r_0}{r}\right)^{s-1}\right]. \tag{4.11}$$

Since $\phi(r)$ is the energy of two ions, $N\phi_z$ is the total energy of a crystal of N ion-pairs. This is the work required to dissociate the crystal into $2N$ separate ions, N positive and N negative. It is not the same as the energy of vaporization, although related to it, because vaporization is dissociation into neutral particles. Since each ion bonds with six nearest neighbours, $\phi_z/6$ is the energy per nearest-neighbour bond, which we shall write as ϕ. The elastic constant of such a bond is, from Equations (4.6) and (4.11), given by

$$\left(\frac{d^2\phi}{dr^2}\right)_{r=r_0} = \frac{(s-1)A\,e^2}{6r_0^3}. \tag{4.12}$$

Suppose that the crystal is subjected to a hydrostatic pressure p which produces the Hookeian volume change $\Delta V/V$. Then K, defined by

$$p = -K\frac{\Delta V}{V},\qquad(4.13)$$

is the *bulk modulus of elasticity* of the material. Consider a cubic face of unit area on the crystal. The number of nearest-neighbour bonds acting through this face, perpendicularly to it, is equal to r_0^{-2}. Hence, the force

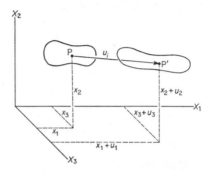

Fig. 4.11. Displacement of a point P in a body.

f supported by one such bond is pr_0^2. If the deformation u is small, then $\Delta V/V = 3u/r_0$. Hence, from Equations (4.6), (4.12) and (4.13),

$$K = \frac{1}{3r_0}\left(\frac{d^2\phi}{dr^2}\right)_{r=r_0} = \frac{(s-1)A\,e^2}{18r_0^4}.\qquad(4.14)$$

This agrees well with observed bulk moduli. In KCl, for example, it gives $K = 1\cdot88 \times 10^{11}$ dyn cm^{-2} ($\simeq 2\cdot8 \times 10^5$ psi), whereas the observed value (extrapolated to $0°K$) is 2×10^{11}.

By further calculation we could find the resistance of the crystal to shear forces. The corresponding calculations of elastic constants of metallic and covalent crystals are much more difficult, because the laws of force can no longer be expressed by such simple formulae as Equation 4.7.

4.6 STRAIN

We must now make our ideas of strain and stress more precise. Strain in a body is caused by the displacement of its particles from one position to another. We define the position of a particle P by its co-ordinates x_1, x_2, x_3, in a set of axes which are fixed and independent of the body (Fig. 4.11). We shall often write these co-ordinates and axes as x_i and X_i, where $i = 1, 2, 3$. Suppose that the deformation and movement of the body displaces P to P′, with co-ordinates $x_1 + u_1, x_2 + u_2, x_3 + u_3$. Then $u_i(=u_1, u_2, u_3)$ is the *displacement* of P. If u_i = constant for all particles, there is no strain, only a rigid translation of the whole body. An essential condition for strain is that u_i should differ from one particle to another, as, for example, when opposite ends of a stretched piece of rubber move in opposite directions. This means that u_i has to be a function $f(x_i)$ of x_i.

The simplest such function is the *linear* one, $u_i \propto x_i$, which gives *uniform* strain. In a one-dimensional problem, such as the uniform stretching of a wire along its axis, the displacement u of a point along that axis is simply $u = ex$, where x is the initial co-ordinate of the point along the axis and where e, the coefficient of proportionality, is the *nominal tensile strain*, i.e. $e = du/dx$. To generalize this to three-dimensional uniform strain, we have to take each of the three u_i components in turn and make each a linear function of each of the three initial co-ordinates of the point. We thus need, in place of e, nine different coefficients of proportionality, which we denote as $e_{11}, \dots, e_{33}$, i.e.

$$u_1 = e_{11}x_1 + e_{12}x_2 + e_{13}x_3,$$

$$u_2 = e_{21}x_1 + e_{22}x_2 + e_{23}x_3, \qquad (4.15)$$

$$u_3 = e_{31}x_1 + e_{32}x_2 + e_{33}x_3.$$

These equations can be written more neatly as

$$u_i = e_{ij}x_j, \qquad (4.16)$$

where $i, j = 1, 2, 3$, and where we have used the *repeated suffix convention*, i.e. because a suffix (in this case j) occurs twice in a term (i.e. $e_{ij}x_j$) that term represents the sum of all terms (i.e. $e_{i1}x_1 + e_{i2}x_2 + e_{i3}x_3$) formed by inserting all the numerical values of that suffix.

The analysis of *non-uniform* strains, in which the u_i are no longer simply proportional to the x_i, is of course more complicated, but this does not affect the *definition* of strain, since we can regard a non-uniform strain field as composed of elementary regions, each small enough to deform uniformly to a first approximation. This can easily be seen by, for example, bending a piece of rubber with a grid of small squares marked on it. The complication lies in the analysis of the *variation* of strain from one element to another, not in the definition of strain.

What do the nine e_{ij} mean? Clearly,

$$e_{11} = \frac{\partial u_1}{\partial x_1}, \qquad e_{22} = \frac{\partial u_2}{\partial x_2}, \qquad e_{33} = \frac{\partial u_3}{\partial x_3}, \qquad (4.17)$$

and these are the *tensile strains* or *unit elongations* (positive for tension, negative for compression) along X_1, X_2, and X_3, respectively. (We have used *partial* differentials here to identify the variations of displacement along particular directions through the material.) Sometimes e_{11}, e_{22}, e_{33}, are the only non-zero strain components. They are then called *principal*

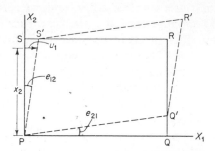

Fig. 4.12. Angular distortion of a prism.

strains and the axes X_i are *principal* axes. Such a deformation would, for example, change a cube with edges along the principal axes into the shape of a brick.

To understand components such as e_{12}, consider in Fig. 4.12 a deformation in the $X_1 X_2$ plane of an initially rectangular prism PQRS. For simplicity we take the origin of co-ordinates at P and take $e_{11} = e_{22} = 0$. The prism has suffered an angular distortion. The displacement of points on the line PS is a u_1, since it takes place in the direction of X_1, but this u_1 increases in proportion to the x_2 co-ordinate along PS and hence, from Equation (4.15), involves the e_{12} strain component. Thus,

$$e_{12} = \frac{SS'}{PS} = \frac{\partial u_1}{\partial x_2} (= \hat{SPS'}, \text{ when this angle is small}). \qquad (4.18)$$

By a similar argument,

$$e_{21} = \frac{QQ'}{PQ} = \frac{\partial u_2}{\partial x_1} (= \hat{QPQ'}, \text{ when this angle is small}). \qquad (4.19)$$

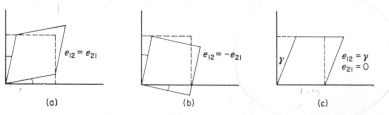

Fig. 4.13. Examples of shear and rotation: (a) pure shear without rotation; (b) pure rotation without shear; (c) simple shear.

These are *positive* when they rotate a line from one positive axis towards the other positive axis, as in Fig. 4.12, and *negative* in the opposite case.

All displacements of the type ($e_{12} + e_{21}$ = constant) produce the same change in the angle $S\hat{P}Q$, i.e. produce the same *shear*, but differ in the amounts by which they *rotate* the prism about P. Figure 4.13 shows examples. We see that, in general, e_{12} and e_{21} produce both *shear strain* and *rigid-body rotation*. To separate these, we write

$$e_{12} = \varepsilon_{12} + \omega_{12},$$
$$(4.20)$$
$$e_{21} = \varepsilon_{21} + \omega_{21},$$

where

$$\varepsilon_{12} = \tfrac{1}{2}(e_{12} + e_{21}), \qquad \omega_{12} = \tfrac{1}{2}(e_{12} - e_{21}),$$
$$(4.21)$$
$$\varepsilon_{21} = \tfrac{1}{2}(e_{21} + e_{12}), \qquad \omega_{21} = \tfrac{1}{2}(e_{21} - e_{12}).$$

We then have

$$\varepsilon_{12} = \varepsilon_{21},$$
$$(4.22)$$

which represent the shear strain produced by e_{12} and e_{21}, as in Fig. 4.13(a); and

$$\omega_{12} = -\omega_{21},$$
$$(4.23)$$

which represent the rigid-body rotation, as in Fig. 4.13(b). For example, *simple shear*, as in Fig. 4.13(c), is obtained when $\varepsilon_{12} = \omega_{12}$.

Generalizing and using Equations (4.17), (4.18) and (4.19), we can express *all* the components by the equations

$$\varepsilon_{ij} = \frac{1}{2}\left(\frac{\partial u_i}{\partial x_j} + \frac{\partial u_j}{\partial x_i}\right),$$
$$(4.24)$$
$$\omega_{ij} = \frac{1}{2}\left(\frac{\partial u_i}{\partial x_j} - \frac{\partial u_j}{\partial x_i}\right),$$

where $i,j = 1, 2, 3$. The ε_{ij} are called *strain components* and the ω_{ij} *rotation components*. The ε_{ij} represent normal strains when $i = j$, and pure shear strains when $i \neq j$. Since $\omega_{ij} = 0$ when $i = j$ and since $\omega_{ij} = -\omega_{ji}$, there are only three rotation components, one for each axis. The symbol ε_{ij} stands for a set of nine distinct terms, called the *strain tensor*, which can be written as a table, thus,

$$\begin{matrix} \varepsilon_{11} & \varepsilon_{12} & \varepsilon_{13} \\ \varepsilon_{21} & \varepsilon_{22} & \varepsilon_{23} \\ \varepsilon_{31} & \varepsilon_{32} & \varepsilon_{33}. \end{matrix}$$
$$(4.25)$$

Because

$$\varepsilon_{ij} = \varepsilon_{ji},$$
$$(4.26)$$

only six of these terms are independent, and the strain tensor is said to be *symmetrical*.

We have defined the strain components in terms of individual deformations of a prism. What interpretation do they have when applied all together? If they are large (i.e. *finite strain*), the problem becomes difficult, because the displacement produced by one strain component can alter the magnitude of another component. For example, if a body were given a simple shear e_{12}, as in Fig. 4.13(c), and then, with the shear displacement u_1 held constant, given a large extension e_{22}, the angle of shear γ would decrease because of the increase in the x_2 co-ordinate. Provided all the strains are sufficiently small (i.e. *infinitesimal strain*), however, these difficulties disappear and the various strain components can be superposed, each as if the others were absent, without ambiguity. Problems of elastic solids are usually handled in this simplified way.

It should be noticed in Figs. 4.12 and 4.13 that the total change of angle is $e_{12} + e_{21} = \varepsilon_{12} + \varepsilon_{21} = 2\varepsilon_{12} = 2\varepsilon_{21}$. Hence, the *angle of shear*, e.g. γ in Fig. 4.13(c), is equal to twice the corresponding component of the strain tensor. To avoid confusion, we shall always use γ for shear strains defined as angles of shear and ε_{ij} for shear strains defined as in Equation (4.24).

The following problem often has to be solved. Given the strains ε_{ij} measured in co-ordinate axes X_i, what are the strains ε'_{ij} measured in a differently oriented set of axes X'_i?. It is solved by working through the following steps: (1) find the displacements u_i from the strains ε_{ij}; (2) find the displacements u'_i in the new axes from the u_i; (3) find the ε'_{ij} from the u'_i. Using Equation (4.16), and neglecting the ω_{ij} because these contribute to neither ε_{ij} nor ε'_{ij}, we have

$$u_i = \varepsilon_{ij} x_j. \tag{4.27}$$

Before taking the second step, we construct a table,

$$
\begin{array}{c|ccc}
 & X_1 & X_2 & X_3 \\
\hline
X'_1 & l_{11} & l_{12} & l_{13} \\
X'_2 & l_{21} & l_{22} & l_{23} \\
X'_3 & l_{31} & l_{32} & l_{33}
\end{array}
\tag{4.28}
$$

to connect the orientations of the X'_i with those of the X_i. The terms l_{ij} are *direction cosines*. Thus, l_{12} is the cosine of the angle turned by the positive arm of X_2 in rotating about the origin to the positive arm of X'_1. These cosines are of course negative for angles between $\pi/2$ and

$3\pi/2$. To show the use of this table, we consider in Fig. 4.14 the co-ordinates of the point P in two dimensions. To find x'_1 from x_1 and x_2, we construct the various perpendiculars which meet the X'_1 axis at A, B

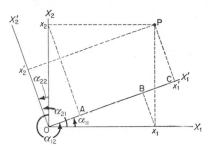

Fig. 4.14.   Rotation of axes.

and C. Then $x'_1 = OC = OB + BC = OB + OA = x_1 \cos \alpha_{11} + x_2 \cos \alpha_{12}$ $= l_{11}x_1 + l_{12}x_2$. The method can be applied to other co-ordinates and generalized to three dimensions. We obtain

$$x'_i = l_{i\beta}x_\beta,$$
$$x_i = l_{\beta i}x'_\beta,$$

(4.29)

where i, $\beta = 1, 2, 3$, and β is a repeated suffix.

We can now take the second step. Consider the displacement along X'_1. The contribution to this from u_1 is $l_{11}u_1$. Similarly, u_2 and u_3 contribute $l_{12}u_2$ and $l_{13}u_3$. Hence,

$$u'_1 = l_{1i}u_i = l_{1i}\varepsilon_{ij}x_j.$$

(4.30)

Here the suffixes i and j are both repeated, so that this is the sum of nine terms, $l_{11}\varepsilon_{11}x_1 + \dots + l_{13}\varepsilon_{33}x_3$. The advantage of the compact suffix notation is now obvious. For any axis X'_α, where $\alpha = 1, 2, 3$, we then have

$$u'_\alpha = l_{\alpha i}\varepsilon_{ij}x_j,$$

(4.31)

and, in view of Equation (4.29), this can be written

$$u'_\alpha = l_{\alpha i}l_{\beta j}\varepsilon_{ij}x'_\beta.$$

(4.32)

For the third step, we write

$$u'_\alpha = \varepsilon'_{\alpha\beta}x'_\beta$$

(4.33)

and so, eliminating u'_α between these two relations, obtain

$$\varepsilon'_{\alpha\beta} = l_{\alpha i}l_{\beta j}\varepsilon_{ij}$$

(4.34)

as the answer to our problem.

As a simple example, we consider the uniform elongation of a bar through a shear of angle γ on planes inclined to its axis, as in Fig. 4.15.

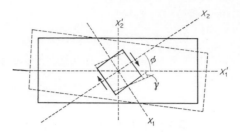

Fig. 4.15. Elongation due to shear.

Labelling the axes and angle of the plane as shown, we have $\gamma = \gamma_{12} = 2\varepsilon_{12} = 2\varepsilon_{21}$. All the ε_{ij} are zero except ε_{12} and ε_{21}; hence,

$$\varepsilon'_{11} = l_{11}l_{12}\varepsilon_{12} + l_{12}l_{11}\varepsilon_{21}$$

$$= \gamma \cos\left(\frac{\pi}{2} - \phi\right)\cos(2\pi - \phi) = \gamma \sin\phi\cos\phi. \qquad (4.35)$$

We see that the greatest elongation, $\varepsilon'_{11} = \varepsilon_{12} = \gamma/2$, is produced when $\phi = 45°$.

As a second example, we consider a square which is elongated along the X_1 axis and contracted along X_2, as in Fig. 4.16, and find the defor-

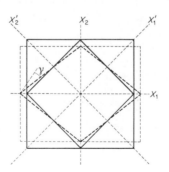

Fig. 4.16. Shear of a diamond inscribed in a square.

mation of a diamond inscribed at 45° in the square. We have $l_{11} = l_{22} = l_{12} = -l_{21} = 1/\sqrt{2}$ and

$$\varepsilon'_{\alpha\beta} = l_{\alpha 1}l_{\beta 1}\varepsilon_{11} + l_{\alpha 2}l_{\beta 2}\varepsilon_{22}. \qquad (4.36)$$

The shears are then

$$\varepsilon'_{12} = \varepsilon'_{21} = -\tfrac{1}{2}(\varepsilon_{11} - \varepsilon_{22}), \tag{4.37}$$

so that the angle of shear is given by

$$\gamma = -(\varepsilon_{11} - \varepsilon_{22}). \tag{4.38}$$

If the contraction along X_2 is equal and opposite to the extension along X_1, i.e. $\varepsilon_{11} = -\varepsilon_{22} = \varepsilon$, then

$$\gamma = -2\varepsilon. \tag{4.39}$$

In this case $\varepsilon'_{11} = \varepsilon'_{22} = 0$ and the deformation is a pure shear.

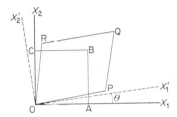

Fig. 4.17. Two-dimensional strain.

We have seen that the relative magnitudes of the strain components vary with the axes in which they are measured. There is, in fact, always a particular set of axes, i.e. *principal axes*, in which all components vanish except ε_{11}, ε_{22} and ε_{33}, and in which the deformation is particularly easy to visualize (e.g. cube → brick). Consider in Fig. 4.17 the two-dimensional problem in which a unit square OABC deforms into OPQR through the strains ε_{11}, $\varepsilon_{12}(=\varepsilon_{21})$ and ε_{22}. Refer these strains to the axes X'_1 and X'_2, rotated through the angle θ. We have $l_{11} = l_{22} = \cos\theta, l_{12} = -l_{21} = \sin\theta$. The shear strain ε'_{12} is then found to be

$$\varepsilon'_{12} = l_{11}l_{21}\varepsilon_{11} + l_{11}l_{22}\varepsilon_{12} + l_{12}l_{21}\varepsilon_{21} + l_{12}l_{22}\varepsilon_{22}$$
$$= -\tfrac{1}{2}(\varepsilon_{11} - \varepsilon_{22})\sin 2\theta + \varepsilon_{12}\cos 2\theta. \tag{4.40}$$

This vanishes when

$$\tan 2\theta = \frac{2\varepsilon_{12}}{\varepsilon_{11} - \varepsilon_{22}} = q, \tag{4.41}$$

and the values of θ that satisfy this are

$$\tfrac{1}{2}\tan^{-1} q \quad \text{and} \quad \frac{\pi}{2} + \tfrac{1}{2}\tan^{-1} q. \tag{4.42}$$

These define the orientations of the principal axes. A similar derivation can be given for three-dimensional strain.

The fact that $\varepsilon_{ij} = 0$ ($i \neq j$) in principal axes does not of course mean that there is no shear strain in the material. Figure 4.16 showed the shear strain in such a case. Since the same state of deformation can appear as different degrees of elongation and shear, according to the axes chosen, we need a means of recognizing how much of the strain is shear. The method is to split the total strain into *dilatation* and *deviatoric* strain. The first changes the volume but not the shape. The second changes the shape but not the volume.

Deformation changes the volume of a unit cube to $(1 + \varepsilon_{11})(1 + \varepsilon_{22})$ $(1 + \varepsilon_{33})$. For small strains this is $1 + \varepsilon_{11} + \varepsilon_{22} + \varepsilon_{33}$.   Hence,

$$\Theta = \frac{\Delta V}{V} = \varepsilon_{11} + \varepsilon_{22} + \varepsilon_{33} = \varepsilon_{ii} \tag{4.43}$$

is the dilatation or fractional volume change. To obtain the deviatoric strains, we subtract $\frac{1}{3}\Theta$ from each tensile strain component. Thus, the deviatoric strain tensor is

$$\begin{array}{ccc} \varepsilon_{11} - \frac{1}{3}\Theta & \varepsilon_{12} & \varepsilon_{13} \\[2mm] \varepsilon_{21} & \varepsilon_{22} - \frac{1}{3}\Theta & \varepsilon_{23} \\[2mm] \varepsilon_{31} & \varepsilon_{32} & \varepsilon_{33} - \frac{1}{3}\Theta. \end{array} \tag{4.44}$$

By defining a symbol δ_{ij} with the properties

$$\begin{aligned} \delta_{ij} &= 1, \quad \text{when } i = j, \\ &= 0, \quad \text{when } i \neq j, \end{aligned} \tag{4.45}$$

the division into dilatational and deviatoric strains can be written more neatly as

$$\varepsilon_{ij} = \tfrac{1}{3}\Theta\, \delta_{ij} + (\varepsilon_{ij} - \tfrac{1}{3}\Theta\, \delta_{ij}), \tag{4.46}$$

where the first term on the right is the dilatation and the second is the deviatoric strain. When the ε_{ij} are principal strains, their deviatoric parts are simply $\varepsilon_{11} - \frac{1}{3}\Theta$, $\varepsilon_{22} - \frac{1}{3}\Theta$, and $\varepsilon_{33} - \frac{1}{3}\Theta$. These are elongations and contractions, along principal axes, that change the shape at constant volume. We saw an example of this in Fig. 4.16 and Equation (4.39).

The problem of transforming strains (or stresses) from one set of axes to another is much easier when the new axes are simply rotated about one of the old axes. It can be solved graphically by *Mohr's circle construction*. Let us consider Fig. 4.14 and let $\alpha_{11} = \theta$. We then have $l_{11} = l_{22} = \cos\theta$, $l_{12} = -l_{21} = \sin\theta$, $l_{33} = 1$, and all the other direction cosines are zero. Suppose that X_1 and X_2 are principal axes, so that $\varepsilon_{ij} = 0$ when $i \neq j$.

The only strain components which are changed, when referred to the new axes X'_1 and X'_2, are then, from Equations 4.34, as follows,

$$\varepsilon'_{11} = \varepsilon_{11} \cos^2 \theta + \varepsilon_{22} \sin^2 \theta = \tfrac{1}{2}(\varepsilon_{11} + \varepsilon_{22}) - \tfrac{1}{2}(\varepsilon_{22} - \varepsilon_{11}) \cos 2\theta,$$

$$\varepsilon'_{22} = \varepsilon_{11} \sin^2 \theta + \varepsilon_{22} \cos^2 \theta = \tfrac{1}{2}(\varepsilon_{11} + \varepsilon_{22}) + \tfrac{1}{2}(\varepsilon_{22} - \varepsilon_{11}) \cos 2\theta,$$

$$\varepsilon'_{12} = -\varepsilon_{11} \sin \theta \cos \theta + \varepsilon_{22} \sin \theta \cos \theta = \tfrac{1}{2}(\varepsilon_{22} - \varepsilon_{11}) \sin 2\theta.$$

$$(4.47)$$

We can represent these relations in a diagram in which ε'_{11} and ε'_{22} are measured along the horizontal axis and ε'_{12} along the vertical axis, as in Fig. 4.18. The points P and Q are marked on the horizontal axis at the

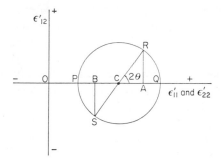

Fig. 4.18. Mohr's circle construction.

strains ε_{11} and ε_{22}, respectively (using the negative part of the axis if one or both of these is negative), and a circle with centre C is drawn with PQ as diameter. A line RCS is then drawn at an angle 2θ, measured anti-clockwise about C, from CQ, and perpendiculars are drawn from R and S to meet OQ at A and B. Then OC $= \tfrac{1}{2}(\varepsilon_{11} + \varepsilon_{22})$, CA $=$ BC $= \tfrac{1}{2}(\varepsilon_{22} - \varepsilon_{11})$ cos 2θ and AR $=$ SB $= \tfrac{1}{2}(\varepsilon_{22} - \varepsilon_{11})$ sin 2θ. Hence, OA $= \varepsilon'_{22}$, OB $= \varepsilon'_{11}$ and AR $= \varepsilon'_{12}$. This is Mohr's circle construction, which enables us to visualize and determine easily the variation of strain components as the axes of reference are rotated. We notice that the problem can be inverted, i.e. if we are given ε'_{11}, ε'_{22} and ε'_{12}, we can then from the same con-struction find ε_{11} and ε_{22}. Knowing these, we can then go on to find the components ε''_{11}, ε''_{22} and ε''_{12} for some other angle of rotation in terms of ε'_{11}, ε'_{22} and ε'_{12}, so that the method is not limited to problems in which the principal axes are known beforehand.

4.7 THERMAL STRAIN

The properties of matter are studied by observing the response (e.g. strain, flow of heat or electricity) of a specimen to a stimulus (e.g. force, thermal or electrical gradient). The relation between a stimulus and its response is a property of the material, i.e. *response = property × stimulus*. Some stimuli and responses are non-directional (e.g. heat content, volume, temperature) and are called *scalars* or *zero-order tensors*. Next in complexity are the *vectors* or *first-order tensors*, e.g. force, displacement, momentum, temperature gradient, which require three numbers (e.g. the components along the co-ordinate axes) to specify their value at a point in the body. After these come the *second-order tensors* such as strain, which in principle require nine numbers to specify them, although this reduces to six when, like strain, they are symmetrical. Nine numbers are needed, because second-order tensors are quantities that connect one vector with another. For example, strain connects the displacement vector (u_i) with the position vector (x_j). Symmetrical second-order tensors can always be represented by means of three principal components referred to principal axes.

Thermal strains provide a simple example of these points. Suppose that the temperature of a body is uniformly increased by ΔT. This is a scalar stimulus, but the response, the strain of thermal expansion, is a second-order tensor ε_{ij}. Hence, the property connecting them, i.e. the *coefficient of thermal expansion*, is also a second-order tensor, α_{ij}; thus,

$$\varepsilon_{ij} = \alpha_{ij}\,\Delta T. \tag{4.48}$$

Being symmetrical, this can be referred to its principal axes, with *principal coefficients of expansion*, α_1, α_2, α_3, along these axes.

A sphere of unit radius inscribed in a crystal is changed by thermal strain into an ellipsoid with semi-axes of lengths $1 + \alpha_1\Delta T$, $1 + \alpha_2\Delta T$, $1 + \alpha_3\Delta T$. Unlike strain, the thermal expansion tensor is a property of the material. Therefore its principal axes must be fixed in relation to the crystal axes. Cubic crystals, by symmetry, must expand equally along their three cubic axes. The ellipsoid with three equal axes is the sphere, $\alpha_1 = \alpha_2 = \alpha_3 = \alpha$. Hence, thermal expansion is *isotropic* in cubic crystals, and only one coefficient, α, is needed to describe it. In other crystals thermal expansion is generally *anisotropic*. Hexagonal crystals, for example, expand equally in directions lying in the hexagonal plane but usually differently along the hexagonal axis.

These effects are important in polycrystals. If a homogeneous polycrystal of cubic structure (e.g. Fe, Cu, Al, SiC, MgO) is uniformly heated or cooled, the thermal strains in its grains produce no intergranular stresses, because each grain expands or contracts isotropically. But if the crystals are non-cubic and thermally anisotropic (e.g. Sn, Zn, U, graphite, quartz, many minerals), each grain strains anisotropically and presses in certain directions against its neighbours, so creating intergranular stresses. These intergranular stresses often distort and sometimes even disintegrate materials. They contribute to the weathering of rocks and to the spalling (crumbling fractures) of building-stones. They cause distortion and loss of plastic strength in uranium-metal fuel elements in nuclear reactors. Tin-based alloys used for sliding bearings show an effect called *thermal fatigue*, in which repeated temperature changes cause the grains to deform plastically, leading to roughness and cracking on external surfaces. The thermal expansion coefficients along the two principal axes in a tetragonal tin crystal are 30·5 and 15·5 ($\times 10^{-6}\,°K^{-1}$), respectively. While the material

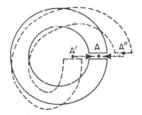

Fig. 4.19. A dislocation in an elastic ring.

is elastic, the strains produced between the grains of a random polycrystal are of order $\Delta\varepsilon \simeq \Delta\alpha\Delta T$, where $\Delta\alpha$ is the deviatoric part of the expansion coefficient. For example, if $\Delta\alpha \simeq 15 \times 10^{-6}\,°K^{-1}$ and $\Delta T = 200(°K)$, then $\Delta\varepsilon \simeq 3 \times 10^{-3}$. This is well beyond the elastic limit of tin, so that plastic deformation must occur. Lead-based bearing alloys, which are f.c.c., are immune to the effect.

4.8 COMPATIBILITY

Intergranular thermal stresses are an example of *self-stresses*, i.e. stresses that can exist in a body free from externally applied forces. Self-stresses are produced whenever two or more parts of the body cannot be joined together without elastic strain. An example is the *elastic dislocation* which is formed when the open ends of a badly made ring are forced together and then welded to make a closed ring, as in Fig. 4.19. The displacement in the ring is *multivalued*, since, for example, the point A has moved in opposite directions according as it has come from A' or A".

In many linear elastic problems we are simply interested in how the body responds to external forces and are not concerned with self-stresses. Even if there are self-stresses, the principle of superposition allows us usually (assuming that there are no elastic instabilities, such as that of Fig. 4.10) to ignore them when calculating the response to external forces. The absence of self-stresses means that, when the body is free from external forces, it is then free from elastic strain. If it were cut into small parts, these could be fitted together again without strain to re-create it perfectly, i.e. these parts are mutually *compatible* in their dimensions. This mutual compatibility persists when the body is elastically deformed by the external forces. The displacements must then vary smoothly throughout the body, i.e. the displacement components u_i must be continuous, single-valued functions of the co-ordinates x_i. This places a restriction on the strain components, because there are only three u_i from which the six ε_{ij} come. For example, the components

$$\varepsilon_{11} = \frac{\partial u_1}{\partial x_1}, \qquad \varepsilon_{22} = \frac{\partial u_2}{\partial x_2}, \qquad \varepsilon_{12} = \frac{1}{2}\left(\frac{\partial u_1}{\partial x_2} + \frac{\partial u_2}{\partial x_1}\right) \qquad (4.49)$$

are all obtained by differentiating u_1 and u_2 with respect to x_1 and x_2. However, if these displacements vary continuously with x_1 and x_2, there must exist relations of the type

$$\frac{\partial}{\partial x_2}\left(\frac{\partial u_1}{\partial x_1}\right) = \frac{\partial}{\partial x_1}\left(\frac{\partial u_1}{\partial x_2}\right), \qquad (4.50)$$

so that certain differential coefficients of ε_{11}, ε_{22} and ε_{12}, must be linked together in a *compatibility relation*. Applied to all the strain components, this argument leads to *six* compatibility equations, i.e.

$$\frac{\partial^2 \varepsilon_{11}}{\partial x_2\, \partial x_3} = \frac{\partial}{\partial x_1}\left(-\frac{\partial \varepsilon_{23}}{\partial x_1} + \frac{\partial \varepsilon_{31}}{\partial x_2} + \frac{\partial \varepsilon_{12}}{\partial x_3}\right),$$

$$2\frac{\partial^2 \varepsilon_{12}}{\partial x_1\, \partial x_2} = \frac{\partial^2 \varepsilon_{11}}{\partial x_2{}^2} + \frac{\partial^2 \varepsilon_{22}}{\partial x_1{}^2}, \qquad (4.51)$$

and four others obtained by permuting the subscripts 1, 2, 3. These equations are important for solving many problems of elasticity (cf. Section 5.7).

The analysis of self-stresses naturally raises more difficult questions. In a state of self-stress the *total* strains undergone by the elements of the body are compatible, but part of these strains are, for example, thermal strains. The remaining part, i.e. the elastic strains, do not satisfy the compatibility equations. In a *multiply connected* body such as the ring of Fig. 4.19 it is possible to have elastic states in which the displacement varies *continuously* everywhere, but is *multivalued* in the sense that if we go completely round the ring, we obtain, at the end-point, a different displacement from that with which we started. The continuity of the displacements, however, enables the compatibility equations to be used. Some simple dislocations have strain fields of this type.

4.9 STRESS

External forces can be applied to a body in two ways: either, as with the forces of gravity and inertia, as *body forces* which act directly on the particles of the body; or as *surface* or *contact forces* which act directly on the particles at the surface but only indirectly, through the forces transmitted along the network of bonds, on those inside the body. Body forces give rise also to contact forces, e.g. the force of gravity is responsible for the pressure on the soles of our feet.

Stress is the state of force produced throughout the body by the mutual interactions of the particles in their displaced positions. Because these interactions are short-range forces, i.e. act over distances of molecular magnitude, we can in a macroscopic theory of stress regard the interaction

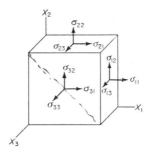

Fig. 4.20.   Stress components.

of one part of a body with another as a contact force exerted across their common boundary. Stress is thus regarded as a distribution of contact forces between successive layers of the body. To define the stress at a point in a body, we consider the contact forces applied to the faces of a small cube of material round that point. We divide each of these forces by the area on which it acts and then resolve it into components parallel to the co-ordinate axes, as in Fig. 4.20. This gives nine *stress components*

$$
\begin{array}{ccc}
\sigma_{11} & \sigma_{12} & \sigma_{13} \\[4pt]
\sigma_{21} & \sigma_{22} & \sigma_{23} \\[4pt]
\sigma_{31} & \sigma_{32} & \sigma_{33}
\end{array}
\qquad (4.52)
$$

to specify the state of stress at the point enclosed by the cube. These components collectively form the *stress tensor* σ_{ij}. The components σ_{11}, σ_{22} and σ_{33} are normal to their faces, and so represent uniaxial tension (positive) or compression (negative) stresses. The other components are shears. The component σ_{ij} acts on the face normal to X_i in the direction of

X_j. A shear stress is positive if it is directed along a positive axis when it acts on a face whose outward normal points along a positive axis. The shears in Fig. 4.20 are all positive.

Only six of the nine stress components are independent, because stress, like strain, is a symmetrical second-order tensor. Consider, for example, the equilibrium of a prism shown in Fig. 4.21 against rotation about the

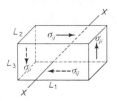

Fig. 4.21.   Equilibrium of shear stresses.

axis X. The force on a face due to σ_{ij} is $\sigma_{ij}L_1L_2$. We take the prism to be sufficiently small that equal and opposite forces act on opposite faces. The total couple exerted by the two forces $\sigma_{ij}L_1L_2$ about the X axis is then $\sigma_{ij}L_1L_2L_3$. Similarly, that due to σ_{ji} is $\sigma_{ji}L_1L_2L_3$. In equilibrium these couples balance. Hence,

$$\sigma_{ij} = \sigma_{ji}, \qquad (4.53)$$

which compares with Equation (4.26).

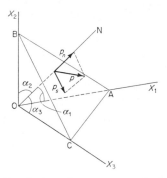

Fig. 4.22.   Resolution of stresses on an inclined plane.

To know the state of stress at a point, we need enough information to be able to determine the contact force locally on any arbitrary plane through that point. The stress tensor gives this information. Consider in Fig. 4.22 a tetrahedron OABC with an arbitrary face ABC on which the force is to be found. We make the tetrahedron very small, so as to be able to disregard any body forces acting on it, since these are proportional to its volume V, whereas the contact forces are proportional to the areas of its faces, i.e. to $V^{2/3}$, and so predominate at very small volumes. From the stress tensor we can find the contact forces on the three faces OBC, OCA and OAB, and so, from the equilibrium of forces, we can find the forces on the remaining face ABC. Let the outward normal ON on this face be inclined at angles α_1, α_2 and α_3 to the co-ordinate axes, and let the direction cosines be

$$l_1 = \cos \alpha_1, \qquad l_2 = \cos \alpha_2, \qquad l_3 = \cos \alpha_3. \qquad (4.54)$$

If the area of ABC is unity, the areas of OBC, OCA and OAB are l_1, l_2 and l_3, respectively. Let the contact stress p applied to ABC have components p_1, p_2 and p_3 along the co-ordinate axes; since ABC has unit area, these are also the forces on this face.

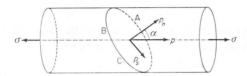

Fig. 4.23. Resolution of tensile stress on an inclined plane.

Consider now the equilibrium of forces acting on the tetrahedron along the X_1 axis. The forces in this direction applied to OBC, OCA and OAB, respectively, are $\sigma_{11}l_1$, $\sigma_{21}l_2$ and $\sigma_{31}l_3$. Hence, $p_1 = \sigma_{11}l_1 + \sigma_{21}l_2 + \sigma_{31}l_3$. Generalizing this to equilibrium along any axis X_i, and, using Equation (4.53), we have

$$p_i = \sigma_{ij}l_j, \qquad (4.55)$$

where $i,j = 1, 2, 3$, and j is a repeated suffix. We thus have the components of p. To find the normal stress p_n on ABC, we resolve the p_i along ON,

$$p_n = p_1l_1 + p_2l_2 + p_3l_3 = p_il_i,$$

and hence, from Equations (4.55)

$$p_n = \sigma_{ij}l_il_j. \qquad (4.56)$$

The formula for resolving stresses is thus essentially the same as that for resolving strains. The shear stress p_s on ABC acts in the direction determined by the parallelogram of forces, as shown in Fig. 4.22, and its magnitude is given by

$$p_s^2 = p^2 - p_n^2 = p_1^2 + p_2^2 + p_3^2 - p_n^2. \qquad (4.57)$$

As a simple example, we consider in Fig. 4.23 the stresses on a plane ABC with a normal at an angle α to the tensile axis of a bar subjected to uniaxial tension $\left(\sigma_{11} = \sigma; \text{ all other } \sigma_{ij} = 0; \alpha_1 = \alpha; \alpha_2 = \dfrac{\pi}{2} - \alpha; \alpha_3 = \dfrac{\pi}{2}\right)$. We then have

$$p_n = \sigma_{11}l_1^2 = \sigma \cos^2 \alpha; \tag{4.58}$$

$$p_1 = \sigma_{11}l_1 = \sigma \cos \alpha; \qquad p_2 = p_3 = 0; \tag{4.59}$$

$$p = (p_1^2 + p_2^2 + p_3^2)^{\frac{1}{2}} = \sigma \cos \alpha; \tag{4.60}$$

$$p_s = (p^2 - p_n^2)^{\frac{1}{2}} = \sigma \cos \alpha(1 - \cos^2 \alpha)^{\frac{1}{2}} = \sigma \cos \alpha \sin \alpha. \tag{4.61}$$

We see that the shear stress has its maximum value, $\sigma/2$, on planes at 45° to the tensile axis.

Like strain, the stress at a point can be expressed as three *principal stresses* which act along *principal axes*. In isotropic materials the axes of principal stress and strain coincide. Often the geometry of a body and arrangement of applied forces enable the principal axes to be found by simple inspection. For example, a free surface always has two principal axes in its plane. In Fig. 4.23, one principal axis lies along the axis of the bar and the others are any two perpendicular axes lying in a transverse plane.

Figure 4.23 shows that a uniaxial tensile stress produces shear stress on suitably inclined planes. As with strain, we split up stress tensors into *hydrostatic* and *deviatoric* parts, to ascertain their general nature. When the principal stresses are all equal, i.e.

$$\sigma_{11} = \sigma_{22} = \sigma_{33} = \tfrac{1}{3}\sigma_{ii} = -p,$$
$$\sigma_{12} = \sigma_{23} = \sigma_{31} = 0, \tag{4.62}$$

the stress is purely hydrostatic, and p is the *hydrostatic* pressure. When the principal stresses are unequal, we have

$$p = -\tfrac{1}{3}\sigma_{ii} = -\tfrac{1}{3}(\sigma_{11} + \sigma_{22} + \sigma_{33}) \tag{4.63}$$

for the pressure. The *deviatoric stresses*, which indicate shear stress, are the remaining part of the stress tensor, i.e. the second term in the relation

$$\sigma_{ij} = -p\,\delta_{ij} + (\sigma_{ij} + p\,\delta_{ij}). \tag{4.64}$$

This is directly analogous to Equation (4.46), the change in signs appearing because p is a negative stress whereas Θ is a sum of positive strains.

4.10   EQUATIONS OF EQUILIBRIUM

There are no compatibility conditions for the stresses, but there are *equilibrium conditions* instead. Consider in Fig. 4.24 a small rectangular prism whose sides δx_i are just large enough to give a significant small variation of stress through the prism. The condition of equilibrium along X_1 is that

$$A_1 + A_2 + B_1 + B_2 + C_1 + C_2 + D = 0, \qquad (4.65)$$

where $A_1 \ldots C_2$ are the six forces acting on the faces in this direction and D is the component of body force in this direction. If the prism is in accelerated motion, D includes the force of inertia. Let F_1, F_2 and F_3 be the re-

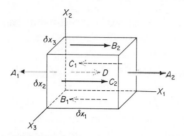

Fig. 4.24. Forces on a prism.

solved components along the co-ordinate axes of total body force per unit mass and let ρ be the density of the material. Then

$$D = F_1 \rho \, \delta x_1 \, \delta x_2 \, \delta x_3. \qquad (4.66)$$

If the normal stress component acting on the face at $X_1 - 0$ is σ_{11}, then, provided the prism is sufficiently small, the force acting on the face at $X_1 = \delta x_1$ is

$$\left(\sigma_{11} + \frac{\partial \sigma_{11}}{\partial x_1} \right) \delta x_1.$$

Since the area of this face is $\delta x_2 \, \delta x_3$, we have

$$A_1 + A_2 = \left[-\sigma_{11} + \left(\sigma_{11} + \frac{\partial \sigma_{11}}{\partial x_1} \right) \right] \delta x_1 \, \delta x_2 \, \delta x_3 = \frac{\partial \sigma_{11}}{\partial x_1} \delta x_1 \, \delta x_2 \, \delta x_3. \qquad (4.67)$$

Similarly for $B_1 + B_2$ and $C_1 + C_2$. Hence, Equation (4.65) becomes

$$\frac{\partial \sigma_{11}}{\partial x_1} + \frac{\partial \sigma_{21}}{\partial x_2} + \frac{\partial \sigma_{31}}{\partial x_3} + F_1 \rho = 0. \qquad (4.68)$$

Similar equations may be derived for equilibrium along the X_2 and X_3 axes. Generalizing, the three equations of stress equilibrium along the three axes are

$$\frac{\partial \sigma_{ij}}{\partial x_j} + F_i \rho = 0, \qquad (4.69)$$

where the suffix i refers to the axis considered and where j is a repeated suffix.

The importance of these equations is that they deal with the *variation* of stress from one point to another. They are thus the starting point for the analysis of *non-uniform* stresses in bodies.

4.11 GENERALIZED HOOKE'S LAW

In Sections 4.6 and 4.9 we have analysed strain and stress as independent entities. The results we have so far established are thus valid for any kind of body that is deformable and capable of transmitting force. Only when we come to choose some relation between σ_{ij} and ε_{ij} do we bring the detailed mechanical properties of the material into the problem. Depending upon the particular relation chosen, we may deal with various types of bodies, e.g. rigid, linear elastic, non-linear elastic, plastic, fluid, viscous. In this section we shall deal with *linear elastic* (i.e. Hookeian) solids, in which the σ_{ij} are linearly proportional to the ε_{ij}.

To be completely general, we must make each stress component depend linearly on every strain component; i.e.

$$\sigma_{11} = c_{1111}\varepsilon_{11} + c_{1211}\varepsilon_{12} + c_{1311}\varepsilon_{13}$$
$$+ c_{2111}\varepsilon_{21} + c_{2211}\varepsilon_{22} + c_{2311}\varepsilon_{23} \qquad (4.70)$$
$$+ c_{3111}\varepsilon_{31} + c_{3211}\varepsilon_{32} + c_{3311}\varepsilon_{33},$$

together with similar equations for the other stress components. The coefficients c_{pqrs} are *elastic constants*. They form a fourth-order tensor, and in principle there are $9 \times 9 = 81$ of them. The symmetry of the stress and strain tensors, however, reduces this number to $6 \times 6 = 36$. It can be shown from considerations of strain energy that only 21 of these are independent. Finally, since the elastic constant tensor is a property of the material, symmetries of structure in the material reduce the number of independent components still further.

It is usual to simplify the notation in equations such as Equation (4.70) as follows:

$$\sigma_{11} \rightarrow \sigma_1, \qquad \sigma_{22} \rightarrow \sigma_2, \qquad \sigma_{33} \rightarrow \sigma_3,$$
$$\sigma_{23} \rightarrow \sigma_4, \qquad \sigma_{31} \rightarrow \sigma_5, \qquad \sigma_{12} \rightarrow \sigma_6,$$
$$\varepsilon_{11} \rightarrow \varepsilon_1, \qquad \varepsilon_{22} \rightarrow \varepsilon_2, \qquad \varepsilon_{33} \rightarrow \varepsilon_3, \qquad (4.71)$$
$$2\varepsilon_{23} \rightarrow \gamma_4, \qquad 2\varepsilon_{31} \rightarrow \gamma_5, \qquad 2\varepsilon_{12} \rightarrow \gamma_6.$$

Thus, Equation (4.70) becomes

$$\sigma_1 = c_{11}\varepsilon_1 + c_{12}\varepsilon_2 + c_{13}\varepsilon_3 + c_{14}\gamma_4 + c_{15}\gamma_5 + c_{16}\gamma_6. \qquad (4.72)$$

We obviously choose co-ordinate axes parallel to cube edges in *cubic crystals*. It can then be proved from the cubic symmetry that

$$c_{11} = c_{22} = c_{33},$$

$$c_{12} = c_{23} = c_{31}, \tag{4.73}$$

$$c_{44} = c_{55} = c_{66},$$

and that all other elastic contents are zero. There are thus only three independent elastic constants in cubic crystals, c_{11}, c_{12} and c_{44}. This is because there are only three basic, independent modes of deformation. The first is dilatation by hydrostatic stress. Here we have $\varepsilon_1 = \varepsilon_2 = \varepsilon_3 = \frac{1}{3}\Theta$, with all other $\varepsilon_{ij} = 0$, and $\sigma_1 = \sigma_2 = \sigma_3 = -p$, with all other $\sigma_{ij} = 0$. Hence,

$$\sqrt{\quad} \quad p = -\tfrac{1}{3}(\sigma_1 + \sigma_2 + \sigma_3) = -\tfrac{1}{3}(\varepsilon_1 + \varepsilon_2 + \varepsilon_3)(c_{11} + 2c_{12})$$

$$= -\tfrac{1}{3}(c_{11} + 2c_{12})\Theta, \tag{4.74}$$

and the *bulk modulus of elasticity* (Equation (4.13)) is given by

$$K = \tfrac{1}{3}(c_{11} + 2c_{12}). \tag{4.75}$$

The second deformation mode is shear on a cubic face of the crystal in the direction of a cubic crystal axis, as shown in Fig. 4.25(a). The elastic

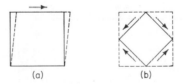

(a) (b)

Fig. 4.25. Two modes of shear in a cubic crystal cell.

constant in this case is a *shear modulus* or *modulus of rigidity*, μ_0, where

$$\sigma_{23} = 2\mu_0\varepsilon_{23}, \tag{4.76}$$

i.e.

$$\mu_0 - c_{44}. \tag{4.77}$$

In the third mode, shown in Fig. 4.25(b), planes rotated about a cubic axis by $45°$ from the cubic faces are sheared in the direction perpendicular to this axis. The shear modulus μ_1 in this case is given by

$$\mu_1 = \tfrac{1}{2}(c_{11} - c_{12}). \tag{4.78}$$

The ratio

$$\frac{\mu_0}{\mu_1} = \frac{2c_{44}}{c_{11} - c_{12}} \tag{4.79}$$

is the *elastic anisotropy factor*. When it is unity, the crystal is elastically isotropic. Values of elastic constants and anisotropy factors are given in Table 4.1. We notice that most crystals are highly anisotropic. In many

TABLE 4.1
Elastic Constants of Cubic Crystals (10^{12} dyn cm^{-2})

Substance	Structure	c_{11}	c_{12}	c_{44}	μ_0/μ_1
Na	b.c.c.	0·055	0·042	0·049	7·5
K	b.c.c.	0·046	0·037	0·026	5·7
Fe	b.c.c.	2·37	1·41	1·16	2·4
W	b.c.c.	5·01	1·98	1·51	1·0
Al	f.c.c.	1·08	0·62	0·28	1·2
Cu	f.c.c.	1·684	1·214	0·754	3·2
Ni	f.c.c.	2·50	1·60	1·185	2·6
Pb	f.c.c.	0·48	0·41	0·14	4·0
Diamond	diamond	9·2	3·9	4·3	1·6
Si	diamond	1·66	0·64	0·79	1·55
Ge	diamond	1·29	0·48	0·67	1·7
NaCl	NaCl	0·486	0·127	0·128	0·7
NaBr	NaCl	0·33	0·13	0·13	1·3
KCl	NaCl	0·40	0·062	0·062	0·36
KBr	NaCl	0·35	0·058	0·050	0·33
KI	NaCl	0·27	0·043	0·042	0·36
LiF	NaCl	1·19	0·54	0·53	1·6

non-cubic crystals, such as graphite, for which Equations (4.73) no longer hold, the anisotropy is even more pronounced.

When the atoms are at centres of symmetry in cubic crystals, the *Cauchy relation*, $c_{12} = c_{44}$, can be proved to hold, provided the interatomic forces act along the lines of centres between atoms. We see that this is roughly true of the ionic crystals but fails badly for most of the metals. This is not surprising, since the main forces in ionic crystals are of the required type, whereas in metals the atoms are not bonded directly to one another but are held together by an electron gas. The weakness of some b.c.c. metals against μ_1 shear should be noticed; this is similar to the instability of Fig. 4.1(c).

4.12   ELASTICALLY ISOTROPIC MATERIALS

Except where the material has a pronounced fibrous texture (e.g. wood) or a preferred orientation of its crystals, a polycrystalline solid is usually elastically isotropic or nearly so. This is because the individual anisotropies of the randomly oriented grains cancel out and an average value of elastic constants is obtained. There are only two independent elastic constants (e.g. c_{11} and c_{12}, since $2c_{44} = c_{11} - c_{12}$) for an elastically isotropic solid. Actually, several constants are used for convenience in various problems. These are *Young's modulus E*, the *bulk modulus K*, the *shear modulus μ*, *Poisson's ratio ν*, and *Lamé's constants λ* and μ. The latter are defined by

$$\lambda = c_{12}, \qquad \mu = c_{44}, \qquad \lambda + 2\mu = c_{11}. \qquad (4.80)$$

The relations between these various constants can be found by considering particular types of deformation. Thus, Hooke's law for *uniaxial tension σ_1* (all other $\sigma_{ij} = 0$) is

$$\begin{aligned} \sigma_1 &= c_{11}\varepsilon_1 + c_{12}\varepsilon_2 + c_{13}\varepsilon_3 \\ &= c_{12}(\varepsilon_1 + \varepsilon_2 + \varepsilon_3) + (c_{11} - c_{12})\varepsilon_1 \\ &= \lambda\Theta + 2\mu\varepsilon_1. \end{aligned} \qquad (4.81)$$

Under a *shear stress σ_4*, the angle of shear γ_4 is given by

$$\sigma_4 = c_{44}\gamma_4 = \mu\gamma_4. \qquad (4.82)$$

Thus, c_{44}, $\frac{1}{2}(c_{11} - c_{12})$, Lamé's μ and the shear modulus μ, are all the same. For *hydrostatic stress* we have

$$p = -K\Theta = -\tfrac{1}{3}(\sigma_1 + \sigma_2 + \sigma_3) = -(\lambda + \tfrac{2}{3}\mu)\Theta, \qquad (4.83)$$

so that

$$K = \lambda + \tfrac{2}{3}\mu. \qquad (4.84)$$

Young's modulus relates the tensile stress and strain in uniaxial tension, i.e.

$$\sigma_1 = E\varepsilon_1. \qquad (4.85)$$

Since $p = -\tfrac{1}{3}\sigma_1$ for uniaxial tension, we have

$$\sigma_1 = -3p = (3\lambda + 2\mu)\Theta. \qquad (4.86)$$

Substituting for σ_1 from Equation (4.81) and rearranging,

$$\varepsilon_1 = \frac{\lambda + \mu}{\mu}\,\Theta, \qquad (4.87)$$

so that Young's modulus is given by

$$E = \frac{\sigma_1}{\varepsilon_1} = \frac{\mu(3\lambda + 2\mu)}{\lambda + \mu}. \qquad (4.88)$$

Poisson's ratio is the ratio of lateral contraction to longitudinal extension in a uniaxial tensile test. For isotropic materials it is given by

$$v = -\frac{\varepsilon_2}{\varepsilon_1} = -\frac{\varepsilon_3}{\varepsilon_1} = \frac{1}{2}\left(1 - \frac{\Theta}{\varepsilon_1}\right) = \frac{\lambda}{2(\lambda + \mu)}. \tag{4.89}$$

Many other relations between the elastic constants of isotropic solids can be obtained. For example,

$$K = \frac{E}{3(1 - 2v)}, \tag{4.90}$$

$$\mu = \frac{E}{2(1 + v)} \tag{4.91}$$

$$E = \frac{3\mu}{1 + \dfrac{\mu}{3K}}. \tag{4.92}$$

Since $3K$ is usually much larger than μ, we can write

$$E = 3\mu\left(1 + \frac{\mu}{3K}\right)^{-1} \simeq 3\mu\left(1 - \frac{\mu}{3K}\right), \tag{4.93}$$

which shows that Young's modulus is contributed mainly by the shear modulus and only slightly by the bulk modulus. This is because in uniaxial stressing the stress-free sides of the material are free to move inwards or outwards, so that the resistance to dilatation is not tested to any great extent. Uniaxial strength is thus a quite different property from hydrostatic strength. As $\mu/K \to 0$ (e.g. as the material becomes *fluid*), we have $v \to 0{\cdot}5$ and $E \to 3\mu$. On the other hand, if $v = 0$, we have $E = 2\mu = 3K$.

We see that Poisson's ratio measures the relative resistance of the material to dilatation and shearing. If uniaxial extension produced no lateral contraction, i.e. $v = 0$, the stress σ_1 would give the strains ε_1, 0, 0, the dilatation $\Theta = \varepsilon_1$ and the angle of shear $\gamma = \varepsilon_1$. In practice, however, the bulk modulus opposes this large dilatation. The atoms are pulled together, sideways, to reduce the volume, i.e. $\Theta = \varepsilon_1(1 - 2v)$. But these lateral strains, $\varepsilon_2 = \varepsilon_3 = -v\varepsilon_1$, increase the angle of shear to the value $\gamma = \varepsilon_1(1 + v)$, which is resisted by the shear modulus. Atoms which slide over each other easily but resist being pulled apart thus prefer to increase γ so as to reduce Θ; Poisson's ratio is then large, and, in the limit of complete fluidity, reaches its maximum value $0{\cdot}5$. Conversely, a small value of v indicates a relatively strong resistance to shear.

Average values of elastic constants of elements are given in Table 1.1 and of some other substances in Table 4.2. In materials used for engineering structures such as aircraft, where lightness is important, the actual mechanical properties are often less important than the "specific" properties, i.e. the actual properties divided by the specific gravity. Some of the

TABLE 4.2
Average Elastic Properties and Densities

Substance	Young's modulus, 10^6 psi	Shear modulus, 10^6 psi	Poisson's ratio	Density, g cm^{-3}
Tungsten carbide	90			14·9
Silicon carbide	60			3·17
Alumina	50			3·8
Titanium carbide	50			4·5
Magnesium oxide	40			3·6
Asbestos fibre	25			2·4
Cellulose fibre	15			1·5
Glass	10	3·5	0·23	2·5
Concrete	2		0·2	2·4
Wood (dry along grain)	1·5	0·08		0·4
Polymethylmethacrylate	0·5		0·4	1·2
Rubber (vulcanized)	0·5		0·4	1·2
Nylon	0·4		0·4	1·15
Polystyrene	0·4		0·4	1·05
Polytetrafluorethylene	0·06		0·4	2·2
Polyethylene	0·02		0·4	0·9
Rubber (soft)	10^{-3} to 10^{-2}		0·49	1·5

highest specific Young's moduli (in 10^6 psi) are 34 (diamond), 23·5 (beryllium), 19 (silicon carbide), 13 (alumina), 11 (asbestos fibre) and 10 (cellulose fibre). Among lower values we note 4 (glass), 3·8 (steel), 3·7 (aluminium and wood) and 3·6 (magnesium).

4.13 ELASTIC ENERGY AND RESILIENCE

To calculate the strain energy in a Hookeian solid, consider a cube with edges of length l_0 strained in uniaxial tension by a force which is slowly raised from zero to F. The loaded faces of the cube move apart a distance δl through the elastic strain, and the work done by the applied force is $\frac{1}{2}F\delta l$, the $\frac{1}{2}$ entering because δl increases linearly with F. In terms of tensile stress, $\sigma = F/l_0^2$, and strain, $\varepsilon = \delta l/l_0$, the *work done per unit volume* or *density of strain energy* in the body is given by

$$\tfrac{1}{2}\sigma\varepsilon = \frac{\sigma^2}{2E} = \tfrac{1}{2}\varepsilon^2 E. \tag{4.94}$$

Similarly, it is

$$-\tfrac{1}{2}p\Theta = \frac{p^2}{2K} = \tfrac{1}{2}\Theta^2 K \tag{4.95}$$

for Hookeian hydrostatic deformation and

$$\tfrac{1}{2}\sigma\gamma = \frac{\sigma^2}{2\mu} = \tfrac{1}{2}\gamma^2\mu \tag{4.96}$$

for Hookeian shear.

This elastic energy is not converted to heat (apart from the thermo-elastic effect discussed in Section 6.5) but is stored in the strained atomic bonds, and can be fully recovered from the body as mechanical work by slowly reducing the applied force to zero. For truly elastic material the path followed in the stress–strain diagram on unloading retraces the loading path identically, so that the work done during unloading is equal to the work received during loading. The forces in the body which produce this behaviour are *conservative* in the sense that they conserve energy available for mechanical work (i.e. free energy). They contrast with frictional forces, which are non-conservative and convert work into heat.

The density of strain energy stored when a body is on the point of failure (e.g. by fracture or plastic flow) is called the *modulus of elastic resilience*. In strong steel this is about $2 \cdot 5 \times 10^7$ erg cm^{-3}, which is approximately 2×10^{-4} eV per atom. It is thus very small compared with the atomic bond energy. We shall see the reason for this in Chapters 9 and 11.

Elastic resilience is a measure of the ability of a body to absorb the energy of a blow elastically, without failure, and is thus important for shock absorbers. A high resilience depends on either a high failure stress, as in a steel buffer spring, or a low elastic constant, as in rubber. Crystalline solids such as metals are in general too stiff, elastically, to have much resilience in bulk form. The advantage of rubber is that, because of its exceptional molecular structure, it is resilient even in large lumps. The resilience of strong metals is increased by fashioning them into slender shapes, such as spiral or leaf springs, so as to obtain large deflexions from small elastic strains. Because of its high failure stress, hard steel can be used in very slender and flexible springs.

For greatest resilience in, say, a tensile bolt which is to be subjected to shock loads, the sectional area should be the same in all transverse cross-sections. Consider a tensile rod of length l, uniform cross-sectional area A, and failure stress σ_f. It can absorb a tensile blow of energy $\sigma_f^2 Al/2E$ elastically without failing. Suppose now that over a part of its length, l_1, its cross-sectional area is *increased* to A_1, the rest remaining unchanged. At failure, the energy absorbed in the unchanged part is $\sigma_f^2 A(l - l_1)/2E$ and that in the thickened part is $(\sigma_f A/A_1)^2 A_1 l_1/2E$, so that the total energy absorbed at failure is then

$$\frac{\sigma_f^2 Al}{2E}\left[1 - \frac{l_1}{l}\left(1 - \frac{A}{A_1}\right)\right]. \tag{4.97}$$

Since $l_1 < l$ and $A_1 > A$, the resilience of the rod is *decreased* by *increasing* the cross-sectional area over part of its length.

FURTHER READING

KITTEL, C., *Introduction to Solid State Physics*, Wiley, New York, 1956.

BORN, M. and HUANG, K., *Dynamical Theory of Crystal Lattices*, Oxford University Press, London, 1954.

NYE, J. F., *Physical Properties of Crystals*, Oxford University Press, London, 1957.

WOOSTER, W. A., *A Textbook on Crystal Physics*, Cambridge University Press, London, 1938.

ZENER, C., *Elasticity and Anelasticity of Metals*, Chicago University Press, 1948.

LOVE, A. E. H., *A Treatise on the Mathematical Theory of Elasticity*, Dover, New York, 1944.

TIMOSHENKO, S., *Theory of Elasticity*, McGraw-Hill, New York, 1934.

PROBLEMS

4.1 A collection of equal, hard spheres in a flexible container is subjected to an external hydrostatic pressure. Prove that it is mechanically stable if packed in an f.c.c. or c.p. hex. structure but not if packed in a b.c.c. or simple cubic structure. To what extent is this a good model for illustrating the mechanical properties of a metal such as copper, silver and gold?

4.2 Using the definition $K = -V(dp/dV)$, find the bulk modulus of a perfect gas under (a) isothermal and (b) adiabatic conditions.   [p and γp]

4.3 An isotropic body is subjected to principal stresses along the axes X_1, X_2, X_3, of such magnitudes that the elongations along X_2 and X_3 are zero. It is then stressed along these same axes in such a way that the elongations along X_2 and X_3 are equal and that along X_1 is zero. Compare the stress–strain ratio along X_1 in the first case with that along X_2 or X_3 in the second.   [$(1 - v):1$]

4.4 An isotropic rod is stretched longitudinally in tension while its sides are prevented from moving inwards. Show that the appropriate elastic constant for this is not E but $K + 4\mu/3$.

4.5 Prove from first principles that $-1 \leqslant v \leqslant +0.5$ for an isotropic elastic solid.

4.6 A wire is stretched elastically by various loads and simultaneous measurements are made of its length l and electrical resistance ρ. Assuming that the resistivity (resistance per unit dimensions) is unchanged by elastic strain, show how Poisson's ratio can be determined from these measurements.   [$d\rho/\rho = (dl/l)(1 - 2v)$]

4.7 A body is subjected to principal stresses of magnitudes 5, 0 and -3, respectively, along the axes X_1, X_2 and X_3. What are the normal and shearing stresses on a plane perpendicular to the X_1X_3 plane and inclined at $30°$ to the X_1X_2 plane?   [0.5, 3.46]

4.8 Prove that the dilatation of a cubic crystal under a uniaxial tension σ is independent of the direction of the tension and is given by $\sigma/(c_{11} + 2c_{12})$.

chapter **5**

Elastic Stress Distributions

In this chapter we shall examine, by solving practical problems, how elastic bodies react to applied forces and deformations. Applied elasticity is a large subject, and we shall for the most part limit ourselves to the simplest examples of problems important in engineering design. First we shall consider a number of situations—thin-walled vessels, torsion of shafts, bending of beams, deflexion of springs, buckling of compressed struts—in which there is some simplifying feature or some simplifying assumption, so that the general equations of elasticity theory can be bypassed and a solution obtained by elementary arguments. Then we shall discuss two-dimensional problems, such as stresses in thick-walled cylinders and stresses round holes in plates, that can be solved from the general theory by using the Airy stress function. Finally, we shall briefly discuss stresses produced by dislocations, grain boundaries and cracks.

5.1 SIMPLIFICATIONS

When solving problems of stress and strain distribution, we must often make simplifications, even when dealing with elastically isotropic Hookcian solids. The equations of elasticity do not even possess unique solutions unless self-stresses are ignored. Fortunately, the principle of superposition permits us usually to deal with self-stresses as a separate problem and to ignore them when considering externally applied forces. This principle is also useful for building up solutions of complicated problems by super-posing the stress fields of simpler problems.

External forces are often applied non-uniformly, through bolts and other fastenings. The stresses vary sharply across such loading points, as also across grooves and other sharp changes in the contours of bodies. These complicated regions of the stress field are usually highly localized and hardly affect the overall elastic deformations of bodies. This is expressed by *St. Venant's principle*, which states that, if the forces acting on a small

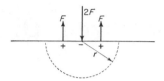

Fig. 5.1.   Example of St. Venant's principle.

part of the surface of a body are replaced by *statically equivalent forces* (i.e. with same resultant force and couple) acting on the same part, the stress state is negligibly changed at large distances compared with the dimensions of that part. Figure 5.1 illustrates this. The two tensile forces F act symmetrically about the compressive force $2F$. Since their resultant force and couple are zero, they produce negligible stresses in regions well beyond the radius r. When studying the long-range effects of applied forces, we may therefore replace the actual conditions at the loading points by the simplest statically equivalent conditions.

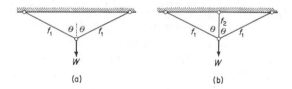

Fig. 5.2.   Pin-jointed frames: (a) perfect;   (b) redundant.

When the above simplifications are made, some problems of external loading become *statically determinate*, i.e. the stresses can be found from the laws of statics alone, without considering deformation and stress–strain relations.   By contrast, in *statically indeterminate problems* the equations of stress-equilibrium alone are not sufficient, and the stress–strain relations and strain-compatibility conditions are also needed to obtain solutions. Figure 5.2 shows an example of this. Here we have two simple *pin-jointed frames*, i.e. structures made from straight bars linked by flexible joints which can transmit only uniaxial tensile or compressive forces along the bars. The analysis of such frames is important in the design of steel bridges and other engineering frameworks. The frame in diagram (a) is *simply stiff* or *perfect*. It has just sufficient linkages to prevent any part of it from rotating about its joints (we are dealing here with *kinematic constraint* in the plane of the diagram only; to prevent rotation out of this plane at least one more bar would be needed, as in Fig. 4.1(a). A frame with less than the minimum number of linkages cannot hold its shape and is said to be *imperfect*. In Fig. 5.2(b) one extra or *redundant* linked bar has been added.

Suppose that a load W is suspended from each of the frames in Fig. 5.2 and that the inclined bars make an angle θ with the vertical axis. Then in diagram (a) we have

$$W = 2f_1 \cos \theta \qquad (5.1)$$

as a condition of static equilibrium, from which the force f_1, and therefore the stress, in the bars can be determined. Hence, this is a statically determinate problem. The redundant frame is, by contrast, statically indeterminate. It could, for example, support self-stresses between the central bar and the others. Even without self-stresses, there is only one equation of equilibrium

$$W = 2f_1 \cos \theta + f_2 \qquad (5.2)$$

for the two forces f_1 and f_2. The displacements of the bars at the loading point must, however, be *compatible*. If each inclined bar elongates from l_1 to $l_1 + \delta l_1$ and the central bar from l_2 to δl_2, where $l_2 = l_1 \cos \theta$, the vertical displacements must be equal, which gives

$$\delta l_1 = \delta l_2 \cos \theta. \qquad (5.3)$$

Suppose all the bars have unit cross-sectional areas and Young's modulus E. Then

$$f_1 = E \, \delta l_1 / l_1 \quad \text{and} \quad f_2 = E \, \delta l_2 / l_2, \qquad (5.4)$$

from which we can obtain a second relation between f_1 and f_2,

$$f_1 = f_2 \cos^2 \theta \qquad (5.5)$$

which, with Equation (5.2), provides the solution

$$f_1 = \frac{W \cos^2 \theta}{1 + 2 \cos^3 \theta}, \quad f_2 = \frac{W}{1 + 2 \cos^3 \theta}. \qquad (5.6)$$

The solution of many problems in elasticity is eased by exploiting simplifying geometrical features which enable the general three-dimensional problem to be reduced to one- or two-dimensional problems. In some cases a solution, either exact or approximate, can be guessed intuitively from the geometry of the problem. We shall see examples of this in Sections 5.3 and 5.4. In others there are general methods of solution which we shall discuss in Section 5.7 and later sections.

5.2 THIN-WALLED PRESSURE VESSELS

As a simple example of a statically determinate problem, we consider the stresses in a thin-walled cylindrical vessel or shell which contains a fluid under pressure p, as in Fig. 5.3(a). Let the cylinder have length l,

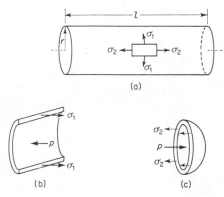

(a)

(b)   (c)

Fig. 5.3. (a) Cylindrical pressure vessel; (b) equilibrium of hoop stress; (c) equilibrium of longitudinal stress.

radius r and thickness δr, where $l \gg r \gg \delta r$. The axes of principal stress are obvious. We denote σ_1 as the circumferential tension or *hoop stress* and σ_2 as the *longitudinal stress*, as shown. Since the shell is thin, we suppose these stresses to be constant through the wall thickness. The third principal stress σ_3 varies from the value $-p$ on the inner face to zero on the outer face. We shall see that p and σ_3 are very small compared with σ_1 and σ_2. We can evaluate σ_1 by balancing forces acting on one half-cylinder as in diagram (b). The pressure provides an outward force $2rlp$. The hoop stress in the two walls provides an opposite force $2 \delta r \, l\sigma_1$. Hence,

$$\sigma_1 = pr/\delta r. \tag{5.7}$$

For the longitudinal stress we consider the equilibrium of an end-cap of the vessel, as in diagram (c). The outward force is $\pi r^2 p$ and the inward force $2\pi r \, \delta r \sigma_2$. Hence,

$$\sigma_2 = pr/2 \, \delta r, \tag{5.8}$$

so that the longitudinal stress is one-half of the hoop stress. This result is independent of the shape of the end-cap. Equation (5.8) also gives the tensile stress in a *spherical* pressure vessel.

To find the strains, we note that a tensile stress σ produces a (tensile) strain $+\sigma/E$ along its own axis and a (compressive) strain $-v\sigma/E$ along a perpendicular axis, where E and v are Young's modulus and Poisson's ratio; and that, from the principle of superposition, the strain along, for example, the axis of σ_1 is $[\sigma_1 - v(\sigma_2 + \sigma_3)]/E$. Hence,

$$\frac{\varepsilon_1}{\sigma_1 - v(\sigma_2 + \sigma_3)} = \frac{\varepsilon_2}{\sigma_2 - v(\sigma_3 + \sigma_1)} = \frac{\varepsilon_3}{\sigma_3 - v(\sigma_1 + \sigma_2)} = \frac{1}{E}. \quad (5.9)$$

5.3 TORSION OF ROUND SHAFTS

An example in which the geometry of the problem makes the elastic state obvious is the torsion of a round bar by couples applied about its axis. We consider first a cylindrical shell of length l, radius r and thickness dr, as in Fig. 5.4(a), the ends of which are twisted through the angle θ.

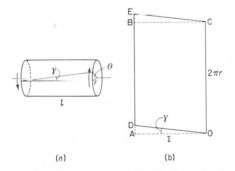

Fig. 5.4. (a) Torsion of round cylinder; (b) equivalent shear.

We suppose the ensuing strain to be distributed uniformly, and see, by imagining the cylinder to be opened out flat, that it is equivalent to a shear OABC → ODEC. For small strains we have $AD = \theta r = \gamma l$ (θ and γ in radians), so that the *angle of shear* is given by

$$\gamma = \theta \frac{r}{l}. \quad (5.10)$$

A tangential force df is applied along DE and CO to maintain this strained condition, with similar forces along DO and CE to balance the couple. In the twisted cylinder df is a circumferential force round each circular end-face. The area of an end-face is $2\pi r\, dr$, so that the *shear stress* σ in the shell is given by

$$\sigma = \frac{df}{2\pi r\, dr} = \mu\gamma,\qquad (5.11)$$

where μ is the shear modulus. The force

$$df = 2\pi r\sigma\, dr = 2\pi r^2\mu\theta\, dr/l\qquad (5.12)$$

acts tangentially at a distance r from the axis of the cylinder, and so produces a *couple* or *torque* of moment $r\, df$. We regard the solid round bar as a coaxial set of cylindrical shells and use Equations (5.10) and (5.11) to describe the strain and stress at a distance r from its axis. The total torque T on the bar is the sum of $r\, df$ over all shells, up to the outer radius r_1 of the cylinder, i.e.

$$T = \int_0^{r_1} \frac{2\pi r^3\mu\theta}{l}\, dr = \frac{\pi r_1^4\mu\theta}{2l},\qquad (5.13)$$

where $\pi r_1^4/2$ is the *polar second moment of area* of the cylinder. It is thus possible to determine μ from the angle of twist produced by a known torque acting on a round bar or wire.

The torsion of non-circular bars cannot be analysed in such a simple way, because transverse planes do not remain plane (cf. section 8.5). It can be shown that, for an elliptical cross-section of semi-axes a and b, the $r_1^4/2$ in Equation (5.13) is replaced by $a^3b^3/(a^2 + b^2)$. Since this is a maximum at $a = b$, the circle has the greatest torsional rigidity among the family of ellipses. For a ribbon-shaped specimen $(a \gg b)$ we have $a^3b^3/(a^2 + b^2) \simeq ab^3$. The torsional rigidity is then small and sensitive to the thickness $2b$ of the ribbon. This flexibility of thin ribbons is utilized in belts, torsion balances and galvanometer suspensions, where a maximum resistance to tension with a minimum resistance to twisting is required.

5.4 BENDING OF BEAMS

Another problem which can be solved (at least approximately) by simple intuitive methods is the bending of a uniform beam. In the simplest treatment an elastic beam of thickness h is bent to a radius ρ, where $\rho \gg h$, and the deformation is assumed to consist solely of the extension or contraction of longitudinal fibres of the beam, in proportion to their distance from a central *neutral surface* which retains its original length (Fig. 5.5). A fibre at a distance $\rho + y$ from the centre of curvature alters its length by the ratio $(\rho + y)/\rho$, so that the longitudinal stress at a distance $\pm y$

from the neutral surface is $\pm Ey/\rho$. If the width of the beam at this distance from the neutral surface is $b(y)$, the longitudinal force df in the layer between y and $y + dy$ is $Eyb\,dy/\rho$. In *pure bending* the total longitudinal force exerted over any cross-section of the beam is zero, i.e.

$$\frac{E}{\rho} \int yb(y)\,dy = 0, \qquad (5.14)$$

which means that the neutral surface passes through the centre of area of the cross-section. Each layer of force df, however, also exerts a moment

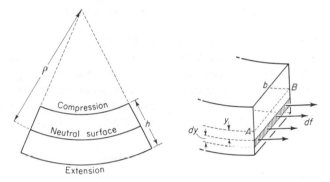

Fig. 5.5. Bending of a beam.

$y\,df$ about an axis (AB in Fig. 5.5) in the neutral surface. The moments from successive layers above and below this surface co-operate in tending to rotate the cross-section about this axis. They produce a total *bending moment* M,

$$M = \frac{E}{\rho} \int y^2 b(y)\,dy = \frac{EI}{\rho}, \qquad (5.15)$$

where $\int y^2 b(y)\,dy$ is the *second moment of area* I of the cross-section. For a rectangular beam, where $b(y) = b = $ constant,

$$I = b \int_{-h/2}^{+h/2} y^2\,dy = \frac{bh^3}{12}. \qquad (5.16)$$

This bending moment due to the longitudinal stresses in the beam must be balanced by the external forces which produce the bent state. Consider a beam subjected to *symmetrical four-point loading*, as in Fig. 5.6. The section ABCD is subjected to the external forces $\pm F$, which exert no resultant force but apply a moment Fd to the beam, and to the bending moment M applied by the longitudinal stresses to the face CD where it joins the rest of the beam. If this element is in static equilibrium, these moments balance, i.e.

$$M = Fd. \tag{5.17}$$

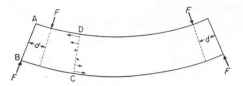

Fig. 5.6. Symmetrical four-point loading.

The argument holds for any position of CD between the central loading points. Hence, symmetrical four-point loading produces uniform pure bending in the region between the central loading points. Using Equations (5.15), (5.16) and (5.17), it is thus possible to measure Young's modulus from the radius of curvature produced in a beam by this type of loading.

The deflexion of such a beam can be found from Fig. 5.7. We have

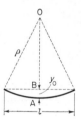

Fig. 5.7. Deflexion of a uniformly bent beam.

$\rho = OA$, $\rho^2 = (OB)^2 + (l/2)^2$, and thus

$$y_0 = AB = \rho - \left(\rho^2 - \frac{l^2}{4}\right)^{\frac{1}{2}} = \rho\left[1 - \left(1 - \frac{l^2}{4\rho^2}\right)^{\frac{1}{2}}\right]. \tag{5.18}$$

When $\rho \gg l$, we can use a series expansion, $(1 - a)^{\frac{1}{2}} = 1 - \frac{1}{2}a - \frac{1}{8}a^2 \ldots$, to simplify this expression. It is usually sufficient to neglect all terms beyond $\frac{1}{2}a$, i.e.

$$y_0 = \frac{l^2}{8\rho} = \frac{Ml^2}{8EI}. \tag{5.19}$$

The resistance of the beam to bending depends on its *flexural rigidity EI*. Beams used in structural engineering are designed to obtain maximum rigidity from minimum material. Flexural rigidity increases rapidly with increasing h, but if h/b is too large, the beam becomes ribbon-shaped and liable to collapse by twisting (cf. Section 5.3).

When the failure stress of the material is the same in both tension and compression (e.g. structural steel and most ductile metals), beams symmetrical about the neutral surface are used. Since fibres near the neutral surface are relatively understressed, this part of a beam is usually reduced to a thin web (but not so thin as to buckle), as in the familiar I-shaped engineering beam. Materials which break more easily in tension than compression (e.g. cast-iron and concrete) are best used in beams in which $b(y)$ is large on the tensile side of the neutral surface and small on the compressive side, so as to match the longitudinal stresses to the strengths of the material. In *prestressed* concrete beams the problem of the tensile weakness of concrete is overcome by introducing longitudinal

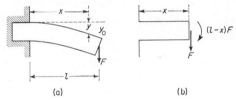

Fig. 5.8. (a) Deformation of a cantilever by an end load; (b) shearing force and bending moment applied to fixed part of beam through the vertical section at x.

compressive stresses into the concrete component of the beam, by means of steel tension bars which run longitudinally through the concrete and pull on its ends.

The stress state produced by symmetrical four-point loading is particularly simple. More commonly, beams are subjected to *non-uniform* bending moments and to *shearing forces*. Figure 5.8 shows a simple example, a *cantilever* of length l with one horizontally *built-in* or *encastré* end and with a vertical load F on its free end. Consider the forces exerted across that section of the beam at a distance x from the fixed end. The force F is transmitted to this section by the material beyond it. There is thus a vertical force F applied on the section at x; since the section is vertical, this is a *shearing force* which acts as shown in diagram (b). The end load F also exerts, through the leverage of the $(l - x)$ part of the beam, a bending moment

$$M = (l - x)F \qquad (5.20)$$

on the cross-section at x. The shearing force is thus constant along the beam, but the bending moment increases linearly with distance from the free end. More complicated distributions of load, which can also produce non-uniform shearing forces, can be analysed by an obvious extension of these same principles of static equilibrium.

The condition for equilibrium of bending moments on the section at x is, from Equations (5.15) and (5.20)

$$\frac{EI}{\rho} = (l - x)F. \tag{5.21}$$

We assume that the beam is only slightly bent out of the horizontal position, so that the relation

$$\frac{d^2y}{dx^2} = \frac{1}{\rho} = \frac{F}{EI}(l - x) \tag{5.22}$$

holds to a good approximation, where y is the vertical deflexion at x. Integrating twice,

$$y = \frac{F}{EI}\left(\frac{lx^2}{2} - \frac{x^3}{6}\right) + c_1 x + c_2. \tag{5.23}$$

Since $y = dy/dx = 0$ at $x = 0$, the integration constants c_1 and c_2 are zero. We thus know the deflexion of the beam. Its maximum value, at $x = l$, is

$$y_0 = \frac{F}{EI}\frac{l^3}{3}. \tag{5.24}$$

As a similar problem, we consider a beam freely supported at its ends and loaded in the middle, as in Fig. 5.9. At its central section $dy/dx = 0$ by

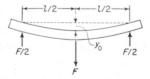

Fig. 5.9.   Symmetrical three-point loading.

symmetry. The problem is the same as that of Fig. 5.8, if we regard this as two encastré beams of length $l/2$ joined end-to-end and deflected upwards at their free ends by forces $F/2$. Hence, replacing F by $F/2$ and l by $l/2$ in Equation (5.24), the central deflexion of this beam is given by

$$y_0 = \frac{F}{EI}\frac{l^3}{48}. \tag{5.25}$$

A slightly more complicated problem is the centrally loaded beam with built-in ends, as shown in Fig. 5.10. We may, however, regard this as four quarter-length beams, each like that of Fig. 5.8, loaded with a force $F/2$ at one end and held horizontally at the other. By symmetry the points of inflexion occur half-way between the ends and the centre. We

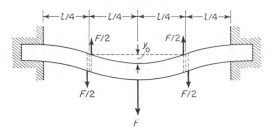

Fig. 5.10.   Beam fixed at ends and loaded in middle.

thus replace F by $F/2$ and l by $l/4$ in Equation (5.24), and double y_0, since two quarter-beams contribute to the central deflexion. Hence,

$$y_0 = \frac{F}{EI}\frac{l^3}{192}.$$ (5.26)

Comparing this with Equation (5.25), we see that unclamping the ends increases the deflexion of a centrally loaded beam fourfold.

Shearing forces also deflect beams, although usually by insignificant amounts compared with the bending deflexions. Consider an I-beam, the shearing deflexion of which can be analysed fairly simply, loaded as in Fig. 5.9. Section and side views of the beam, and the shearing deflexion, are shown in Fig. 5.11. It follows from Fig. 5.9 that the shear force carried on each half of the beam is $F/2$. This force is mainly supported by the resistance of the central web to shear deformation, as in Fig. 5.11(c)

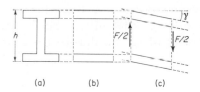

Fig. 5.11.   An I-beam: (a) section;   (b) side view;   (c) shear deflexion.

Hence, if A_w is the cross-sectional area of the web and μ the shear modulus, the angle of shear γ is given by

$$\gamma = \frac{F}{2A_w\mu}$$ (5.27)

approximately. The *extra* deflexion of the beam at the centre, y'_0, due to this shear is then given by

$$y'_0 = \frac{F}{2A_w\mu}\frac{l}{2}.$$ (5.28)

The total deflexion at the centre is then $y_0 + y'_0$, where y_0 is the deflexion due to bending, as given by Equation (5.25). If the area of each flange is A_f, then $I \simeq \frac{1}{2}A_f h^2$ and $y_0 \simeq Fl^3/24EA_f h^2$. Thus,

$$\frac{y'_0}{y_0} \simeq \frac{6E}{\mu}\frac{A_f}{A_w}\frac{h^2}{l^2}.$$ (5.29)

For materials such as steel, $6E/\mu \simeq 15$. The shearing deflexion is then appreciable, if l is not very long compared with h or if the web is very thin.

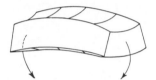

Fig. 5.12.   Anticlastic bending.

In the above examples we have analysed only the simplest bending problems by the simplest methods. One important effect we have neglected is *anticlastic bending*. If we bend a thick piece of rubber sharply, it takes a shape such as that of Fig. 5.12. This is an effect of Poisson's ratio, which causes the longitudinal fibres of the beam to become thin when stretched and thick when compressed. These lateral contractions and expansions distort the cross-section in the manner shown. The ratio of curvatures on such a doubly bent surface can be shown, in fact, to be equal to Poisson's ratio, which can thus be determined from anticlastic bending.

5.5   DEFLEXION OF COILED SPRINGS

A close-coiled helical spring, as shown in Fig. 5.13(a), deflects mainly by torsion. The second diagram shows one arm of the spring, of wire diameter d and coil radius R, under an axial load F. The *twisting moment* has the same value FR on every cross-section and is balanced by the torsional resistance of the wire. Replacing r_1 by $d/2$ in Equation (5.13), and $\mu\theta r_1/l$ by σ_s (the maximum shear stress produced in the wire by the torsion), we obtain $\pi\sigma_s d^3/16$ as the moment of torsional resistance. Hence,

$$F = \frac{\pi\sigma_s d^3}{16R} . \tag{5.30}$$

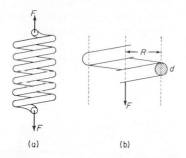

(a) (b)

Fig. 5.13. A close-coiled helical spring.

The angle of twist $\delta\theta$ of a short length δl of wire is given by

$$\delta\theta = \frac{2\sigma_s}{\mu d} \delta l, \tag{5.31}$$

and this rotation causes F to fall a distance $R\,\delta\theta$. If the total length of wire in the coil is l, the total elongation u of the spring is then given by

$$u = \frac{2\sigma_s Rl}{\mu d} = \frac{32FR^2 l}{\pi\mu d^4} = \frac{64FR^3 n}{\mu d^4}, \tag{5.32}$$

where $l = 2\pi Rn$ and n is the number of coils. The coefficient of proportionality, $\mu d^4/64R^3 n$, between force and deflexion is the *spring constant*. The *elastic resilience* of the spring is given by

$$\tfrac{1}{2}uF = \frac{\sigma_s^2 V}{4\mu}, \tag{5.33}$$

where V ($= \pi l d^2/4$) is the volume of wire and σ_s now refers to the failure stress of the material.

5.6 ELASTIC STABILITY AND BUCKLING

Suppose that a body, elastically stressed in equilibrium, is given a small additional deflexion. Does the body then tend to return to its original equilibrium point or does it tend to depart from it even further?

This is the general problem of *elastic stability* and is important in engineering, since structures in unstable equilibrium are unreliable; a small disturbance can lead to total elastic collapse (cf. Fig. 4.10). The equilibrium is stable if every small and geometrically possible distortion of the structure sets up restoring forces which tend to return the structure to its initial form. Many forms of instability can occur, e.g. combined twisting and bending of beams, buckling of compressed columns and struts, and whirling of rotating shafts. The investigation of the limiting conditions for stable equilibrium in such cases is an important branch of engineering. We shall consider only the buckling of struts under compression.

Let ABC in Fig. 5.14(a) represent a long, initially straight strut of length

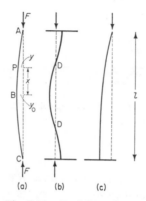

Fig. 5.14.   Buckling of struts.

l, uniform in cross-section and properties. Suppose that the compressive force F is applied at the centres of area of the end-faces and acts along the initial centre line of the strut. Suppose also that the strut is then very slightly bent as shown. We have, for the curvature at a general point P,

$$\frac{d^2y}{dx^2} = -\frac{1}{\rho} = -\frac{M}{EI} = -\frac{Fy}{EI},\qquad(5.34)$$

the minus sign being used because the curve is concave to the x axis. We write $\alpha^2 = F/EI$ and obtain

$$\frac{d^2y}{dx^2} + \alpha^2 y = 0\qquad(5.35)$$

as the equation of the bent strut, with the solution $y = m \sin \alpha x + n \cos \alpha x$. The strut is symmetrical about the point $x = 0$, which is in accordance with the cosine function, since $\cos(-x) = \cos x$, but not with the sine function, since $\sin(-x) = -\sin x$. Hence, we put $m = 0$ and obtain

$$y = y_0 \cos \alpha x. \qquad (5.36)$$

We have $y = 0$ at $x = \pm l/2$. Thus, $y_0 \cos(\alpha l/2) = 0$. This is satisfied (with $y_0 \neq 0$) if $\alpha l = \pi$; i.e.

$$F = \alpha^2 EI = \frac{\pi^2 EI}{l^2}, \qquad (5.37)$$

which is *Euler's formula* for the buckling load.

The value $\alpha l = \pi$ implies that $\cos(\alpha l/2) = 0$ and, hence, that y_0 is *indeterminate* from Equation (5.36). What this means is that at this critical value of F the strut is in *neutral equilibrium*; the buckling forces and elastic restoring forces are precisely balanced for every (small) value of y_0, so that the strut can equally well take up any amplitude of deflexion. For end loads smaller than this, the restoring forces predominate and the strut can return to its initially straight form by pushing back the end loads; this is stable equilibrium. For larger loads the converse is true; the slightest disturbance from the exactly straight form will grow catastrophically into a complete folding of the strut. In the later stages of collapse the curvature becomes too sharp for the above equations to apply, and eventually the fibre stresses in the strut may exceed the elastic limit of the material.

The above formula applies only to struts loaded through pin-joints at their ends. If the ends are forced to remain parallel to the initial line of the strut, as in Fig. 5.14(b), the line of thrust of the applied force passes through the points of inflexion D, thus producing by its bending moment the opposite curvatures of the strut outside and inside these points. By symmetry these points occur at $l/4$ from each end (cf. Fig. 5.10). Hence, the central half of the strut, between the points of inflexion, behaves like a pin-jointed strut of length $l/2$. The failure load is thus four times that of Equation (5.37). Similar analyses can be made of Fig. 5.14(c) and other arrangements of struts. The general result is

$$F = \frac{n\pi^2 EI}{l^2}, \qquad (5.38)$$

where n has the following values:

(a) Both ends pin-jointed (Fig. 5.14(a)), $n = 1$.
(b) Both ends fixed in direction and position (Fig. 5.14(b)), $n = 4$.
(c) Both ends fixed in direction, one end free to move sideways, $n = 1$.
(d) Both ends fixed in position, one end pin-jointed, $n = 9/4$.

(e) One end fixed in direction, other end pin-jointed and free to move sideways (Fig. 5.14(c)), $n = 1/4$.

To avoid buckling, a strut needs to be short and stocky, with a large moment of area. In the limit, the Euler collapse load then exceeds the simple crushing strength of the strut, which is $\sigma_f A$, where σ_f is the failure stress (by yielding or fracture) and A the cross-sectional area. This is shown in Fig. 5.15. Very short struts (e.g. cylindrical columns not more than

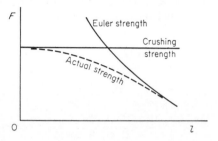

Fig. 5.15.   Failure load in relation to the length of a strut.

about twice their diameter in length) fail by simple crushing, and long ones fail by elastic buckling. Struts of intermediate length fail by a mixture of buckling and crushing, and their failure loads fall below those given by the curves of buckling and crushing alone.

5.7   PLANE STRESS AND STRAIN

We now turn to the general problem of solving the equations of Chapter 4 which govern the deformation and state of stress in an isotropic Hookeian solid. In some problems, particularly those of *plane stress* and *plane strain*,

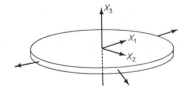

Fig. 5.16.   Plane stress.

there are geometrical simplifications which enable a general method of solution to be found.

When a flat sheet or membrane is stressed by forces in its plane, as in Fig. 5.16, its stress-free surfaces (normal to X_3) are principal planes and stress components (σ_{13}, σ_{23}, σ_{33}) directed out of this plane are zero. If the sheet is very thin, this stress state continues throughout its thickness and is then two-dimensional; i.e. it is a *state of plane stress* in which σ_{11}, σ_{12} and σ_{22} are constant through the thickness and all other stress components are zero. Because of the Poisson's ratio effect, the strains may not be zero through the thickness.

In *plane-strain* problems the *displacements*, not the stresses, are all parallel to a plane. These are mostly problems of *thick plates*. As an example, we show in Fig. 5.17 a thick block indented by a knife ($l \gg \rho$).

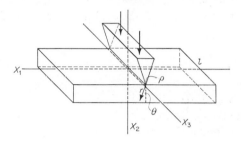

Fig. 5.17. Plane strain produced by a knife.

Particles are obviously pushed downwards and sideways by the knife but not *along* the knife (except possibly at the ends, where they are free to fall outwards along the X_3 axis). Over practically the whole strain field, the displacement u_3 along X_3 is zero and u_1 and u_2 are independent of X_3. The strains are two-dimensional; $\varepsilon_{13} = \varepsilon_{23} = \varepsilon_{33} = 0$. Also $\sigma_{13} = \sigma_{23} = 0$. The stress σ_{33} may exist because of the Poisson's ratio effect,

$$\sigma_{33} = v(\sigma_{11} + \sigma_{22}). \tag{5.39}$$

In both plane stress and plane strain we thus have to determine σ_{11}, σ_{12} and σ_{22}.

The vanishing of various stresses and strains greatly simplifies the equations of Chapter 4. The only *compatibility equation* required in plane stress and strain is

$$2\frac{\partial^2 \varepsilon_{12}}{\partial x_1 \, \partial x_2} = \frac{\partial^2 \varepsilon_{11}}{\partial x_2^2} + \frac{\partial^2 \varepsilon_{22}}{\partial x_1^2}. \tag{5.40}$$

The equations of *stress equilibrium* (assuming no body forces) become

$$\frac{\partial \sigma_{11}}{\partial x_1} + \frac{\partial \sigma_{12}}{\partial x_2} = \frac{\partial \sigma_{22}}{\partial x_2} + \frac{\partial \sigma_{12}}{\partial x_1} = 0, \tag{5.41}$$

and *Hooke's law* becomes

$$\varepsilon_{11} = \frac{1}{E}(\alpha\sigma_{11} - \beta\nu\sigma_{22}),$$

$$\varepsilon_{22} = \frac{1}{E}(\alpha\sigma_{22} - \beta\nu\sigma_{11}), \qquad (5.42)$$

$$2\varepsilon_{12} = \frac{\sigma_{12}}{\mu} = \frac{2(1+\nu)}{E}\sigma_{12},$$

where $\alpha = \beta = 1$ for plane stress; and $\alpha = 1 - \nu^2$ and $\beta = 1 + \nu$ for plane strain. Equations (5.41) and (5.42) enable us to write Equation (5.40) in the form

$$\frac{\partial^2(\sigma_{11} + \sigma_{22})}{\partial x_1^2} + \frac{\partial^2(\sigma_{11} + \sigma_{22})}{\partial x_2^2} = \left(\frac{\partial^2}{\partial x_1^2} + \frac{\partial^2}{\partial x_2^2}\right)(\sigma_{11} + \sigma_{22}) = 0. \quad (5.43)$$

We thus have three differential equations ((5.41) and (5.43)) for the three stresses σ_{11}, σ_{12} and σ_{22}. Many useful solutions of these equations can be found by constructing various mathematical functions which satisfy them and then ascertaining which of these functions fit the problems in question. This is done by means of the *Airy stress function* χ, which is defined by the relations

$$\sigma_{11} = \frac{\partial^2\chi}{\partial x_2^2}, \qquad \sigma_{22} = \frac{\partial^2\chi}{\partial x_1^2}, \qquad \sigma_{12} = -\frac{\partial^2\chi}{\partial x_1\,\partial x_2}, \qquad (5.44)$$

and which gives the compatibility condition (Equation (5.43)) in the form

$$\frac{\partial^4\chi}{\partial x_1^4} + 2\frac{\partial^4\chi}{\partial x_1^2\,\partial x_2^2} + \frac{\partial^4\chi}{\partial x_2^4} = \left(\frac{\partial^2}{\partial x_1^2} + \frac{\partial^2}{\partial x_2^2}\right)^2\chi = 0. \qquad (5.45)$$

This is called the *biharmonic equation*, and the forms of χ which are solutions of it are *biharmonic functions*. From the definition of χ in Equations (5.44), any biharmonic function automatically satisfies Equations (5.41). We notice that the elastic constants do not appear in Equations (5.41) and (5.45). The mechanical state given by a biharmonic function is thus valid for any isotropic, Hookeian solid. The elastic constants only enter after the problem has been solved, when we find the strains from the stresses.

The simplest useful solutions are of the form $\chi = ax_1^2 + bx_1x_2 + cx_2^2$. They represent *constant* stresses in the body. Such a polynomial solution must be of at least second degree if it is not to vanish when differentiated twice to give the stresses. The simplest solution which gives non-uniform stresses is of the form $\chi = ax_2^3$. It gives $\sigma_{11} = 6ax_2$, $\sigma_{12} = \sigma_{22} = 0$, and so represents pure bending of a beam about the X_3 axis with X_1 as the longitudinal fibre axis and with $6a = E/\rho$ (cf. Section 5.4). There are various ways of arriving at other solutions, some of which we shall discuss later. The principle of superposition is often helpful, since it enables a complicated solution χ to be obtained as a sum

$$\chi = \chi_1 + \chi_2 + \cdots \tag{5.46}$$

of simpler solutions $\chi_1, \chi_2 \ldots$ etc.

5.8 CYLINDRICAL POLAR CO-ORDINATES

Cartesian co-ordinates are used for setting up the basic equations of elasticity, but the solution of actual problems often goes more easily in

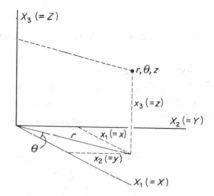

Fig. 5.18. Cylindrical polar co-ordinates.

other co-ordinate systems. *Cylindrical polar co-ordinates*, r, θ, z, where

$$r^2 = x^2 + y^2 = x_1^2 + x_2^2,$$

$$x = r \cos \theta, \qquad y = r \sin \theta, \tag{5.47}$$

$$z = x_3,$$

and which are shown in Fig. 5.18, are particularly useful. We note the following differential relations between these co-ordinates,

$$\frac{\partial r}{\partial x} = \frac{x}{r} = \cos \theta, \qquad \frac{\partial r}{\partial y} = \frac{y}{r} = \sin \theta,$$

$$\frac{\partial \theta}{\partial x} = -\frac{y}{r^2} = -\frac{\sin \theta}{r}, \qquad \frac{\partial \theta}{\partial y} = \frac{x}{r^2} = \frac{\cos \theta}{r}. \tag{5.48}$$

To express the basic equations in these co-ordinates, we consider a volume element PQRS in Fig. 5.19 with sides of lengths $dr(=PQ)$, $r\,d\theta(=PR)$ and

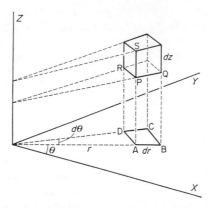

Fig. 5.19.   A volume element in polar co-ordinates.

$dz(=PS)$. The definitions of the displacement $u_3(=u_z)$ and strain $\varepsilon_{33}(=\varepsilon_{zz})$ remain unchanged. To define the others, we consider the projection ABCD of the element on the XY plane in both its unstrained (ABCD) and strained (A'B'C'D') forms, as in Fig. 5.20. Both the dimensions

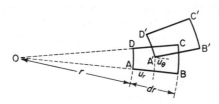

Fig. 5.20.   Displacement and strain of volume element.

and the deformation of the element are taken as infinitesimal. If A moves a distance u_r in the *radial* direction, the corresponding radial component of the movement of B is $u_r + (\partial u_r/\partial r)dr$. Since $AB = dr$, the *radial strain* is given by

$$\varepsilon_{rr} = \frac{\partial u_r}{\partial r}. \tag{5.49}$$

The unit elongation $\varepsilon_{\theta\theta}$ in the circumferential direction (along AD) is made up of two parts. First, even if the volume element is merely moved along the radial axes OA and OD, as in Fig. 5.21, there would be a circumferential strain, since AD would increase in length from $r\,d\theta$ to $(r + u_r)\,d\theta$. The unit elongation of AD due to this is given by

$$\frac{(r + u_r)\,d\theta - r\,d\theta}{r\,d\theta} = \frac{u_r}{r}. \tag{5.50}$$

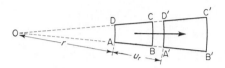

Fig. 5.21. Circumferential strain produced by radial displacement.

The second contribution comes from the *circumferential* component u_θ of the displacement A → A′ in Fig. 5.20. The corresponding value for D → D′ is $u_\theta + (\partial u_\theta/\partial\theta)\,d\theta$. Since AD $= r\,d\theta$, this gives a strain $(1/r)\,\partial u_\theta/\partial\theta$. The total circumferential strain is then

$$\varepsilon_{\theta\theta} = \frac{1}{r}\frac{\partial u_\theta}{\partial\theta} + \frac{u_r}{r}. \tag{5.51}$$

As regards shear, the radial components of the displacements D → D′ and A → A′ differ by the amount $(\partial u_r/\partial\theta)\,d\theta$. Since AD $= r\,d\theta$, the angle betwen AD and A′D′ is $(1/r)\,\partial u_r/\partial\theta$. Similarly, the angle between AB and A′B′ is $\partial u_\theta/\partial_r$. Some of this second angle is, however, contributed by rotation of the whole element through the angle AÔA′ ($=u_\theta/r$). Hence, the shear strain $\varepsilon_{r\theta}$ ($=\varepsilon_{\theta r}$) is given by

$$2\varepsilon_{r\theta} = \frac{\partial u_\theta}{\partial r} - \frac{u_\theta}{r} + \frac{1}{r}\frac{\partial u_r}{\partial\theta}. \tag{5.52}$$

The shears along the z axis are easily derived. The strain components are then

$$\varepsilon_{rr} = \frac{\partial u_r}{\partial r}, \qquad \varepsilon_{\theta\theta} = \frac{1}{r}\frac{\partial u_\theta}{\partial\theta} + \frac{u_r}{r}, \qquad \varepsilon_{zz} = \frac{\partial u_z}{\partial z},$$

$$2\varepsilon_{r\theta} = \frac{\partial u_\theta}{\partial r} - \frac{u_\theta}{r} + \frac{1}{r}\frac{\partial u_r}{\partial\theta}, \tag{5.53}$$

$$2\varepsilon_{\theta z} = \frac{1}{r}\frac{\partial u_z}{\partial\theta} + \frac{\partial u_\theta}{\partial z}, \qquad 2\varepsilon_{rz} = \frac{\partial u_r}{\partial z} + \frac{\partial u_z}{\partial r}.$$

The positive stress components are defined as in Fig. 5.22. The equations of stress equilibrium are found by the same method as before, although

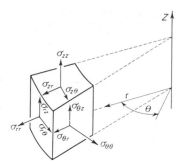

Fig. 5.22.   Stress components.

the non-rectangular shape of the volume element brings in two new features. First, the area of the cylindrical face at $r + dr$ exceeds that at r by the amount $(r + dr)\, d\theta\, dz - r\, d\theta\, dz = dr\, d\theta\, dz$, so that there is a difference of radial force $\sigma_{rr}\, dr\, d\theta\, dz$ on these two faces, quite apart from the contribution due to the difference $(\partial\sigma_{rr}/\partial r)\, dr$ in σ_{rr} on these two faces. Hence, the total difference in radial force on these faces is given by

$$\left(\frac{\partial\sigma_{rr}}{\partial r} + \frac{\sigma_{rr}}{r}\right) r\, dr\, d\theta\, dz. \tag{5.54}$$

A similar argument applies to the other stress components acting on the cylindrical faces. The second effect is that, because AB and CD are not quite parallel, the circumferential and radial forces acting on these faces have small components acting in each other's directions. For example, to the first order of infinitesimal quantities, the circumferential stresses $\sigma_{\theta\theta}$

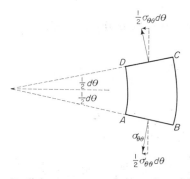

Fig. 5.23.   Radial components of circumferential stresses.

and $\sigma_{\theta\theta} + (\partial\sigma_{\theta\theta}/\partial\theta)\, d\theta$ on these two faces each produce a radial component $\frac{1}{2}\sigma_{\theta\theta}\, d\theta$, as shown in Fig. 5.23. They thus contribute $\sigma_{\theta\theta}\, dr\, d\theta\, dz$ to the total radial force. When these effects are taken into account, the equations of stress equilibrium become

$$\frac{\partial\sigma_{rr}}{\partial r} + \frac{1}{r}\frac{\partial\sigma_{r\theta}}{\partial\theta} + \frac{\partial\sigma_{rz}}{\partial z} + \frac{\sigma_{rr} - \sigma_{\theta\theta}}{r} + F_r\rho = 0,$$

$$\frac{\partial\sigma_{r\theta}}{\partial r} + \frac{1}{r}\frac{\partial\sigma_{\theta\theta}}{\partial\theta} + \frac{\partial\sigma_{\theta z}}{\partial z} + \frac{2\sigma_{r\theta}}{r} + F_\theta\rho = 0, \qquad (5.55)$$

$$\frac{\partial\sigma_{rz}}{\partial r} + \frac{1}{r}\frac{\partial\sigma_{\theta z}}{\partial\theta} + \frac{\partial\sigma_{zz}}{\partial z} + \frac{\sigma_{rz}}{r} + F_z\rho = 0.$$

In plane stress and strain the stresses σ_{rr}, $\sigma_{r\theta}$ and $\sigma_{\theta\theta}$ are constant with respect to z, σ_{zz} is either zero or $\nu(\sigma_{rr} + \sigma_{\theta\theta})$, and σ_{rz} and $\sigma_{\theta z}$ are zero. Equations (5.55) then reduce (omitting the body forces) to

$$\frac{\partial\sigma_{rr}}{\partial r} + \frac{1}{r}\frac{\partial\sigma_{r\theta}}{\partial\theta} + \frac{\sigma_{rr} - \sigma_{\theta\theta}}{r} = 0,$$

$$\frac{\partial\sigma_{r\theta}}{\partial r} + \frac{1}{r}\frac{\partial\sigma_{\theta\theta}}{\partial\theta} + \frac{2\sigma_{r\theta}}{r} = 0. \qquad (5.56)$$

These are satisfied by the stress function χ, where

$$\sigma_{rr} = \frac{1}{r}\frac{\partial\chi}{\partial r} + \frac{1}{r^2}\frac{\partial^2\chi}{\partial\theta^2},$$

$$\sigma_{r\theta} = \frac{1}{r^2}\frac{\partial\chi}{\partial\theta} - \frac{1}{r}\frac{\partial^2\chi}{\partial r\,\partial\theta}, \qquad (5.57)$$

$$\sigma_{\theta\theta} = \frac{\partial^2\chi}{\partial r^2},$$

and in terms of which the compatibility equation is

$$\left(\frac{\partial^2}{\partial r^2} + \frac{1}{r}\frac{\partial}{\partial r} + \frac{1}{r^2}\frac{\partial^2}{\partial\theta^2}\right)^2 \chi = 0. \qquad (5.58)$$

5.9 THICK-WALLED CYLINDERS

To show the use of the above formulae, we shall consider problems in which there is circular symmetry about the z axis, i.e. in which the stresses are constant with respect to θ. Many important problems are of this type, e.g. stresses in thick-walled pressure tubes and in rotating shafts and turbine rotors. Because of this symmetry, $\sigma_{r\theta}$ and all $\partial/\partial\theta$ terms vanish. Thus, Equations (5.56) and (5.58) reduce to the ordinary differential equations

$$\frac{d\sigma_{rr}}{dr} + \frac{\sigma_{rr} - \sigma_{\theta\theta}}{r} = 0, \tag{5.59}$$

$$\frac{d^4\chi}{dr^4} + \frac{2}{r}\frac{d^3\chi}{dr^3} - \frac{1}{r^2}\frac{d^2\chi}{dr^2} + \frac{1}{r^3}\frac{d\chi}{dr} = 0. \tag{5.60}$$

The general solution is of the form

$$\chi = A \ln r + Br^2 \ln r + Cr^2 + D,$$

$$\sigma_{rr} = \frac{A}{r^2} + B(1 + 2\ln r) + 2C,$$

$$\sigma_{\theta\theta} = -\frac{A}{r^2} + B(3 + 2\ln r) + 2C, \tag{5.61}$$

$$\sigma_{r\theta} = 0.$$

If the cylinder is solid, then $A = B = 0$, since the stresses otherwise become infinite at $r = 0$. The only solution is then $\sigma_{rr} = \sigma_{\theta\theta} = $ constant, i.e. uniform stress. If there is a hole along the axis of the cylinder, A and B need not be zero. By evaluating the displacements contributed by it, the B term can be shown to represent a state of self-stress due to a type of *dislocation* (with a wedge-shaped half-plane) in the cylinder. We shall suppose $B = 0$. Then

$$\sigma_{rr} = \frac{A}{r^2} + 2C,$$

$$\sigma_{\theta\theta} = -\frac{A}{r^2} + 2C. \tag{5.62}$$

Consider a cylinder of inner and outer radii r_0 and r_1, respectively, under

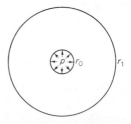

Fig. 5.24. Thick-walled cylinder under internal pressure.

an internal pressure p, as in Fig. 5.24. We then have the boundary conditions $\sigma_{rr} = -p$ at $r = r_0$ and $\sigma_{rr} = 0$ at $r = r_1$, i.e.

$$\frac{A}{r_0^2} + 2C = -p, \qquad \frac{A}{r_1^2} + 2C = 0, \tag{5.63}$$

so that $A = -pr_0^2 r_1^2/(r_1^2 - r_0^2)$ and $2C = pr_0^2/(r_1^2 - r_0^2)$. The stresses are thus

$$\sigma_{rr} = \frac{pr_0^2}{r_1^2 - r_0^2}\left(1 - \frac{r_1^2}{r^2}\right),$$

$$\sigma_{\theta\theta} = \frac{pr_0^2}{r_1^2 - r_0^2}\left(1 + \frac{r_1^2}{r^2}\right). \tag{5.64}$$

We see that σ_{rr} is compressive and $\sigma_{\theta\theta}$ tensile. The greatest tensile stress occurs at the inner surface and its value is

$$\sigma_{\theta\theta} = p\frac{(r_0^2 + r_1^2)}{(r_1^2 - r_0^2)} \quad \text{at} \quad r = r_0. \tag{5.65}$$

Thus, $\sigma_{\theta\theta}$ is never smaller than p, however thick the tube. As $r_1/r_0 \to \infty$, then $\sigma_{\theta\theta} \to p$, but we notice that $\sigma_{\theta\theta} = 1\cdot 25p$ when $r_1 = 3r_0$, so that, from the practical point of view, there is no great gain from making the tube much thicker than this. Since the outer layers of such a tube are relatively understressed (because of the r^{-2} terms in Equations (5.64)), a much stronger tube can be made by introducing *self-stresses* into it by shrinking hoops or tubes on to the central cylinder, or by tightly binding this cylinder with tape or wire. The superposition of these self-stresses on those from the internal pressure produces a more uniform circumferential stress through the thickness of the wall.

5.10 STRESS CONCENTRATION DUE TO A CIRCULAR HOLE

As a second example, we shall consider the way in which a small circular hole disturbs an otherwise uniform tensile stress σ in a large thin plate. Figure 5.25 shows the arrangement. Since the surface of the hole is free from applied forces,

$$\sigma_{rr} = \sigma_{\theta r} = 0 \quad \text{at} \quad r = r_0. \tag{5.66}$$

From St. Venant's principle, the disturbance to the stress field vanishes at large distances, i.e.

$$\sigma_{xx} \to \sigma, \qquad \sigma_{xy} \to \sigma_{yy} \to 0 \quad \text{as} \quad r \to \infty. \tag{5.67}$$

With the aid of Equations (5.47) and (5.48), we may resolve this into the stresses

$$\sigma_{rr} = \sigma \cos^2 \theta = \tfrac{1}{2}\sigma(1 + \cos 2\theta),$$

$$\sigma_{r\theta} = -\sigma \sin \theta \cos \theta = -\tfrac{1}{2}\sigma \sin 2\theta, \qquad (5.68)$$

$$\sigma_{\theta\theta} = \sigma \sin^2 \theta = \tfrac{1}{2}\sigma(1 - \cos 2\theta).$$

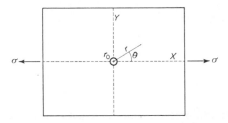

Fig. 5.25. Small hole in a thin plate under uniaxial tension.

We may thus regard this as two superposed stress fields σ' and σ'', where

$$\sigma'_{rr} = \tfrac{1}{2}\sigma, \qquad\qquad \sigma'_{r\theta} = 0, \qquad\qquad \sigma'_{\theta\theta} = \tfrac{1}{2}\sigma,$$

$$\sigma''_{rr} = \tfrac{1}{2}\sigma \cos 2\theta, \quad \sigma''_{r\theta} = -\tfrac{1}{2}\sigma \sin 2\theta, \quad \sigma''_{\theta\theta} = -\tfrac{1}{2}\sigma \cos 2\theta, \qquad (5.69)$$

when $r \to \infty$. Since σ'' depends on θ, whereas σ' does not, the way to satisfy the boundary conditions at $r = r_0$ is to make each satisfy them separately, i.e.

$$\sigma'_{rr} = \sigma'_{\theta r} = \sigma''_{rr} = \sigma''_{\theta r} = 0 \quad \text{at} \quad r = r_0. \qquad (5.70)$$

Since σ' is independent of θ, we may use Equations (5.62) for it. The boundary conditions, given by Equations (5.69) and (5.70), then require that $A = -2Cr_0{}^2$ and $C = \sigma/4$. Thus,

$$\sigma'_{rr} = \frac{\sigma}{2}\left(1 - \frac{r_0{}^2}{r^2}\right), \qquad \sigma'_{\theta\theta} = \frac{\sigma}{2}\left(1 + \frac{r_0{}^2}{r^2}\right). \qquad (5.71)$$

Since simple trigonometric functions of θ occur in the boundary conditions for σ'', we may try the stress function

$$\chi'' = f(r) \cos 2\theta \qquad (5.72)$$

for this stress, where $f(r)$ depends on r only. Substituting into Equations (5.57) shows that this gives the required angular terms in the stresses. The compatibility relation, Equation (5.58), then reduces to

$$\left(\frac{d^2}{dr^2} + \frac{1}{r}\frac{d}{dr} - \frac{4}{r^2}\right)^2 f(r) = 0, \qquad (5.73)$$

which has the general solution

$$f(r) = Er^2 + Fr^4 + \frac{G}{r^2} + H. \tag{5.74}$$

By expressing the stresses in terms of this solution, using Equations (5.57) and (5.72), and satisfying the boundary conditions, we find that

$$E = -\frac{\sigma}{4}, \qquad F = 0, \qquad G = -\frac{r_0^4}{4}\sigma, \qquad H = \frac{r_0^2}{2}\sigma. \tag{5.75}$$

Adding σ' and σ'', we then obtain the total state of stress round the hole as

$$\sigma_{rr} = \frac{\sigma}{2}\left[1 - \frac{r_0^2}{r^2} + \left(1 + \frac{3r_0^4}{r^4} - \frac{4r_0^2}{r^2}\right)\cos 2\theta\right],$$

$$\sigma_{r\theta} = -\frac{\sigma}{2}\left(1 - \frac{3r_0^4}{r^4} + \frac{2r_0^2}{r^2}\right)\sin 2\theta, \tag{5.76}$$

$$\sigma_{\theta\theta} = \frac{\sigma}{2}\left[1 + \frac{r_0^2}{r^2} - \left(1 + \frac{3r_0^4}{r^4}\right)\cos 2\theta\right].$$

The only non-vanishing stress at the surface of the hole is $\sigma_{\theta\theta}$, with the value

$$\sigma_{\theta\theta} = \sigma(1 - 2\cos 2\theta). \tag{5.77}$$

This is a maximum, $\sigma_{\theta\theta} = 3\sigma$, at the points $\theta = \pm\pi/2$.

The uniaxial stress is thus increased locally by a factor of three at the sides of a circular hole. This is important in engineering, because of the risk of local failure due to overstressing in the neighbourhood of bolt-holes, etc. To reduce this risk the material is often made thicker in the neighbourhood of such a hole. If σ is a uniaxial *compressive* stress, the greatest *tensile* value of $\sigma_{\theta\theta}(=-\sigma)$ is found at those points on the hole where $\theta = 0$ or π. Thus, if a glass plate containing such a hole is compressed, cracks form first along the axis of compression at the hole.

Since the stress disturbance varies as r^{-2}, its effect is very small at distances beyond about $3r_0$ from the hole. The above results can thus be extended to plates containing several holes, or to holes near free surfaces, provided the distances between them are not too small. By contrast, if a very small hole is drilled very near a large one, at $\theta = \pi/2$, the maximum stress at the edge of this small hole is then approximately $3^2\sigma$. We can extend this argument to a nest of n successively smaller holes, each in the maximum stress of the preceding one. We then have a stress of order $3^n\sigma$ at the edge of the smallest. If the radii decrease by a factor 10 from one to the next, and those of the largest and smallest are c and ρ, respectively, then $c = 10^n \rho$ and the concentrated stress σ_c at the smallest is very roughly given by $\log_{10} (\sigma_c/\sigma) \simeq n \log_{10} 3 \simeq \frac{1}{2}n \simeq \frac{1}{2} \log_{10} (c/\rho)$ i.e. by

$$\sigma_c \simeq \sigma \sqrt{\frac{c}{\rho}}. \qquad (5.78)$$

This is an extremely crude calculation, but it does indicate that a large stress concentration is to be expected at the sharp end of a notch or corner of a square hole. Proper calculations of the stress concentrations formed in such cases do, in fact, show that the stress increases as the square root of the ratio of width of the hole (across the uniaxial stress axis) to radius of the corner. Such large stress concentrations can be dangerous in engineering components, and are avoided as far as possible, e.g. by rounding off the corners to increase ρ.

5.11 DISLOCATIONS, BOUNDARIES AND CRACKS

We saw in Section 3.4 that the crystal dislocations of Fig. 3.13 could in principle be made by cutting into the crystal and then sliding one cut face over the other in a certain direction by a certain constant distance b (the Burgers vector). In Fig. 4.19 we also made a dislocation in a ring-shaped elastic body in the same way. The distortion of structure round a crystal dislocation can, in fact, be described by elasticity theory. Only at the centre of the dislocation does this approach break down. In an ideal Hookeian solid the stress would become infinite at the centre, if the dislocation were not hollow, but in the crystal there is no problem, since stress cannot vary across the space between neighbouring atoms. Although the displacement is multivalued across the cut face, in Fig. 4.19, it varies continuously round the dislocated ring, and we can apply all the equations of elasticity to it, provided the displacement increases by one Burgers vector each time we go round the dislocation.

Consider a screw dislocation in a cylindrical shell of length l and radius r, as in Fig. 5.26. The elastic strain has to accommodate a displacement $u_z = b$ round a length $2\pi r$. In an elastically isotropic material the accommodation must occur equally all round the shell, which suggests the simple result

$$u_z = \frac{\theta}{2\pi} b, \qquad u_r = u_\theta = 0, \qquad (5.79)$$

giving a (shear) strain and stress in the shell

$$\gamma_{\theta z} = (\gamma_{z\theta}) = \frac{b}{2\pi r}, \qquad \sigma_{\theta z} = (\sigma_{z\theta}) = \frac{\mu b}{2\pi r}, \qquad (5.80)$$

when substituted into Equations (5.53). We see from these equations that all the other strain and stress components are zero. In the absence of body forces, Equations (5.55) of stress equilibrium are satisfied by this state of stress. The compatibility relations are satisfied, since the strains are derived from a continuous displacement field. Hence, Equations (5.79) and (5.80) satisfy all the elastic equations.

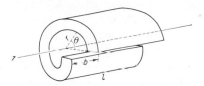

Fig. 5.26. A screw dislocation in a cylindrical shell.

We notice that there is no stress in the radial direction. This same stress field thus describes the elastic state round a screw dislocation in a thick cylinder whose cylindrical walls are free from external forces. The shear stress $\sigma_{z\theta}$ acting on the end-faces of the cylinder produces, however, a torque about the z axis. If the dislocated cylinder is entirely free from external forces and couples, it will be twisted slightly about its axis by this torque. Such twists have been detected experimentally in very thin crystals. The effect, which is readily calculated, superposes a stress of order $-\mu b r / r_1^2$ on that given by Equation (5.80), where r_1 is the outer radius of the cylinder. When the dislocation produces a step on the end-face of the cylinder, the stress field is further complicated by the stress concentration at the base of the step. This is important, however, only within a distance of order b from the base of the step.

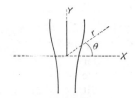

Fig. 5.27. An edge dislocation.

An edge dislocation has a more complicated stress field than a screw. We see from Fig. 4.19 that a displacement b has to be accommodated round a ring of length $2\pi r$, so that the strains and stresses must contain a term of type $b/2\pi r$. However, there is no longer circular symmetry about the dislocation axis, since the atoms in the half-plane are compressed and those beneath are expanded; cf. Fig. 5.27. The strain field is therefore of the form $(b/2\pi r)f(\theta)$, where $f(\theta)$ is a function, such as $\sin\theta$, that changes sign once as θ increases from 0 to 2π. It can be shown, in fact, that the field is described by the stress function,

$$\chi = -Dr\ln r\sin\theta, \tag{5.81}$$

where

$$D = \mu b/2\pi(1-v), \tag{5.82}$$

which gives the stresses

$$\sigma_{rr} = \sigma_{\theta\theta} = -D\frac{\sin\theta}{r}, \qquad \sigma_{r\theta} = D\frac{\cos\theta}{r},$$

$$\sigma_{xx} = -D\frac{y(3x^2+y^2)}{(x^2+y^2)^2}, \qquad \sigma_{yy} = D\frac{y(x^2-y^2)}{(x^2+y^2)^2}, \qquad \sigma_{xy} = D\frac{x(x^2-y^2)}{(x^2+y^2)^2}. \tag{5.83}$$

We see that the normal stress σ_{xx} in the direction of the Burgers vector is compressive on the half-plane side of the dislocation and tensile on the other side. The existence of the radial stress σ_{rr} means that this solution represents a dislocated cylinder with *stressed* cylindrical faces. Correction for this introduces additional stresses of order Dr/r_1^2, where r_1 is the outer radius of the cylinder. If the cylinder is hollow, with a stress-free inner boundary of radius r_0, a further stress field of order Dr_0^2/r^3 has also to be added.

The elastic strain energy of a dislocation is of the order of magnitude

$$U \simeq \int_{r_0}^{r_1} \frac{1}{2}\frac{b}{2\pi r}\frac{\mu b}{2\pi r}2\pi rl\,dr \simeq \frac{\mu b^2 l}{4\pi}\ln\left(\frac{r_1}{r_0}\right), \tag{5.84}$$

where we have integrated the energy in successive shells of the type of Fig. 5.26. For a unit dislocation in a typical crystal, we take $r_1 = 1$ cm, $r_0 \simeq 10^{-7}$ cm, $b \simeq 2\cdot5 \times 10^{-8}$ cm, and obtain

$$U \simeq \tfrac{4}{3}\mu b^3 n \tag{5.85}$$

for a dislocation line n ($\simeq l/b$) atoms long. For many crystalline solids $\mu b^3 \simeq 5$ eV, so that $u \simeq 7$ eV per atom length of the dislocation. This value is not greatly altered by the corrections for stress-free boundaries, mentioned above, or by the addition of the strain energy in the *core* of the dislocation, i.e. inside r_0, which is of order 1 eV per atom length. It is also insensitive to the shape of the crystal, provided most of the dislocation lies at a distance of order r_0 from any free surface, because the energy depends only logarithmically on r_0.

We see that even a short dislocation line has an energy orders of magnitude larger than kT ($\simeq 1/40$ eV at room temperature), so that, from the type of argument developed in Section 3.3, a dislocation is not thermodynamically stable in a crystal. If a crystal were annealed for a sufficiently long time at a high temperature, all its dislocations would disappear. Normally, however, some dislocations survive such annealing treatments, because they arrange themselves into small-angle boundaries, networks and other metastable substructures in crystals. Crystals free from dislocations have, however, been prepared, in silicon, by very slow crystallization from the melt.

Equation (5.84) shows that the energy of a dislocation depends on b^2. This means that a dislocation with a large Burgers vector $m\mathbf{b}$, where $m \gg 1$, is unstable, since it can reduce energy from m^2b^2 to mb^2 by splitting up into m separate and widely spaced dislocations each with a Burgers vector $\mathbf{b}$. Crystal dislocations, in fact, mostly have Burgers vectors equal to one lattice spacing. Since two parallel dislocation lines of the same sign (i.e. with parallel Burgers vectors $+\mathbf{b}$ and $+\mathbf{b}$) halve their energy when they move apart, there exists a force of repulsion between them. Conversely, since two dislocations of opposite signs ($+\mathbf{b}$ and $-\mathbf{b}$) reduce their energy to zero when they come together and mutually annihilate, there is a force of attraction between them.

A force is exerted on a dislocation line whenever the energy varies with the position of the dislocation, since, by definition, force is equal to change of energy with position. We see from Equation (5.84) that most of the energy lies at large distances from the centre of the dislocation; more than half of it lies outside the radius 10^{-4} cm, when $r_1 = 1$ cm and $r_0 = 10^{-7}$ cm. This is because the dislocation has a *long-range* elastic field, the stresses falling off only as r^{-1}. It follows that really large changes in this energy can be made only by reducing or "cutting-off" these long-range stresses. The most obvious way to do this is to place a second dislocation line, with a Burgers vector equal but opposite in sign, parallel to the first dislocation at a short distance r from it, forming a *dislocation dipole*. By the principle of superposition, the stress σ at any point due to this dipole is given by $\sigma = \sigma_1 + \sigma_2$, where σ_1 and σ_2 are the separate stresses of the two dislocations at that point. Beyond a distance of order r from the dipole we have $\sigma_1 + \sigma_2 \simeq 0$, so that there is little strain energy from the dipole at this range. At most points inside this order of distance we have either $\sigma_1{}^2 \gg \sigma_2{}^2$ or $\sigma_2{}^2 \gg \sigma_1{}^2$, so that the energy of the dipole is roughly equal to that of two dislocations each of length l in a cylinder with a radius of order r, i.e.

$$U \simeq \frac{\mu b^2 l}{2\pi} \ln\left(\frac{r}{r_0}\right). \qquad (5.86)$$

This varies with the distance r between the dislocations in the dipole, and, hence, is equivalent to an attractive force, F, per unit length of dislocation line, between them which is

$$F = \frac{1}{l}\frac{dU}{dr} \simeq \frac{\mu b^2}{2\pi r}.$$ (5.87)

Another example is provided by a dislocation near a free surface. The stresses from the dislocation are cut off at the surface, and this leads to energy and force formulae very similar to those of Equations (5.86) and (5.87), with r as the distance from the line to the surface. A dislocation is thus attracted towards a free surface.

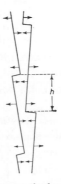

Fig. 5.28.   Forces across the boundary of Fig. 3.19.

The forces between parallel edge dislocations are complicated by the angular term in the stress field (Equations (5.83)). As well as the force F directed along the line between centres, there is also a force tending to alter the θ co-ordinate of one dislocation relative to another. In particular, this force tends to pull edge dislocations of the same sign into positions vertically above one another, so forming small-angle boundaries of the type shown in Fig. 3.19. This process of alignment occurs when severely bent crystals are annealed, and is called *polygonization*. The elastic energy of a small-angle boundary can be deduced by a simple argument, as follows. In Fig. 5.28 we consider the boundary of Fig. 3.19 to be made by bringing together two stepped crystal faces. Compressive forces are exerted between the crystals at places where a step presses into the other crystal face; and tensile forces across the widely spaced regions between steps. In equilibrium, these compressive and tensile forces balance. They produce no resultant force and no couple. They alternate with a periodicity h along the boundary. It then follows from St. Venant's principle that the resultant stresses from them do not appreciably penetrate into the crystals beyond a distance of order h from the boundary. We thus have a cut-off, and can estimate the energy of the boundary approximately as the sum of separate dislocation energies, each of order $(\mu b^2 l/4\pi) \ln (h/r_0)$. In terms of the boundary angle $\theta = b/h$, the energy γ per unit area of the boundary is then given by

$$\gamma = \frac{\mu b}{4\pi} \theta(A - \ln \theta), \tag{5.88}$$

where $A \simeq \ln (b/r_0)$. This formula, which we shall find useful when we come to discuss energies of boundaries in Chapter 8, has been confirmed experimentally.

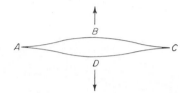

Fig. 5.29. A tensile crack.

Elastic *cracks* are closely related to dislocations. Suppose that such a crack is formed in a body under tensile stress, as shown in Fig. 5.29. If the original atomic bonds between the faces ABC and ADC remained unbroken and Hookeian at the large spacing which now separates these faces, a very large force would be exerted by them. The actual force is, however, zero, since Hooke's law, indeed all elastic resistance, completely fails at large spacings. The same zero force could, however, also be obtained, without any failure of Hooke's law, by inserting into the cavity ABCD a perfectly fitting piece of unstressed material. All the bonds in this piece and those that join it to the crack faces could then have their unstrained lengths. This lens-shaped insert can be regarded as the half-plane of a more general kind of edge dislocation whose Burgers vector (i.e. thickness of the lens) varies in length from zero at A and C to a maximum between B and D. We can regard this general dislocation as a lens-shaped group of half-planes of simple (i.e. constant h) edge dislocations, as in Fig. 5.30(a). If, after making the insert, we then removed the applied stress, the body would be in a state of self-stress produced by this group of dislocations. The stress field of the crack when the applied stress is present can thus be regarded as the superposition of the stress field of the dislocations together with the external stress.

Figure 5.30 shows a *tensile crack* and two forms of *shear cracks* or *shear faults* (displacement parallel to plane of fault) regarded as groups of simple dislocations. The tensile crack consists of edge dislocations with half-planes in the plane of the crack. The first shear fault consists of edge dislocations with half-planes perpendicular to the crack and to the applied stress; the second consists of screw dislocations parallel to the applied stress. Clearly, the dislocations are piled up more closely at the heads of these cracks. Since cracks have non-coherent surfaces, the Burgers vectors of these dislocations are no longer restricted to crystal vectors. In fact, we can use the same picture to describe cracks and faults in non-crystalline materials, such as the shear faults in the earth's crust which cause earthquakes when the stress due to the piled-up dislocations in them exceeds the strength of the material ahead. It is useful to regard such piled-up groups

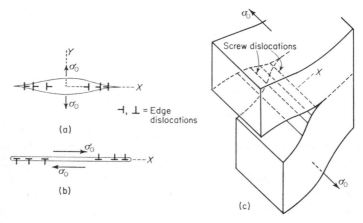

Fig. 5.30. Cracks regarded as groups of dislocations: (a) tensile crack; (b) shear crack; (c) shear crack (one half shown).

as *continuous* sheets of dislocations with infinitesimal Burgers vectors; we then define $b(x)\,dx$ as the *total* Burgers vector of all the dislocations lying at a distance between x and $x + dx$ from the centre of the crack.

Such a sheet represents a crack when the total stress (applied plus dislocation stresses) acting on any part of the crack face is zero. We have to add, at any point x_0 along the crack, all the stresses acting there from the dislocations at all other points x along the crack. For a crack of length $2c$, this sum extends from $x = -c$ to $x = +c$. We use Equations (5.80) and (5.83) to find the stresses at x_0 due to dislocations at x, and so deduce that the dislocations of Fig. 5.30 represent cracks with stress-free faces, provided that the equation

$$\sigma_0 + A \int_{-c}^{+c} \frac{b(x)\,dx}{x_0 - x} = 0 \qquad (5.89)$$

is satisfied over the range between $x_0 = -c$ and $x_0 = +c$, where σ_0 is the applied stress, as shown in Fig. 5.30, and where A is equal to $\mu/2\pi(1 - v)$

for edge dislocations and $\mu/2\pi$ for screw dislocations. The solution of this equation can be shown to be

$$b(x) = \frac{\sigma}{\pi A} \frac{x}{\sqrt{c^2 - x^2}}, \qquad (5.90)$$

which gives an elliptical shape of crack and

$$\sigma = \sigma_0 \frac{x}{\sqrt{x^2 - c^2}} \qquad (5.91)$$

for the stress σ on the x axis at points beyond the crack front. Let $r \, (=x - c)$ be the distance ahead of this front. Then, in the region $r \ll c$ the stress approximates to

$$\sigma \simeq \sigma_0 \sqrt{\frac{c}{2r}}, \qquad (5.92)$$

which is very similar to Equation (5.78), although here we are dealing with the stress at a distance r from the tip of an arbitrarily sharp crack. Equation (5.92) can, in fact, be applied to a notch of finite sharpness, provided $r \gg \rho$, where ρ is the radius of its tip. This characteristic stress concentration factor, with slightly different numerical factors, describes the stresses near the tips of all types of sharp notches, surface steps, keyways and other geometrical features, that cut off the applied stress over a length c and concentrate it into a much smaller length r at their ends.

FURTHER READING

CASE, J. and CHILVER, A. H., *Strength of Materials*, Arnold, London, 1959.

TIMOSHENKO, S., *Theory of Elasticity*, McGraw-Hill, New York, 1934.

SOUTHWELL, R. V., *Introduction to the Theory of Elasticity for Engineers and Physicists*, Oxford University Press, London, 1941.

READ, W. T., *Dislocations in Crystals*, McGraw-Hill, New York, 1953.

COTTRELL, A. H., *Dislocations and Plastic Flow in Crystals*, Oxford University Press, London, 1953.

NABARRO, F. R. N., Mathematical Theory of Stationary Dislocations, *Advan. Phys.*, **1**, 269 (1952).

MUSKHELISHVILI, N. I., *Some Basic Problems of the Mathematical Theory of Elasticity*, Noordhoff, Groningen, 1953.

PROBLEMS

5.1   A circular cylinder of internal bore 12 in. diameter and of wall thickness $\frac{1}{2}$ in. contains liquid under pressure from a piston which slides in it. The tensile stress in the cylinder wall must not exceed 10,000 psi. What is the maximum force which can be applied to the piston? [94·4 lb wt.]

5.2 Show that the circumferential stress in a circular ring rotating about its axis is given by $\rho v^2/g$, where ρ is the density of the material and v is the circumferential velocity of the ring. Assuming that this can be applied to a rotor, 6 ft in diameter, made of steel ($\rho = 0.278$ lb in^{-3}) and that the stress must not exceed 18,000 psi, what is the maximum rotational speed?   [1327 rev/min]

5.3 What percentage of material can be saved if a propeller shaft, of given torsional rigidity, is made hollow instead of solid, with an internal diameter which is half the external diameter?   [22%]

5.4 A beam of length l, carrying a uniformly distributed load w per unit length, is simply supported at its ends by equal vertical forces. Show that the maximum bending moment occurs at the centre of the beam and is equal to $wl^2/8$.

5.5 The maximum loading allowed on a floor is 336 lb/ft^2. The floor joists have a span of 14 ft, are 12 in. deep and 4·5 in. wide. The stress in them must nowhere exceed 1000 psi. What is the greatest distance, from centre to centre, that they can be placed apart?   [13·1 in.]

5.6 A light cantilever of length l carries a uniformly distributed load w per unit length. Show that the vertical deflexion of its free end is $wl^4/8EI$.

5.7 A steel scaffold tube has an external diameter of 2 in. and a wall thickness of 0·25 in. The bending stress in it must not exceed 20,000 psi. What is the greatest allowable bending moment on the tube?   [37,200 lb in.]

5.8 A close-coiled helical spring is required to absorb 2000 ft/lb of energy. It is initially unstressed, the maximum stress in it must not exceed 80,000 psi, and the maximum compression is 13·5 in. The mean diameter of the coils is 10 times that of the wire. The shear modulus is 11,200,000 psi. Calculate the number of coils and their mean diameter.

[5·65, 10·64 in.]

5.9 Show that, if a function $f(x, y)$ satisfies the equation

$$\frac{\partial^2 f}{\partial x^2} + \frac{\partial^2 f}{\partial y^2} = 0,$$

then $xf(x, y)$ and $(x^2 + y^2) f(x, y)$ can be used as stress functions.

5.10 Referring to Fig. 5.17, show that

$$\chi = -\frac{P}{\pi} r\theta \sin \theta$$

is a stress function which describes the state of stress in the plate, when the applied load is P per unit length of knife-edge. Show that the state of stress at the point r, θ is a simple radial stress,

$$\sigma_{rr} = -\frac{2P}{\pi} \frac{\cos \theta}{r}.$$

By a similar analysis show that the radial stress in a sharp knife of angle 2α, uniformly loaded with a force P per unit length is

$$\sigma_{rr} = \frac{P \cos \theta}{r(\alpha - \frac{1}{2} \sin 2\alpha)}.$$

Discuss the significance of this result near the knife-edge.

5.11 A horizontal beam of rectangular cross-section is loaded at one end, $x = 0$, with a vertical force F. Its other end, at $x = l$, is built into a rigid vertical wall, so that it forms a cantilever. The top and bottom faces of the beam have co-ordinates $y = +h/2$ and $y = -h/2$, respectively. The thickness b of the beam in the z direction is uniform and $b \ll h \ll l$. Show that

$$\chi = Axy^3 + Bxy,$$

where A and B are adjustable constants, is a stress function for this problem, when A and B are chosen to satisfy the boundary conditions. Then show that the stresses in the beam are

$$\sigma_{xx} = -xyF/I, \quad \sigma_{yy} = 0, \quad \sigma_{xy} = -\tfrac{1}{2}\left(\frac{h^2}{4} - y^2\right)F/I,$$

where I is the second moment of area.

5.12 Prove that, according to linear elasticity theory, there is no force between a straight edge dislocation and a screw dislocation parallel to it.

5.13 Show that the stress state round a screw dislocation, as given by Equation (5.80), produces a couple of magnitude $\frac{1}{2}\mu b(r_1^2 - r_0^2)$ on the end-faces of the cylinder, where r_1 and r_0 are the outer and inner radii of the cylinder. Then show that this couple can be annulled by superposing the stress state represented by Equations (5.10) and (5.11) with a suitable value of θ. In this way show that a cylindrical crystal of radius r with a screw dislocation along its axis, when free from external forces, has a twist of approximately $b/\pi r^2$ radians per unit length.

5.14 It can be proved that, when a small elliptical hole with semi-axes of lengths a and b lies in a plate with its a axis normal to an applied tensile stress σ, the concentrated tensile stress at the end of this axis is $\sigma\left(1 + \dfrac{2a}{b}\right)$.

Compare this formula, for various values of a and b, with Equations (5.77) and (5.78).

chapter 6

Waves and Vibrations in Solids

*In this chapter we shall consider the propagation of stresses
and strains in solids. We start with the velocities of stress
waves under various conditions and look briefly at shock waves
and plastic waves. Then we discuss the characteristic vibrations
of solids and problems of mechanical damping. The discussion
of vibrations then leads us to the theory of the specific heat of
solids. Finally, we outline the ways in which deviations from
Hooke's law and other factors determine thermal expansion
and thermal conductivity in solids.*

6.1 PROPAGATION OF MECHANICAL DISTURBANCES

We turn now to the way that solids behave when the forces acting on
them are no longer in static equilibrium. The unbalanced components of
force f then set up various motions in the body, according to Newton's law

$$f = \frac{d(mv)}{dt}, \tag{6.1}$$

which equates them to the rate of momentum change (m = mass, v = velocity, t = time). Many new topics come within our reach when we make
this transition from statics to dynamics; not only waves, vibrations,
mechanical damping and impact effects, but also thermal properties, such
as specific heat, thermal expansion and conductivity.

Consider a long uniform cylinder of material, with density ρ and cross-
sectional area A, in static equilibrium. Now let an unbalanced force be
applied to one end of it, setting the material there into motion. This motion

is transmitted to the next layer of material by the atomic interactions between layers, and so travels along the cylinder as a mechanical disturbance. Consider the forces which then act on a short element PQ of the cylinder, as in Fig. 6.1. At a given instant of time the force applied to the

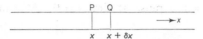

Fig. 6.1. Equilibrium of an element of a cylinder.

face P (co-ordinate x) by the material to its left is $f(x)$, and that applied to Q (co-ordinate $x + \delta x$) by the material to its right is $f(x + \delta x)$. The resultant force applied to the element is then

$$f(x + \delta x) - f(x) = \left[f(x) + \frac{\partial f(x)}{\partial x} \delta x \right] - f(x) = \frac{\partial f(x)}{\partial x} \delta x = A \frac{\partial \sigma}{\partial x} \delta x,$$

(6.2)

where $\sigma \, (= f/A)$ is a stress. The element has mass $\rho A \, \delta x$. Let u be its displacement in the direction of the resultant force. Thus, $mv = \rho A (\partial u / \partial t) \delta x$ and Newton's law becomes, from Equations (6.1) and (6.2),

$$\frac{\partial \sigma}{\partial x} = \frac{1}{A \, \delta x} \frac{d(mv)}{dt} = \rho \frac{\partial^2 u}{\partial t^2}.$$

(6.3)

We assume a stress–strain relation

$$\sigma = \alpha \frac{\partial u}{\partial x},$$

(6.4)

where α is an elastic constant, which enables us to write Newton's law as a *wave equation*,

$$\frac{\partial^2 u}{\partial x^2} = \frac{1}{c^2} \frac{\partial^2 u}{\partial t^2},$$

(6.5)

where

$$c = \sqrt{\frac{\alpha}{\rho}}.$$

(6.6)

If $y = ct$, this becomes

$$\frac{\partial^2 u}{\partial x^2} = \frac{\partial^2 u}{\partial y^2}.$$

(6.7)

Since x and y have equal status in this equation, *any* function of the variable $(x + y)$ is a solution. Similarly, *any* function of $(x - y)$ is also a solution, since the minus sign contributes a factor $(-1)^2$ when differentiated twice.

Hence, all functions of the general form $u = g(x + ct)$ and $u = g(x - ct)$ are solutions, e.g. z^n, z^{-n}, $\log z$, $\exp z$, $\sin z$, etc., where $z = x + ct$ or $z = x - ct$.

These solutions represent *stress pulses, sound waves* or *wave packets*, running along the cylinder as in Fig. 6.2. We can draw an arbitrary curve $g(x)$ to represent a pulse at time $t = 0$. At a later time t the same

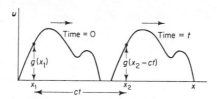

Fig. 6.2.   A propagating stress pulse.

curve is found at a distance ct along the rod, since $g(x_2 - ct) = g(x_1)$ when $x_2 = x_1 + ct$. The size and shape of the pulse depend on the initial disturbance, but the propagation depends only on the wave equation. The pulse thus propagates at the *wave velocity c*, without change of shape or size, i.e. without *dispersion* or *attenuation*. Such arbitrary pulses can be regarded as superposed continuous series of sinusoidal waves, all moving with the same speed c; hence the name *wave packets*.

We now consider some particular examples of this general argument. If the cylinder of Fig. 6.1 is a thin solid rod subjected to a uniaxial tensile force f, then $\alpha = E$ (Young's modulus) and the wave velocity is then

$$c_0 = \sqrt{\frac{E}{\rho}}.$$ (6.8)

If the rod is not thin (compared with the wavelength of the sinusoidal waves which form the pulse), the Poisson's ratio effect sets up lateral stresses and the wave packet then no longer consists of *plane waves*, i.e. waves with plane fronts perpendicular to the direction of propagation.

Some wave velocities are given in Table 6.1. We notice that aluminium and glass, because of their low densities, transmit waves quickly, despite their low elastic constants. The low value in rubber is important for motor tyres; at 100 mile/h (4500 cm sec^{-1}) a tyre is rolled on to a road at the speed of a longitudinal wave, so that elastic disturbances tend to build up into bulges near the region of contact.

As a second interpretation of Equation (6.5), we consider a wave spreading out spherically from a point of disturbance inside an infinite body. At great distances from the source (compared with the wavelength) any small part of the spherical wave-front is approximately a plane wave. We can therefore regard Fig. 6.1 as representing a radial cylinder of material and

TABLE 6.1

Elastic Wave Velocities, in 10^5 cm sec^{-1} ($=3281$ ft sec^{-1})

	Steel	Copper	Aluminium	Glass	Rubber
c_0	5·19	3·67	5·09	5·30	0·046
c_1	5·94	4·56	6·32	5·80	1·04
c_2	3·22	2·25	3·10	3·35	0·027
c_s	2·98	2·12	2·92	3·08	0·026

NOTE: c_0 = longitudinal waves in a thin rod.
        c_1 = longitudinal waves in an infinite body.
        c_2 = transverse waves in an infinite body.
        c_s = Rayleigh surface waves.

P and Q as traces of the intersection of the wave-front sweeping along this cylinder as the wave expands spherically. The displacement now has components v and w perpendicular to x, as well as u parallel to x. Because of the spherical symmetry, u, v and w can vary only along the (radial) x axis, i.e. the only non-vanishing strains are the normal strain $\partial u/\partial x$ and the shears $\partial v/\partial x$ and $\partial w/\partial x$. The normal strain is not that produced by uniaxial tension, since the sides of the bar can move neither inwards nor outwards. The Poisson's ratio strain is therefore suppressed,

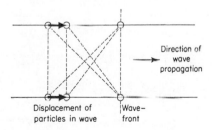

Fig. 6.3. Compression and shear in a longitudinal wave.

i.e. lateral stresses are set up to annul it. This type of strain, in which all particles in one sheet of the wave-front move equally along the x axis without moving sideways, is called *longitudinal* or *irrotational*. It is sometimes inaccurately called *dilatational*, but there is shear as well as dilatation in it, as Fig. 6.3 shows. The elastic constant for this mode of deformation is, from Equation (4.72), c_{11}. Using Equations (4.80) et seq., the longitudinal wave velocity c_1 is then

$$c_1 = \sqrt{\frac{c_{11}}{\rho}} = \sqrt{\frac{K + \frac{4}{3}\mu}{\rho}}, \qquad (6.9)$$

where K and μ are the bulk modulus and shear modulus, respectively. The shears $\partial v/\partial x$ and $\partial w/\partial x$ produce *transverse* or *equivoluminal* waves, in which the particles move perpendicularly to the direction of propagation. For these, $\alpha = \mu$ and the wave velocity is then

$$c_2 = \sqrt{\frac{\mu}{\rho}}. \qquad (6.10)$$

Values of c_1 and c_2 are given in Table 6.1. We notice that c_1 is much the largest of the various wave velocities in rubber, owing to the contribution of the bulk modulus.

These results can also be applied to waves in *fluids*. An ideal liquid cannot support a shear stress; i.e. $\mu = 0$, $c_2 = 0$, and

$$c_1 = \sqrt{\frac{K}{\rho}}. \qquad (6.11)$$

This provides a way of detecting the fluidity of the centre of the earth. Both longitudinal and transverse waves can be detected from earthquakes and other seismic disturbances. Near the surface the longitudinal waves travel at 8×10^5 cm sec^{-1} (26,250 ft sec^{-1}) and transverse waves at $4 \cdot 45 \times 10^5$ cm sec^{-1} (14,600 ft sec^{-1}). These speeds increase with depth, owing to the increased elastic constants of the highly compressed deep layers, until at a depth of 1860 miles there is a discontinuity of properties. The material below does not transmit transverse waves.

At a free surface of an elastic solid the wave velocity is altered, because stress components directed through the surface are zero there. Surface waves, called *Rayleigh waves*, are propagated with the velocity

$$c_s = f(v)\sqrt{\frac{\mu}{\rho}}, \qquad (6.12)$$

where $f(v)$ is a function of Poisson's ratio equal to $0 \cdot 9194$ when $v = 0 \cdot 25$ and to $0 \cdot 9553$ when $v = 0 \cdot 5$. Values of c_s are given in Table 6.1. Through an effect similar to St. Venant's principle, these waves are strong only within a depth of about one wavelength from the surface.

Surface waves are important geologically, because they spread out in two dimensions only, not three, as do spherical waves. They therefore do not lose their amplitude at large distances from the source (e.g. an earthquake) as rapidly as do the spherical waves, and so are the most damaging of earthquake waves. The first shocks to arrive from distant earthquakes are the longitudinal waves. Then come the transverse waves; finally, the surface waves, with much larger amplitudes. The later stages are complicated by the arrival of multiply reflected waves and by extra surface waves (called *Love waves*) due to stratification of the earth's surface.

Returning to Fig. 6.1, we now consider the propagation of some simple non-uniform stresses. *Torsional waves*, due to the twisting of one end of a round bar, are easily analysed. In place of the force f we have the applied torque T. The variation, $(\partial T/\partial x)\,\delta x$, of this along the element δx of the bar has to be balanced against $I(\partial^2\theta/\partial t^2)\,\delta x$, where I is the polar moment of inertia and $\partial^2\theta/\partial t^2$ the angular acceleration. The torsional wave is propagated with the velocity c_2 (Equation (6.10)).

Bending or *flexural* waves are more complicated. Figure 6.4 shows the

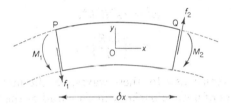

Fig. 6.4. An element of a beam in flexure.

element PQ transmitting a flexural wave. We suppose the bend to be small, disregard the rotation of PQ about O, and consider only the translational motion in the y direction, i.e. we limit the problem to long wavelengths only. The shear forces f_1 and f_2 have a vertical resultant $(\partial f/\partial x)\,\delta x$ which balances the force of inertia, i.e.

$$\rho A \frac{\partial^2 y}{\partial t^2} = \frac{\partial f}{\partial x}. \qquad (6.13)$$

Taking moments about O, we have

$$(M_2 - M_1) + \tfrac{1}{2}(f_1 + f_2)\,\delta x = \frac{\partial M}{\partial x}\,\delta x + f\,\delta x = 0,$$

i.e.

$$f = -\frac{\partial M}{\partial x}, \qquad (6.14)$$

where $f = \tfrac{1}{2}(f_1 + f_2)$. Since $M = EI(d^2y/dx^2)$ when the radius of curvature is large, where I is the second moment of area for this bending mode, Equation (6.13) can be written as

$$\frac{\partial^2 y}{\partial t^2} = -c_0^2 R^2 \frac{\partial^4 y}{\partial x^4}, \qquad (6.15)$$

where $c_0 = \sqrt{E/\rho}$ and where $R\,(= \sqrt{I/A})$ is the *radius of gyration*.

This is not a simple wave equation and, as substitution will prove, it is not satisfied by $y = g(ct - x)$. We try a solution of the form

$$y = y_0 \cos (\omega t - kx), \qquad (6.16)$$

where y_0 is the amplitude and

$$k = \frac{2\pi}{\lambda}, \qquad \omega = \frac{2\pi c}{\lambda}, \qquad (6.17)$$

λ being the wavelength and c the velocity. Substituting Equation (6.16) into Equation (6.15), we find that

$$\omega^2 = c_0^2 R^2 k^4,$$

$$c = \left(\frac{2\pi R}{\lambda}\right) c_0. \qquad (6.18)$$

Thus, c varies inversely as λ for these waves, i.e. the motion is *dispersive*. The shorter waves of a wave packet run on ahead of the longer ones, and the packet becomes increasingly diffuse the further it spreads. This solution is of course valid only for wavelengths long compared with the lateral dimensions of the beam. Since $\lambda \gg R$, then $c \ll c_0$.

In a dispersive wave system the velocity with which a message or signal can be transmitted, called the *group velocity* c_g, differs from the velocity c of individual waves, called the *phase velocity*. It is known from the general theory of wave motion that the relation between these is

$$c_g = c - \lambda_0 \frac{dc}{d\lambda}, \qquad (6.19)$$

where λ_0 is the mean wavelength of the packet which forms the signal. For flexural waves in a beam this becomes

$$c_g = c + \lambda\left(\frac{2\pi c_0 R}{\lambda^2}\right) = 2c, \qquad (6.20)$$

so that the group velocity is twice the phase velocity. This group velocity is of course small compared with the velocity c_0 with which longitudinal waves are transmitted along the beam.

6.2 IMPACT STRESSES

We now consider the stresses produced when a body suffers an impact. In Fig. 6.5 we show a light elastic wire of length l which receives the impact of a weight W dropped from a height h. The work done by W on the

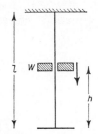

Fig. 6.5.   Impact of a weight on a tensile wire.

wire is $(h + \delta l)W$, where δl is the elastic extension when the weight is first brought to rest. Converted to strain energy, this is equivalent to a stress

$$\sigma_m = \sqrt{\frac{2(h + \delta l)EW}{Al}} \qquad (6.21)$$

in the wire. When $h \gg \delta l$,

$$\sigma_m = v c_0 \rho \sqrt{\frac{M}{m}}, \qquad (6.22)$$

where M and $m\,(= \rho Al)$ are the masses of the weight and wire, A and ρ are the cross-sectional area and density of the wire, $c_0 = \sqrt{E/\rho}$, and v is the striking velocity of the weight $(hW = \frac{1}{2}Mv^2)$. When $h = 0$, which corresponds to a live load W suddenly released on the end of a weightless spring, we have, since $E(\delta l/l) = \sigma_m$,

$$\sigma_m = \frac{2W}{A}. \qquad (6.23)$$

The maximum stress produced in the spring is thus twice that from the same load applied statically. The spring then oscillates in length about the static equilibrium point with an amplitude equal to the static equilibrium strain.

This simple theory neglects effects due to the finite speed of stress propagation along the wire. Suppose that one end R of a rod is suddenly set in motion with constant velocity v, as in Fig. 6.6. A stress wave runs along it, all material overrun by the stress-front S being in motion and under stress, all that before it being at rest and unstressed. An observer at rest in the unstressed part of the rod would see the stress-front approaching with speed c_0. Thus, in time t an initial length $c_0 t$, of mass $\rho A c_0 t$, is

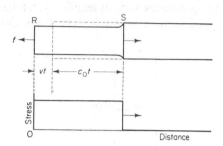

Fig. 6.6. Stress wave in a rod.

overrun by the wave and given a velocity v. From Equation (6.1), the force f required to supply its momentum is

$$f = \frac{d}{dt}(\rho A c_0 t v) = \rho A c_0 v. \qquad (6.24)$$

The stress in the wave is thus

$$\sigma = \frac{f}{A} = v c_0 \rho = E\frac{v}{c_0}, \qquad (6.25)$$

where $E = \rho c_0{}^2$, and the strain, v/c_0, is the change vt in the length $c_0 t$ of the rod. We notice that the wave-front and material move in opposite directions in a tensile wave.

Equation (6.25) shows, in contrast to Equation (6.22), that the stress wave starts with an intensity that depends only on v, not on M. Even a *small* mass (but large compared with the attachment at the struck end) could instantly break the wire of Fig. 6.5 at its struck end, if its striking velocity were large enough. For a steel wire of 200,000 psi tensile strength ($\simeq E/150$) this critical velocity is about $c_0/150$, i.e. 100 ft sec^{-1}. For an ordinary glass rod (strength about $E/1000$) it is about 16 ft sec^{-1}, which is reached when a weight falls 4 ft freely.

If the rod survives this initial impact, the falling weight then decelerates. Thus, from Equation (6.25), the later parts of the stress wave injected into the rod are weaker (Fig. 6.7(a)). A uniform rod will then survive at least until the wave-front reaches its far end. Here the wave is *reflected*, and, if the rod is rigidly fastened, this greatly intensifies the stress and chance of fracture at this end. A small decrease in the initial striking velocity across the critical range of velocities can therefore transfer the fracture from the near to the far end. Figure 6.7 shows the reflection. The general solution of the wave equation gives two waves, $u_a = g_a(ct + x)$ and $u_b = g_b(ct - x)$, moving along the rod in opposite directions. Let $x = 0$ at the fixed end. Then, $u = u_a + u_b = 0$ at $x = 0$ for all values of t, i.e.

$$g_a(c_0 t) = -g_b(c_0 t). \tag{6.26}$$

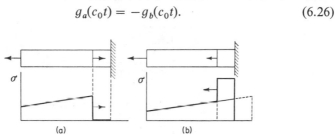

Fig. 6.7. A stress wave (a) due to a decreasing force; (b) after reflection from a fixed boundary.

The displacement produced by the reflected wave is thus *equal and opposite* to that produced by the incident wave. Since the direction of propagation is also reversed, the reflected stress wave, $E(\partial u_b/\partial x)$, has the *same* sign and intensity as the incident wave.

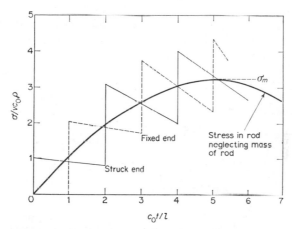

Fig 6.8. Rise in stress at ends of rod after impact, for $M = 10$ m (after G. I. Taylor, *Proc. Inst. Civil Engrs.* (*London*), **26**, 486 (1946)).

If the rod survives the first reflection, a second reflection occurs when the returning front reaches the struck end. Figure 6.8 shows the way the stress in the rod can build up through repeated reflections of this type at its ends. The peak stresses exceed σ_m (Equation (6.22)) by an amount roughly equal to the initial height $vc_0\rho$ of the stress wave. If $M = 10\ m$, the peak is about $1\cdot3\sigma_m$.

When the far end of the rod is free, reflection at this free surface produces a stress wave of opposite sign, which cancels the incident wave, because the boundary condition now is zero stress (i.e. $\partial u/\partial x = 0$) at the

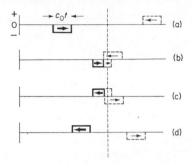

Fig. 6.9. Stages (a)–(d) in the reflection of a short compressive wave at a free boundary.

reflecting surface. An important effect follows from this when the incident wave is a compressive pulse of finite duration t, as shown in Fig. 6.9. Where the reflected tensile pulse overlaps the incident pulse there is a cancellation of stress. Later, as the waves pass each other, a tensile stress appears where the reflected wave is no longer cancelled. This tensile stress appears first at a distance $\tfrac{1}{2}c_0t$ from the free end and subsequently grows into a fully formed tensile wave.

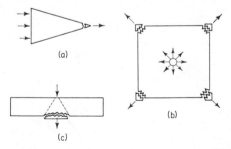

Fig. 6.10. Examples of tensile fractures produced by compressive shocks.

If this tensile stress is large, it may fracture the material. The far end of a free bar can thus be broken off by striking the near end sharply in compression! Figure 6.10 shows some examples of this effect. Diagram (a) shows a conical specimen, the tip of which is broken off by a compressive shock applied to its base. The conical shape causes the stress wave to intensify as it converges towards the tip. A similar effect happens in diagram (b), where the corners of a square plate are thrown off by a compressive shock (e.g. a bullet impact) at its centre. Diagram (c) shows how a cap of material is thrown off the back face of a plate due to a compressive shock on its opposite face (*scabbing fracture*).

The stress waves produced by the impacts of specks of grit blown in the wind and of water drops (e.g. fast aircraft in rainstorms, turbine blades in wet steam) can damage solids, e.g. produce surface cracks in brittle materials and surface plastic deformation in ductile ones. The stress is high during such an impact, because the time of contact t_0, before the particle rebounds, is short. The rebound time is that required for the stress waves to cross the impacting particle a few times, i.e. a small multiple of l/c_0 for a particle of length l. Detailed calculations show, in fact, for a sphere of radius r with initial velocity v, that

$$t_0 \simeq 5\frac{r}{c_0}\left(\frac{c_0}{v}\right)^{1/5}. \qquad (6.27)$$

Consider, for example, a steel ball of 0·5 cm radius and 4 g mass dropped freely on to a thick steel plate from a height of 0·5 cm. Then, $v = 31$ cm sec^{-1}, $(c_0/v)^{1/5} \simeq 7$, and $t_0 \simeq 35$ μsec. The average force exerted between the ball and plate during this time is $2mv/t_0$, i.e. about 8×10^6 dyn, equivalent to 18 lb wt. If the ball were pressed slowly and elastically into the plate by this force the stress under the indentation would be about 150,000 psi ($\simeq E/200$). This exceeds the yield or fracture stress of all but the strongest steel. Quite small heights of fall can thus cause damage. These impulsive effects are important in ball and roller bearings, and are exploited in the processes of *shot-peening* and *surface-hammering* for producing plastically worked surfaces on metals. In *ultrasonic machining*, used for the precision cutting of hard, brittle materials, the cutting tool does not touch the work piece but is oscillated at high-frequency (e.g. 2×10^7 c/s), for example by magnetostriction, to supply impact energy to abrasive particles (e.g. SiC or Al_2O_3) carried in a cutting fluid flooded round it. Ultrasonic oscillators are also used to polish materials by vibrating them in contact with polishing powders, to clean them in agitated cleaning fluids, and to destroy surface oxide films during welding processes.

The reflection of stress waves from free surfaces and other surfaces of mechanical discontinuity is used in methods for detecting *flaws* inside solids. A stream of low-energy ultrasonic pulses, from a magnetostrictive or piezoelectric oscillator, is injected into the specimen through a liquid film and reflected waves from internal cracks and foreign inclusions, as well as from the external surface, are detected by a probe.

6.3 SHOCK-FRONTS

In a large compressive wave, produced, for example, by an explosion, the atoms may be compressed sufficiently to raise their elastic constant significantly. Figure 6.11 shows an effect of this. The high stress at the

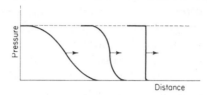

Fig. 6.11.   Formation of a compressive shock-front.

back of the pulse increases the elastic constant and wave velocity, so that this part of the wave travels faster than the front and the wave profile sharpens up into a discontinuous *shock-front*, across which there is a *finite* change of stress and strain. Such fronts have been detected in solids subjected to explosive impacts. Some of the p, V curves of Fig. 4.8 were determined from such experiments. The general effect is complicated by changes produced in the material. We saw in Fig. 6.3 that a longitudinal pulse involves both shear and dilatation. At a certain intensity of wave the shear stress reaches the plastic yield strength of the material. Waves much stronger than this in fact travel only with the speed $\sqrt{K/\rho}$, not $\sqrt{(K + \frac{4}{3}\mu)/\rho}$, (cf. Equations (6.9) and (6.11)), because the failure of the shear stress to rise above the yield strength is equivalent to $\mu \simeq 0$ at high stresses. At even higher pressures, however, the rise in K more than compensates for the loss of the μ term in the wave velocity. In some solids changes of crystal structure are brought about by the high pressures (cf. b.c.c. → f.c.c. iron) and produce additional shock-fronts.

These very strong compressive shocks are *supersonic*. When a supersonic projectile moves through an elastic medium, it loses kinetic energy by *radiation* (in addition to the normal losses due to friction and plastic deformation). Figure 6.12 illustrates the effect. When the projectile is supersonic, the spherical waves which radiate from its tip are left behind and form a conical wave (cf. Section 12.11). In the subsonic case (diagram (a)) there is time for the (elastic) material to close in against the trailing end of the projectile. The pressure exerted on the head of the projectile by the elastic resistance of the material is then compensated by the corresponding pressure on the tail, so that (in the absence of friction and plasticity) the projectile is in *neutral* equilibrium and loses no energy as it moves. However, in the supersonic case there is not time for the material to close in on the trailing end and the pressure on the head is no longer compensated. The work which the projectile does against this pressure provides the kinetic energy of the radiated waves.

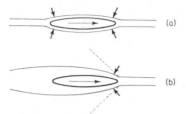

Fig. 6.12. Passage of a projectile (a) slowly, (b) supersonically down a (frictionless) channel in a solid.

Since elastic constants decrease when atomic bonds are finitely stretched, the later parts of a *tensile wave* in a thin rod travel slowly, compared with the leading parts. In ductile materials, however, this elastic effect is overshadowed by the large plastic strains that also occur. Because of the large displacements and momenta they can produce, plastic waves travel slowly. An analysis similar to that of Fig. 6.1 leads to the result

$$c(\varepsilon) = (1 + \varepsilon)\sqrt{\frac{d\sigma/d\varepsilon}{\rho}}, \qquad (6.28)$$

where $c(\varepsilon)$ is the velocity of that part of the plastic wave where the strain is ε, and $d\sigma/d\varepsilon$ is the slope of the *nominal* stress–strain curve at that strain (nominal stress = force divided by *initial* area; nominal strain = elongation divided by *initial* length). Equation (6.28) is clearly analogous to Equation (6.8) and reduces to it in the limit of small, Hookeian strains.* The factor $(1 + \varepsilon)$ appears because, in a large plastic wave, the length and area of the element PQ (Fig. 6.1) are each finitely changed by the strain.

Plastic stress–strain curves often have zero or even negative slopes in parts. Figure 6.13 shows examples of this in mild steel, the *yield drop* AB and the *ultimate tensile stress* D. We shall discuss the physical significance of these features in Chapter 9. Since the stress actually falls at the start of yielding, the initial part of the plastic wave has zero velocity. Only when the *yield elongation* BC (usually $\simeq 0.03$) has been exceeded does $d\sigma/d\varepsilon$ become positive again. These later parts of the plastic wave have

---

* Plastic strain is, however, produced by a process of *flow*, so that, unlike elastic strain, it does not fully appear simultaneously with the stress. Even in the plastic range, therefore, it may be possible for small stress pulses to travel at elastic speeds, owing to delayed plasticity, with the ensuing plastic wave following later.

a finite velocity and so overtake the part representing the yield elongation. Thus, a *finite plastic front* across which there is a strain discontinuity BC forms at the leading edge of the plastic wave. Since the wave velocity in

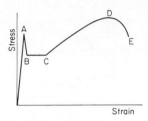

Fig. 6.13.　Nominal stress–strain curve of mild steel.

the BC region is zero, this discontinuous plastic front always forms, no matter how slowly the plastic wave is propagated. It is called a *Lüders front* (Fig. 6.14). The rate of strain $\dot{\varepsilon}$ localized in the front itself can be very high, since

$$\dot{\varepsilon} = \frac{v\Delta\varepsilon}{s}, \tag{6.29}$$

where v is the velocity of propagation of the front, $\Delta\varepsilon$ is the strain discontinuity, and s is the thickness of the front. If, for example, $v = 1$ cm sec^{-1}, $\Delta\varepsilon = 0.03$ and $s = 0.01$ cm (e.g. a typical grain size),

Fig. 6.14.　A Lüders front propagating along a bar.

then $\dot{\varepsilon} = 3$ sec^{-1}. An entire bar deformed simultaneously at this rate would double its length in about one-third of a second.

The plastic wave velocity also becomes zero at the ultimate tensile stress D. The nominal stress–strain curve always falls, beyond this point, until fracture occurs at E (cf. Section 9.6). The element of the bar which first reaches this point simply elongates more and more, becoming smaller in cross-section and forming a narrow *neck* as shown in Fig. 6.15, until it breaks.

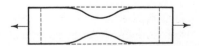

Fig. 6.15.　Formation of a neck at the ultimate tensile stress.

6.4 HARMONIC VIBRATIONS

We turn now from travelling waves to *standing waves* that represent steady modes of elastic vibration in a body. In the elementary theory of harmonic vibrations the displacement u of a mass m is opposed by a Hookeian restoring force, $f = -\alpha u$, from a weightless spring with a spring constant α. Newton's law then becomes

$$m\frac{d^2u}{dt^2} + \alpha u = 0 \tag{6.30}$$

and has two independent solutions obtained by adding together functions of the type

$$u = A\,e^{Bt}. \tag{6.31}$$

By substitution we find

$$B = \pm i\omega, \quad \text{where} \quad \omega = \sqrt{\frac{\alpha}{m}}, \tag{6.32}$$

and $i = \sqrt{-1}$. We add two solutions to obtain a solution of the type

$$u = A_1\,e^{i\omega t} + A_2\,e^{-i\omega t}. \tag{6.33}$$

Since $e^{\pm ix} = \cos x \pm i \sin x$, we can rearrange this as

$$u = C\cos \omega t + D \sin \omega t, \tag{6.34}$$

where $C = A_1 + A_2$ and $D = i(A_1 - A_2)$. The two real constants C and D are determined from the initial conditions. For example, if $u = 0$ at $t = 0$, then $C = 0$ and

$$u = u_0 \sin \omega t, \tag{6.35}$$

where u_0 is the *amplitude* of vibration. The *period* τ is the time for one complete vibration, i.e.

$$\tau = \frac{2\pi}{\omega} = 2\pi\sqrt{\frac{m}{\alpha}}, \tag{6.36}$$

and the *frequency* ω is the number of oscillations in a time 2π.

We have here used two properties of the spring: its weightlessness, which allows us to neglect its inertia, and its Hookeian elasticity, represented by α. This simple analysis can thus be applied fairly well to the motion of heavy loads attached to light elastic beams and rods. Thus, for a tensile rod (length l, cross-sectional area A, Young's modulus E) we have

$$\alpha = EA/l, \tag{6.37}$$

and for a round shaft (length l, radius R, shear modulus μ), oscillating in torsion about its cylindrical axis, we have

$$\alpha = \pi\mu R^4/2l \tag{6.38}$$

and a period

$$\tau = 2\pi\sqrt{\frac{I}{\alpha}}. \tag{6.39}$$

For a beam (length l, flexural rigidity EI) freely supported at its ends and carrying a heavy vibrating load at its centre we have

$$\alpha = 48EI/l^3. \tag{6.40}$$

This result also roughly describes the lowest vibrational mode of an un-loaded beam of the same mass.

Equation (6.30) is modified when a *frictional* force opposes the motion. The vibrations are then *damped*, and the vibrational energy gradually becomes heat. For *linear viscous* damping (i.e. frictional force proportional to velocity) the equation can be written as

$$m\frac{d^2u}{dt^2} + k\frac{du}{dt} + \alpha u = 0, \tag{6.41}$$

where k measures the strength of the damping. Applying Equation (6.31) again, we obtain $mB^2 + kB + \alpha = 0$, i.e.

$$B = -\frac{k}{2m} \pm i\sqrt{\left(\frac{\alpha}{m} - \frac{k^2}{4m^2}\right)}. \tag{6.42}$$

If the damping is small, i.e. $(k^2/4m^2) \ll (\alpha/m)$, the analysis can be developed as before, to give

$$u = e^{-kt/2m}(C\cos\omega t + D\sin\omega t), \tag{6.43}$$

with

$$\omega = \sqrt{\left(\frac{\alpha}{m} - \frac{k^2}{4m^2}\right)}. \tag{6.44}$$

The frequency of the harmonic vibrations is thus slightly reduced by the frictional drag, and the amplitude steadily decays, through the factor $\exp(-kt/2m)$, as shown in Fig. 6.16. The *decay* of the vibrations is usually measured by the *logarithmic decrement* δ,

$$\delta = \ln\frac{u_a}{u_b} = \frac{k\tau}{2m}, \tag{6.45}$$

where u_a, u_b and τ are defined as in Fig. 6.16.

When the damping is large, the motion is no longer even approximately sinusoidal, and, when $(k^2/4m^2) > (\alpha/m)$, it becomes *aperiodic*. Equation (6.42) then becomes

$$B_1 = -\frac{k}{2m} + \sqrt{\left(\frac{k^2}{4m^2} - \frac{\alpha}{m}\right)}, \quad \text{and} \quad B_2 = -\frac{k}{2m} - \sqrt{\left(\frac{k^2}{4m^2} - \frac{\alpha}{m}\right)}, \tag{6.46}$$

and

$$u = A_1 e^{B_1 t} + A_2 e^{B_2 t}. \tag{6.47}$$

This does not represent an oscillation but an exponential *creeping* of the displaced mass back to the equilibrium point $u = 0$. The *critical damping*,

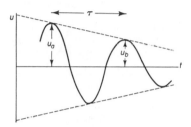

Fig. 6.16. Damped vibrations.

at $k^2/4m^2 = \alpha/m$, is the optimum degree of damping for galvanometer and other instrument suspensions. A greater frictional drag delays the return to the equilibrium point unduly; a smaller one allows the equilibrium point to be overshot repeatedly.

Elastic systems are often maintained in *forced vibration* by rhythmically applied forces, which we shall represent by $f_0 \cos \omega t$. Equation (6.41) then becomes

$$m \frac{d^2 u}{dt^2} + k \frac{du}{dt} + \alpha u = f_0 \cos \omega t. \tag{6.48}$$

We try the function

$$u = A \cos(\omega t - \gamma) = A \cos \gamma \cos \omega t + A \sin \gamma \sin \omega t. \tag{6.49}$$

Substitution shows that this is a solution if

$$A = \frac{f_0}{\sqrt{[m^2(\omega_0^2 - \omega^2)^2 + k^2 \omega^2]}} \quad \text{and} \quad \tan \gamma = \frac{k\omega}{m(\omega_0^2 - \omega^2)}, \tag{6.50}$$

where $\omega_0 = \sqrt{\alpha/m}$ is the frequency of free oscillations. The forced vibrations are maintained without decay, despite the damping, by the supply of energy to the oscillator from the driving force.

When the *driving frequency* approaches the *natural frequency*, i.e. $\omega \simeq \omega_0$, *resonance* occurs and the amplitude A reaches a maximum, $A_{max.}$,

$$A_{max.} = \frac{f_0}{k\omega}, \tag{6.51}$$

the height of which is limited only by the strength k of the frictional drag. Figure 6.17 shows how the amplitude A and *phase angle* γ vary across the resonant frequency. At low driving frequencies the oscillator moves in

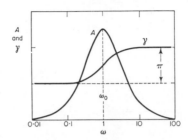

Fig. 6.17. Change of amplitude A and phase angle γ across a resonance.

phase with the driving force, and the phase angle is almost zero. As the frequency is increased, the oscillator lags increasingly behind the force, until at very high frequencies it is completely out of phase, $\gamma = \pi$. The oscillations are large, however, only when the driving frequency is tuned fairly closely to the natural frequency.

We saw in Section 6.1 that the inertial mass of an elastic material causes stress waves to propagate at finite speeds. The step from propagating waves to standing waves in such cases is made by considering surface reflections. As the incident and reflected waves pass through each other, they add together (principle of superposition) to form a pattern of waves which repeats itself periodically. If the waves are sinusoidal and have certain wavelengths related to the dimensions of the body, they form a pattern of standing waves. The displacement of an element of the body is then no longer of the form $u = g(x - ct)$ but of $u = u_0 \sin \omega t \sin kx$. The time and space variables then enter in *separate* periodic terms, which is the characteristic feature of standing waves.

Consider again the rod of Fig. 6.1. We notice that in Equation (6.5), which governs its motion, t and x occur in separate terms. Thus, we can look for solutions of the type

$$u = X(x)T(t), \tag{6.52}$$

where the function $X(x)$ depends on x but not t, and $T(t)$ depends on t

but not x. Equation (6.5) then becomes

$$\frac{1}{c_0{}^2 T}\frac{d^2 T}{dt^2} = \frac{1}{X}\frac{d^2 X}{dx^2}. \tag{6.53}$$

The left side is a function of t only and the right side a function of x only, yet they are equal. They must therefore both be *constant*. Let this constant be $-k^2$. We can then split the equation into two separate ones,

$$\frac{d^2 X}{dx^2} + k^2 X = 0 \quad \text{and} \quad \frac{d^2 T}{dt^2} + k^2 c_0{}^2 T = 0, \tag{6.54}$$

each of which is like Equation (6.30). We can thus use solutions such as

$$X = A \sin kx + B \cos kx. \tag{6.55}$$

Suppose that the rod is of length l and has fixed ends. Then, $u = 0$ and, hence, $X = 0$ at both $x = 0$ and $x = l$ for all values of t. This requires that $B = 0$ and that

$$k = \frac{n\pi}{l}, \tag{6.56}$$

where $n = 1, 2, 3, \dots$ etc. These conditions ensure that *nodes* occur at the ends of the rod, as in Fig. 6.18. These *standing waves* have wavelengths

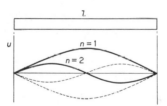

Fig. 6.18. Standing waves in a rod with fixed ends.

given by

$$\lambda = \frac{2l}{n}. \tag{6.57}$$

The longest wave ($n = 1$), i.e. the *fundamental mode*, has a wavelength of twice the length of the rod. The higher modes, i.e. *harmonics*, have nodes within the rod as well as at its ends. A similar analysis can be made for a free-ended bar. The cosine function in Equation (6.55) has then to be used, and *antinodes* occur at the ends.

Such standing waves are fundamental in vibration problems and are known as *natural vibrations, fundamental* and *harmonic oscillations, vibrational modes, proper modes* and *eigenfunctions*. Clearly, we could examine the various other types of progressive waves discussed in Section 6.1 and deduce their equivalent standing waves. We see, for example, from the analysis leading to Equations (6.9) and (6.10) that in three-dimensional solids there are two transverse modes to each longitudinal one.

Once k is known, the solution of Equations (6.54) is easily found. We have

$$\omega = \frac{\pi n c_0}{l} \qquad (6.58)$$

for the number of oscillations of the nth mode in time 2π and

$$u = u_0 \sin\left(\frac{\pi n c_0 t}{l}\right) \sin\left(\frac{\pi n x}{l}\right) \qquad (6.59)$$

for the displacement of a point with co-ordinate x at time t in this vibrational mode.

To find the natural frequencies of flexural vibrations in a slightly bent beam, we assume the displacement y of a point x along the beam to be given by $y = X(x) \sin \omega t$. From Equation (6.15), we then obtain

$$\frac{d^4 X}{dx^4} - \frac{\omega^2}{c_0^2 R^2} X = 0, \qquad (6.60)$$

which has the general solution

$$X(x) = A \sin \gamma x + B \cos \gamma x + \gamma C \cosh \gamma x + \gamma D \sinh \gamma x, \qquad (6.61)$$

where $\gamma^2 = \omega/c_0 R$. For a simply supported beam (e.g. a xylophone) of length l (Fig. 6.19), we have $B = C = D = 0$ and $\gamma = \pi n/l$, where $n = 1, 2,$

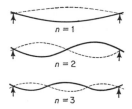

Fig. 6.19.   Flexural vibrations of a simply supported beam.

3, ... etc., so that

$$y = y_0 \sin\left(\frac{\pi n x}{l}\right) \sin \omega t, \qquad (6.62)$$

$$\omega = \gamma^2 c_0 R = \frac{\pi^2 n^2 c_0 R}{l^2} \qquad (6.63)$$

The frequencies of the modes are thus simple harmonics of the fundamental frequency. The measurement of these frequencies provides a simple method for determining Young's modulus under dynamic conditions. If the beam is clamped at one or both ends, the vibrations are complicated by the hyperbolic terms in Equation (6.61), which then enter into the solution. For a cantilever or vibrating reed, clamped at one end and free at the other, as in Fig. 6.20, we have $X = dX/dx = 0$ at the clamped end

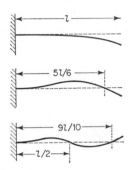

Fig. 6.20. Vibrational modes of a cantilever.

and $d^2X/dx^2 = d^3X/dx^3 = 0$ at the free end (zero couple and shearing force). It follows that $C = -A$ and $D = -B$, so that Equation (6.61) can then be rewritten in terms of A and B alone. The conditions at the free end lead to two independent equations, and, for these to be valid simultaneously, it is necessary that

$$\cos \gamma l \cosh \gamma l = -1. \tag{6.64}$$

This is solved by plotting $\cos \gamma l$ and $1/\cosh \gamma l$ against γl and finding the intersections. The fundamental mode is given by $\gamma = 1\cdot875/l$ and has a frequency $\omega = (1\cdot875)^2 c_0 R/l^2$. The higher modes are not simple multiples of one another (cf. Fig. 6.20). Not forming a harmonic sequence, they produce the harsh musical tone familiar when a clamped bar is struck.

6.5 DAMPING AND INTERNAL FRICTION

Percussive musical instruments are damped as little as possible, to preserve their vibrational energy. Church bells, for example, are made from hard, brittle alloys (e.g. 25% Sn in Cu) and are carefully designed to minimize loss of energy to their supports. In many other cases, e.g. organ pipes and vibrating machinery, the vibrations within the solid are deliberately damped, particularly where there is a risk of resonances. This damping can be produced externally, e.g. by adhesive tape wrapped on vibrating springs or by hydraulic dashpot shock absorbers, but is also often produced by *internal friction* within the vibrating body itself.

In an ideal elastic solid there is a one-to-one relation between stress and strain, and the vibrational energy is conserved. In practice there are usually small deviations, called *anelastic effects*, from this one-to-one relation, and these are sources of internal friction. The stress–strain diagram of a cyclically loaded specimen then forms a closed loop, the area of which gives the energy dissipated per cycle. A loop is formed because the strain cycle lags behind the stress cycle. In the simplest cases both stress and strain vary sinusoidally with time, as $\sigma = \sigma_0 \sin \omega t$ and $\varepsilon = \varepsilon_0 \sin (\omega t - \gamma)$, respectively, where γ is the *phase angle* which measures this lag (cf. Section 6.4). The stress–strain loop is then an ellipse, as in Fig. 6.21.

Fig. 6.21.   A simple stress–strain loop.

Internal friction is measured in several ways. The *specific damping capacity* is $\Delta U/U$, the energy ΔU dissipated per cycle of vibrational energy U. The *logarithmic decrement* δ (Equation (6.45)) is often used for freely vibrating systems. For forced vibrations near a resonance the factor

$$Q^{-1} = \frac{\omega_2 - \omega_1}{\omega_0} \tag{6.65}$$

is often used, where ω_1 and ω_2 are the frequencies on the two sides of the resonant frequency ω_0 at which the amplitude of oscillation is $1/\sqrt{2}$ of the resonant amplitude. Alternatively, $\tan \gamma$ is sometimes used. The relation between these measures is

$$Q^{-1} = \tan \gamma = \frac{\delta}{\pi} = \frac{1}{2\pi} \frac{\Delta U}{U}. \tag{6.66}$$

These measures are usually used for internal frictions below about 0·01. Difficulties of definition become serious at values greater than this.

There are many causes of internal friction. One of the most general is *thermoelasticity*. The volume of a body can be altered by both temperature change and hydrostatic components of applied stress. A suddenly applied tensile stress causes the body to cool slightly as it expands. As the body warms back to its initial temperature, by extracting heat from its surroundings, it expands thermally. The dilatation strain thus continues to increase after the stress has become steady, so that there is a phase lag. We define the *adiabatic Young's modulus* E_S as that measured by the instantaneous response to a suddenly applied stress and the *isothermal Young's modulus* E_T by the response to a stress applied so slowly that the temperature remains constant throughout. The *modulus defect*

$$\Delta_0 = \frac{E_S - E_T}{E_T} \tag{6.67}$$

then measures the intensity of the effect. For metals Δ_0 is usually about 10^{-3}–10^{-2}.

If an elastically anisotropic polycrystal is vibrated, some grains are more highly strained than others and change temperature more. Heat then

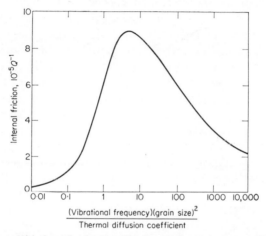

Fig. 6.22. Internal friction of polycrystalline brass, due to intercrystalline heat currents (R. H. Randall, F. C. Rose and C. Zener, *Phys. Rev.* **56**, 343 (1939)).

flows from one grain to another and causes damping. Figure 6.22 shows an example of this effect. Thermoelastic damping also occurs in bent beams. The compressed longitudinal fibres become hotter and the extended ones cooler. A temperature gradient is thus set up across the beam, which alternates with the period of the flexural vibrations. Heat can flow from side to side down this gradient and so dissipate vibrational energy. The time τ to reach thermal equilibrium in a body is of order

$$\tau \simeq x^2/D \tag{6.68}$$

(cf. Equation (1.51)), where x is the length of the heat path and D is the *coefficient of thermal diffusion*,

$$D = \kappa/C, \tag{6.69}$$

κ being the thermal conductivity and C the specific heat per unit volume. The damping is greatest when $\omega\tau \simeq 1$ and diminishes at other frequencies of vibration according to the formula

$$Q^{-1} = \Delta_0 \frac{\omega\tau}{1 + (\omega\tau)^2}. \qquad (6.70)$$

If the vibrations are too quick, there is no time for the heat to flow (adiabatic vibrations); if too slow, the heat flows in a thermodynamically reversible manner down a vanishingly small temperature gradient (isothermal damping). For a 1 mm thick vibrating reed of a typical metal, for which $D \simeq 0.5$ cm^2 sec^{-1}, the optimum frequency is about 50 sec^{-1}.

Thermoelastic damping is an example of *linear damping*, i.e. Q^{-1} is independent of strain amplitude. This is because the heat flow and strain rate $\dot{\varepsilon}$ due to it are both proportional to the temperature gradient and, hence, to the stress. A stress σ gives a strain energy $U \propto \sigma^2$ and a rate of dissipation $\Delta U \propto \dot{\varepsilon}\sigma$. In linear damping the strain rate obeys Newton's law of viscosity, $\dot{\varepsilon} \propto \sigma$, so that $\Delta U \propto \sigma^2 \propto U$. Many other sources of internal friction give linear damping at low strain amplitudes. Examples are: (1) stress-induced ordering of interstitial solute atoms in certain crystals; for example, carbon and nitrogen atoms in b.c.c. steel can jump from one interstitial site to another and so produce slight changes of shape in the crystals; (2) slip of one crystal past another by a small amount along the grain boundary between them, in polycrystals (cf. Section 8.9).

In contrast to linear damping are the *static hysteresis processes*. An example is the hysteresis loop which an elastic–plastic solid shows in its stress–strain curve (cf. Fig. 4.6). Such loops are roughly independent of the frequency of oscillation. The non-elastic strain appears almost instantaneously with the stress, remains roughly constant at constant stress, and is not removed until the stress is greatly reduced or even reversed in sign. This form of damping can appear to some extent, even at low stress amplitudes, in plastically deformed solids or in very heterogeneous materials which contain cracks or other stress concentrators that produce localized plastic deformation at small overall stresses. The usefully high damping capacity of *grey cast-iron* is attributed to the large crack-like flakes of graphite that exist in this material. Some *ferromagnetic materials* also show static hysteresis due to magnetostrictive strains in their grains. There is usually a marked dependence of damping on strain amplitude in all such cases, because the basic mechanism involves the sudden "breaking free" or "yielding" of some feature such as a plastic slip band, or a boundary between differently oriented magnetic domains, at a critical stress. These are discontinuous processes which allow the non-elastic strain to increase suddenly at the critical stress.

The pronounced hysteresis in its stress–strain curve, shown in Fig. 6.23, enables *rubber* to convert large amounts of mechanical energy to heat,

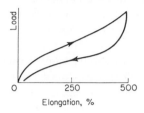

Fig. 6.23. Hysteresis of rubber.

which contributes to the value of this material as a shock absorber. This elastic hysteresis is also a cause of friction when rubber rolls or slides on a (slippery) corrugated surface. The deformation produced by a rolling indenter on rubber is shown in Fig. 6.24. Deformed elements recover

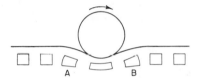

Fig. 6.24. Deformation of elements in rubber under a rolling indenter.

their shape after the indenter has passed, but according to a curve such as that of Fig. 6.23, so that the forward force exerted on the indenter by the element at A in Fig. 6.24 is smaller than the backward force exerted by that at B. There is thus a net backward force on the indenter, irrespective of conditions at the surface itself. This is important when motor tyres skid on slippery roads. Rubbers with large hysteresis losses (butyl rubber) have been developed to take advantage of this effect.

Materials of high damping capacity are used for sound-proofing. The *reverberation* of a closed room is caused by the reflections of sound waves from the walls. If these reflections were perfect, with no absorption, the intensity of sound at the antinodes of the reverberating chamber would build up to an indefinitely high level in the presence of a continuing source of sound. Actually it builds up to the level at which the rate of absorption of acoustical energy in the walls and other absorbers equals the power output from the source. The average *density of acoustical energy*, ρ, in the chamber in this steady state is given by

$$\rho = \frac{4W}{c\alpha S},$$ (6.71)

where W is the power output from the source, c is the velocity of sound in the reverberating medium (1100 ft sec^{-1} in air at room temperature), S the total area of absorbing surfaces, and α the average *absorption coefficient*, i.e. proportion of incident energy not reflected from a surface. Sound-proofing materials mostly consist of loosely woven fibres interspersed with air channels. Sound waves enter these channels and are then dissipated by the friction of the fibres rubbing against one another as they vibrate. Typical values of α are 0·3 (fibrous plaster), 0·5 (rock-wool felt) and 0·7 (unpainted insulation paper board and fibreglass insulating board).

6.6   THERMAL VIBRATIONS OF SOLIDS

Considered as a problem in kinetic theory, the thermal motion of a large number of atoms bound to one another by atomic bonds is obviously complicated. The simplicity of the kinetic theory of gases lies in the fact that the individual gas particles are almost independent of one another. In a solid, however, we have just the opposite situation, a *many-body problem* in which the motion of each particle depends on that of every other particle linked to it by the network of bonds. Fortunately, these strong links produce the elastic properties we have been analysing in the last three chapters. The theory of elasticity thus opens a way to the analysis of the thermal motion of solids. Since *natural vibrations* are steady-state elastic motions of a solid, we suppose that they also represent its thermal motion. In other words we no longer think of atoms as moving individually with thermal energy but as vibrating *collectively*, through the bonds that join them, in the various natural modes of the solid. A great advantage of this approach is that, in so far as Hooke's law is valid, the motion in each vibrational mode is independent of what is happening in the other modes. The modes can be simply superposed, each as if the others did not exist (principle of superposition). The problem is thus reduced basically to something as simple as that of a gas, with the individual, independent vibrational modes taking the place of the individual, independent gas particles.

In how many different modes can the solid vibrate? Clearly (cf. Fig. 6.18) there are few long-wave low-frequency modes and many short-wave high-frequency ones. However, since it is the atoms themselves that vibrate and not the space between them, there are no modes with wavelength λ shorter than $\lambda_{\min.}$, where $\lambda_{\min.}$ is twice the nearest-neighbour atomic spacing, because the mode $\lambda = \lambda_{\min.}$ represents the state in which nearest neighbours vibrate in opposite phases. Figure 6.25 shows this and also shows that shorter wavelengths than $\lambda_{\min.}$ produce the same atomic movements as longer ones. Consider a one-dimensional crystal of length *l*, as in Fig. 6.26, composed of *N* equal atoms spaced at intervals *b*. Then $\lambda_{\min.} = 2b = 2l/N$, assuming $N \gg 1$. For natural vibrations there must exist some definite boundary conditions, e.g. that the end atoms of the chain are fixed or free, with the result that only certain values of λ are

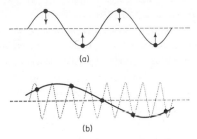

(a)

(b)

Fig. 6.25. Atomic vibrations: (a) shortest real wavelength; (b) shorter wavelengths are equivalent to longer ones.

allowed. We may therefore apply Equation (6.57), i.e. $\lambda = 2l/n$, taking all the values $n = 1, 2, 3, \ldots$ etc., to enumerate all the possible modes up to the limit $\lambda = \lambda_{\min.}$, which is reached when $n = N$. Hence, the number of such modes is equal to the number of atoms. This result can be generalized in various ways. The analysis of Fig. 6.26 by the theory leading

Fig. 6.26. A linear chain of atoms.

to Equation (6.57) deals with only one of the displacement modes, the longitudinal one. Similar analyses can be made of the transverse modes. Similarly, in three dimensions, problems may be analysed in terms of plane waves, two transverse to each longitudinal. The general result, counting both longitudinal and transverse modes, is that there are $3N$ modes in a simple crystal of N equal atoms. Most of these modes have very short wavelengths and high frequencies. In modes such as that of Fig. 6.25(a) we may think of each atom as vibrating against its neighbours, and so evaluate the frequency of an individual atomic vibration in the solid. If the relation $f = \alpha u$ (f = force, u = displacement, α = spring constant) is applied to such an atom, and if m, b, ρ, σ, ε and E are atomic mass, atomic spacing, density, stress, strain and elastic constant, respectively, then $m \simeq \rho b^3$, $f \simeq \sigma b^2$, $u \simeq \varepsilon b$ and $\alpha \simeq Eb$. The vibrational frequency, *v per second*, is then given by

$$v = \frac{\omega}{2\pi} \simeq \frac{1}{2\pi}\sqrt{\frac{\alpha}{m}} \simeq \frac{1}{2\pi b}\sqrt{\frac{E}{\rho}} \simeq \frac{c_0}{2\pi b},$$ (6.72)

where c_0 is given by Equation (6.8). For example, if $c_0 = 5 \times 10^5$ cm sec^{-1} and $b = 2 \cdot 5 \times 10^{-8}$ cm sec^{-1}, then $v \simeq 3 \times 10^{12}$ sec^{-1}. A more precise calculation, which takes account of the motions of the neighbouring atoms, would give about twice this value. Frequencies of this magnitude can be detected in crystals from certain types of X-ray reflections and from the absorption of infra-red light in ionic crystals.

Each of the $3N$ vibrational modes is a simple harmonic motion, the energy of which is expressible as the sum of two squared terms, kinetic (proportional to $m\dot{u}^2$) and potential (proportional to αu^2). Hence, from the principle of equipartition (Section 1.5), if the vibrations are *classical*, a thermal energy kT belongs to each mode at the absolute temperature T and the vibrational specific heat C_v is

$$C_v = 3Nk, \tag{6.73}$$

i.e. $3R$ per mole of atoms. This is the *Dulong–Petit law* and is approximately obeyed by many solids at room temperature and above.

At low temperatures, however, the observed specific heat falls towards zero, as shown in Fig. 6.27. This is an effect of *quantum restrictions* on the motion. A simple harmonic oscillator is not free to take up any level of vibrational energy. Its energy can change only by definite units or

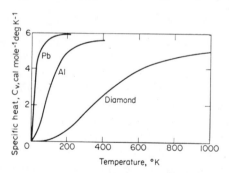

Fig. 6.27. Specific heats of solids at various temperatures (C_p, not C_v, is usually measured, but the difference between them is small).

quanta. At sufficiently low temperatures the thermal energy kT is so much smaller than the smallest allowable quantum that the oscillator is unable to accept any thermal energy at all and its specific heat approaches zero. At high temperatures kT is large compared with the quantum, and the fact that the thermal energy is transferred in discrete units rather than as a continuous flow then makes little difference; the equipartition principle and the Dulong–Petit law are valid in this range.

Analysis of the quantized motion of a simple harmonic oscillator has shown that the allowed energy levels are

$$(n + \tfrac{1}{2})hv, \tag{6.74}$$

where h = Planck's constant, v = frequency per second, and $n = 0, 1, 2, \ldots$ We notice that, even in its lowest state (i.e. at $0°K$), the oscillator is not at rest but vibrates with an energy $\tfrac{1}{2}hv$. This *zero-point motion* is the analogue of the quantized motion of an electron in a confined space (cf. Section 2.7). The fact that the energy quantum hv involves the frequency v means that the transition from quantized to classical behaviour occurs in a temperature range that depends on the frequency of the vibrational mode. We define the *characteristic temperature*, θ, of this transition range by

$$k\theta = hv \tag{6.75}$$

and see that this is of the order $\theta \simeq 300°K$ when $v \simeq 6 \times 10^{12}$ sec^{-1}. Weakly bonded, heavy solids (low E, large ρ) have lower characteristic temperatures than strongly bonded, light solids, as can be seen from Fig. 6.27.

To calculate the average thermal energy of an oscillator, we use the Maxwell–Boltzmann distribution law. The probability that the oscillator has an energy ε or greater is $\exp(-\varepsilon/kT)$. The probability that it has an energy between 0 and hv is thus $1 - \exp(-hv/kT)$, on substituting the two values 0 and hv for ε. In this state the oscillator has the energy $\tfrac{1}{2}hv$ (the zero-point energy). Similarly, the probability that it has an energy between hv and $2hv$ is $\exp(-hv/kT) - \exp(-2hv/kT)$ and in this state it has the energy $(1 + \tfrac{1}{2})hv$. Its average energy over all states is thus, writing $x = hv/kT = \theta/T$ and summing the geometrical series, given by

$$\tfrac{1}{2}hv(1 - e^{-x}) + (1 + \tfrac{1}{2})hv(e^{-x} - e^{-2x}) + (2 + \tfrac{1}{2})hv(e^{-2x} - e^{-3x}) + \ldots$$

$$= \tfrac{1}{2}hv + hv(e^{-x} + e^{-2x} + e^{-3x} + \ldots)$$

$$= \tfrac{1}{2}hv + hv[e^{-x} + (e^{-x})^2 + (e^{-x})^3 + \ldots]$$

$$= \tfrac{1}{2}hv + \frac{hv\, e^{-x}}{1 - e^{-x}} = \tfrac{1}{2}hv + \frac{hv}{e^{hv/kT} - 1}. \tag{6.76}$$

When $T = 0$, this reduces to the zero-point energy, $\tfrac{1}{2}hv$. At high temperatures the second term becomes

$$\frac{xkT}{e^x - 1} \simeq \frac{xkT}{x + \tfrac{1}{2}x^2 + \ldots} \simeq kT\left(1 - \frac{x}{2}\right) \simeq kT - \tfrac{1}{2}hv, \tag{6.77}$$

and, hence, the thermal energy attains its classical value kT.

The differential of this average energy with respect to temperature gives the specific heat of the oscillator. If we multiply this by $3N$ and assume that v refers to the oscillations of highest frequency (cf. Equation (6.72)), we obtain *Einstein's formula*

$$C_v = 3Nk\left(\frac{\theta}{T}\right)^2 \frac{e^{\theta/T}}{[(e^{\theta/T}) - 1]^2} \tag{6.78}$$

for the specific heat. This is the simplest formula that takes the quantum effects into account. Although inaccurate, because it assumes that all the vibrational modes have the same frequency as the maximum, it nevertheless gives curves fairly similar to those of Fig. 6.27.

Many improved theories have been made. One of the best known is that of *Debye*, who worked out the spectrum of vibrational modes for an isotropic elastic solid and obtained the result

$$C_v = 3Nk\left[\frac{12T^3}{\theta^3}\int_0^{\theta/T}\frac{x^3\,dx}{e^x - 1} - \frac{3\theta/T}{(e^{\theta/T}) - 1}\right] \tag{6.79}$$

Some values of the characteristic temperature obtained by fitting Debye's curve to experimental results are given in Table 6.2. They are of the same

TABLE 6.2
Debye Characteristic Temperatures (°K)

Pb	Na	Ag	Cu	Al	Fe	Be	Diamond
88	150	215	315	385	420	1000	2000

order of magnitude as those estimated from atomic bonds and elastic constants.

In layer crystals, very weakly bonded between layers, the particular vibrational modes representing the sliding of those layers across one another become classical at low temperatures. This can be detected in the low-temperature specific heats of such materials, and has been used to determine the shear modulus for sliding the layers. In graphite, for example, it gives the value $\mu = 2\cdot3 \times 10^{10}$ dyn cm^{-2}.

The amplitude of atomic thermal vibrations in solids is fairly large even at low temperatures, e.g. about 0·1 atomic spacings, and increases with rising temperature. This has various effects on the mechanical properties of solids. The *elastic constants* of crystals, for example, are reduced at the melting point to about 0·6 of their value at 0°K. The *mechanical strength* remains finite even at the melting point (cf. ice and water at 0°C), because this is determined by the co-operative movements of large numbers of atoms (cf. Chapter 9). The large amplitude motions produce oscillating stresses over atomic-sized regions but do not greatly assist applied stresses acting uniformly over large regions. The long-wave thermal vibrations, which could assist applied stresses, have very small amplitudes. The energy $3 \, kT$, for example, spread uniformly as energy of elastic tensile strain through 1 cm^3 of iron at room temperature, is equivalent to a stress of about 10^{-5} psi. It has been suggested that *melting* occurs when the amplitude of atomic vibrations is large enough to shear atoms past their neighbours by a critical amount which corresponds to the ideal elastic limit of the crystal (cf. Section 9.3). There is, in fact, a rough correlation between shear moduli and melting points. The critical amplitude, deduced from the observed melting point, compares reasonably well with the value expected from atomic forces.

6.7 THERMAL EXPANSION OF SOLIDS

The amplitudes of atomic vibrations in solids are large enough to produce significant deviations from Hooke's law. This leads to *anharmonicity*, i.e. deviations from simple harmonic motion. The motion can still be described in terms of superposed natural vibrations, but these are no longer entirely independent of one another. For example, the presence of one wave in the solid alters the elastic constants slightly and so alters the motion of other waves. The vibrational specific heat is not much affected by this, but other properties depend profoundly on it. Without anharmonicity there would be no thermal expansion and heat would travel at the speed of sound through the solid. The random exchanges of energy between one vibrational mode and another, by which thermal equilibrium is established, could not occur if the modes were mechanically independent of one another.

Almost all solids expand on heating. As the atomic spacing increases, the spring constant of the atomic bonds (i.e. the curvature of the energy curve in Fig. 2.2) decreases. The atoms then vibrate more slowly, with a lower characteristic temperature. A lower frequency means a smaller energy quantum, hv. Since, at a constant temperature T in the classical region, the total vibrational energy is constant at $3NkT$, the number n of vibrational quanta,

$$n = 3NkT/hv, \qquad (6.80)$$

in the material is increased by the expansion. The more quanta there are, the more ways there are of distributing them randomly throughout the body. It then follows, from the entropy argument of Section 2.3, that the body tends to undergo such an expansion. This tendency is opposed by the bulk modulus, and a balance is struck at a certain (linear) coefficient of thermal expansion

$$\alpha = \frac{1}{l}\frac{dl}{dT}, \tag{6.81}$$

where l is a linear dimension of the body.

The quantitative development of this theory leads to the relation

$$\alpha = \frac{\gamma c_v}{3v_0 K}, \tag{6.82}$$

where c_v = specific heat per gramme, v_0 = volume per gramme at $0°K$, K = bulk modulus at $0°K$, and γ is *Grüneisen's constant* for the change of atomic vibration frequency with atomic volume,

$$\gamma = -\frac{d \ln v}{d \ln V}. \tag{6.83}$$

Table 6.3 gives some values of γ. *Hard* atoms are those whose interatomic forces change rapidly with spacing, i.e. those with large γ. A gold atom, for example, is much more like a hard sphere than a lithium atom; its atomic vibration frequency changes about 9% for a 1% change in atomic spacing (i.e. 3% change in volume).

TABLE 6.3
Values of Grüneisen's Constant

Ag	Al	Au	Cu	Fe	K	Li	Na	Ni	Pt
2·40	2·17	3·03	1·96	1·60	1·34	1·17	1·25	1·88	2·54

To obtain the typical magnitude of α, we take $\gamma \simeq 2$, $c_v = 0·1$ cal g^{-1} $degC^{-1} \simeq 4·2 \times 10^6$ erg g^{-1} $degC^{-1}$, $v_0 \simeq 0·1$ cm^3, and $K \simeq 1·5 \times 10^{12}$ dyn cm^{-2}, to obtain $\alpha \simeq 20 \times 10^{-6}$ degC^{-1}, which compares with the values given in Table 1.1. The expansion coefficient diminishes with the vibrational specific heat, at low temperatures. We have used a single atomic vibrational frequency in Equation (6.83), whereas, strictly, γ is a weighted average over the whole spectrum of vibrational modes. In fact, the value of γ may change at low temperatures (below about $\frac{1}{3}\theta$) where the high-frequency vibrations cease to contribute to the specific heat. The transverse vibrational modes sometimes have negative values of γ, and since these, with their lower frequencies, predominate at low temperatures, the expansion coefficient sometimes becomes *negative*. This is observed in vitreous silica, silicon, zinc blende and indium antimonide.

A low co-ordination number in the crystal structure is expected to favour this effect by providing interstitial space for the atoms to vibrate laterally. Some of the transverse modes in b.c.c. metals are believed to vibrate faster when the atomic volume is increased, which may account for the low values of γ observed in these metals. One crystalline form of plutonium has a negative coefficient of thermal expansion, due to the unusual electronic structure of this metal.

The ability of vitreous silica to withstand sudden changes of temperature without breaking, i.e. its *thermal shock resistance*, is due to its very low expansion coefficient (0.5×10^{-6} degC^{-1}). Values for other glasses (in 10^{-6} degC^{-1}) are 8 (soda-glass), 3 (borosilicate or Pyrex glass) and 0.8 (96% silica, or Vycor). The alloy *Invar* (36% Ni in Fe; $\alpha \simeq 10^{-6}$ degC^{-1} at 50°C) has an atomic spacing slightly below the optimum for ferromagnetism. When heated towards its Curie point ($\simeq 200$°C), the crystal demagnetizes and thus contracts slightly, which annuls the normal thermal expansion of the material. Above the Curie point it expands normally (16×10^{-6} degC^{-1}). By careful adjustment of composition, iron–nickel alloys can be prepared with expansion coefficients that match those of glass, suitable for metal–glass seals. It is also possible in some invars to compensate for the normal change of Young's modulus with temperature, which is useful in precision balance springs.

6.8 THERMAL CONDUCTIVITY IN SOLIDS

Heat can pass through solids in various ways: by thermal vibrations; by the movements of free electrons in metals; and by direct radiative transfer in transparent materials (e.g. passage of infra-red waves through transparent silica). Transfer by thermal vibrations involves the propagation of elastic waves. Heat is injected into the solid at the hot end in the form of elastic wave packets. Each injected quantum of vibrational energy, called a *phonon*, travels along the bar and so propagates the heat.

Although moving elastically, this heat does not travel through the material with the speed of sound, because the phonons, like the particles of a gas, have interrupted journeys. There exists a *mean free path l* ($\simeq 23$ Å at 0°C and 100 Å at -190°C in NaCl) for a phonon between successive "collisions" that change its directions of propagation. By the same kind of analysis as that used to derive Equation (1.41), the coefficient of *thermal conductivity* κ can be derived as

$$\kappa = \tfrac{1}{3}C\bar{c}l, \tag{6.84}$$

where $C\,(=\rho C_p)$ is the specific heat per unit volume and $\bar{c}$ is the average velocity of the elastic waves involved.

The mean free path depends on the "collisions" of phonons with impurities, lattice defects, the free surface, and other phonons. Phonon–phonon collisions could not occur if the principle of superposition were rigorously obeyed. An elastic wave produces a small, periodic change of elastic constants through the material, and this acts like a diffraction grating on other waves passing through it, tending to scatter them.

Anharmonicity by itself does not ensure that the crystal has a thermal resistance to phonon–phonon collisions. The basic process is the coalescence of two phonons to form a third. In this the phonons do not collide with fixed obstacles in the crystal but only with each other. They are like the collisions of particles in a gas flowing down a long smooth tube; the overall momentum of the gas is not altered by such collisions and no resistance to flow is produced by them.

The phonon–phonon collisions that produce thermal resistance are of a special type called *umklapp processes*. The essential point is that shown in Fig. 6.25(b); very short waves ($\lambda < \lambda_{min}$.) are really the same as longer waves ($\lambda > \lambda_{min}$.). If two waves each with $\lambda > \lambda_{min}$. coalesce to form a wave with $\lambda < \lambda_{min}$., then in reality this is still a wave with $\lambda > \lambda_{min}$.· Suppose, for example, that one of these waves, A in Fig. 6.28, is moving to

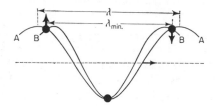

Fig. 6.28.   Movements of atoms in thermal waves.

the right with a wavelength λ slightly larger than λ_{min}., so that initially the atoms are moving as shown, and that it then absorbs another wave that reduces its wavelength to a value just smaller than λ_{min}.. It is then represented by wave B, still moving to the right. But this is a wave with $\lambda < \lambda_{min}$., and is identical with a wave $\lambda > \lambda_{min}$. moving to the *left*. The flow of phonons down the bar is thus not conserved in such processes. Energy is of course conserved, and to fulfil this condition the character of the waves has to change, e.g. from transverse to longitudinal, in such collisions.

Thermal resistance due to umklapp collisions involves phonons with wavelengths of order λ_{min}., i.e. with quanta of order $k\theta$. The frequency of umklapp collisions depends on the number of such phonons in unit volume. From the Maxwell–Boltzmann formula this is proportional to

$$e^{-\beta\theta/T} \tag{6.85}$$

at the absolute temperature T, where $\beta \simeq 1$. At very low temperatures such phonons are rare. Long-wave phonons are present and can collide with one another but cannot produce umklapp processes. If the crystal is large and perfect, the phonon conductivity is then large, since there are very few scattering collisions. For example, at room temperature diamond conducts heat four times better than copper (cf. Table 1.1), which explains why diamonds feel cold to the touch. At $50°K$ a sapphire crystal has a conductivity of 14 cal cm^{-1} sec^{-1} degC^{-1}, which exceeds the best value for copper at any temperature (11 cal cm^{-1} sec^{-1} degC^{-1} at $20°K$). Beryllia is another good non-metallic conductor (0·1 cal cm^{-1} sec^{-1} degC^{-1} at $600°C$).

The low conductivity of glass (about 0·0025 cal cm^{-1} sec^{-1} degC^{-1} at room temperature) is due to the irregular network of silicon–oxygen bonds in this material (cf. Section 7.5); the characteristic length of the molecular linkages, about 10 Å, determines the mean free path of the phonons in this material. This mean free path is practically independent of temperature, but the specific heat increases with temperature, and therefore the thermal conductivity (unlike that of perfect non-metallic crystals) also increases with increasing temperature. A similar increase, by a factor of about two from room temperature to $1000°C$, is observed in other irregular non-metallic structures such as fused silica and fire-brick. The best *thermal insulators* (e.g. insulating fire-brick, cork, insulating wool, diatomaceous earth) are of course highly porous materials and take advantage of the low thermal conductivity of still air (0·00007 cal cm^{-1} sec^{-1} degC^{-1}).

Elastic waves contribute little to the conductivity of pure metals such as copper, silver and aluminium, because the heat is carried rapidly by the free electrons. Equation (6.84) still applies, but with C, $\bar{c}$ and l signifying the thermal capacity, velocity and mean free path of the conducting electrons. Comparing the electron and phonon values at, say, $T \simeq \theta \simeq$ room temperature, the following differences emerge: (1) l is much larger for electrons than phonons (e.g. 10^{-5} cm as against 10^{-6} cm) in a good metallic conductor; (2) $\bar{c}$ is much larger for electrons than phonons (e.g. 10^7 cm sec^{-1} as against 10^5 cm sec^{-1}); (3) C is small for electrons, only about one electron per hundred atoms contributing to thermal conduction at room temperature. The result is that in good metallic conductors the electronic conductivity is 10–100 times larger than the phonon conductivity. However, in highly concentrated solid solution alloys, e.g. austenitic stainless steels, constantan (60 Cu/40 Ni), German silver (52 Cu/

26 Zn/22 Ni), the mean free paths of both electrons and phonons drop to about 10^{-7} cm, owing to scattering by the alloy atoms, and the thermal conductivity is then quite low ($\simeq 0.05$ cal cm^{-1} sec^{-1} degC^{-1}) for a metallic substance. This is useful in cryogenic apparatus.

6.9 PROPAGATION OF SOUND WAVES IN THERMALLY VIBRATING SOLIDS

If elastic energy is injected into a solid as heat waves, it diffuses slowly, by the processes discussed in Section 6.8. If injected as sound waves, it propagates rapidly, by the processes discussed in Section 6.1. Why are sound waves, and macroscopic elastic disturbances generally, immune to the phonon-scattering processes that slow down the flow of heat? The answer lies in their very long wavelengths, compared with atomic dimensions. The phonons with the strong anharmonic scattering properties have wavelengths of order λ_{min}. and represent the fine-scale atomic motion of the solid. Viewed on this fine scale, a sound wave appears like an elastic strain applied *uniformly* over large groups of atoms, which changes *slowly* with time measured by atomic standards ($\simeq 10^{-13}$ sec). The frequencies of the individual atomic vibrations are too far out of resonance with that of the sound wave for them to interfere with it. The atoms in a sound wave vibrate thermally in the same way as in a statically stressed solid, and they interact with it in the same way, i.e. merely through macroscopic effects such as thermal expansion, change of elastic constants with temperature, and thermoelasticity. If ultrasonic vibrations of sufficiently high frequency were injected into the solid, they would be scattered by phonons, but the necessary frequency range ($> 10^9$ sec^{-1}) is barely within range of present experimental techniques.

FURTHER READING

TIMOSHENKO, S., *Theory of Elasticity*, McGraw-Hill, New York, 1934.

LOVE, A. E. H., *A Treatise on the Mathematical Theory of Elasticity*, Dover, New York, 1944.

KOLSKY, H., *Stress Waves in Solids*, Oxford University Press, London, 1953.

LINDSAY, R. B., *Mechanical Radiation*, McGraw-Hill, New York, 1960.

KITTEL, C., *Introduction to Solid State Physics*, Wiley, New York, 1956.

PEIERLS, R., *Quantum Theory of Solids*, Oxford University Press, London, 1955.

BRILLOUIN, L., *Wave Propagation in Periodic Structures*, McGraw-Hill, New York, 1946.

BORN, M., and HUANG, K., *Dynamical Theory of Crystal Lattices*, Oxford University Press, London, 1954.

PROBLEMS

6.1   Prove that a torsional wave in a twisted, round, elastic bar is propagated with the velocity of transverse elastic waves.

6.2   Show that a shear wave, propagating along a close-packed direction with atoms moving in a close-packed direction, in an f.c.c. crystal, has a velocity $\sqrt{[(c_{11} - c_{12})/2\rho]}$.

6.3   Show that the characteristic frequencies of longitudinal vibration of a rod of length l, with one end fixed and the other free, are given by the formula $[(2n + 1)\pi/2l]\sqrt{E/\rho}$.

6.4   Show that a longitudinal wave propagates along a thin, wide sheet with a velocity $\sqrt{[E/\rho(1 - \nu^2)]}$.

6.5   A weight of 1 lb is fixed at the free end of a horizontal steel strip, 12 in. by 1 in. by 0·1 in., the other end of which is rigidly clamped. What is the period of natural vibration, if $E = 30 \times 10^6$ psi?   [0·153 sec]

6.6   Prove the relation between the various measures of internal friction given by Equations (6.66).

6.7   The velocity of longitudinal stress waves along a wire is 10^5 cm sec^{-1}. A 1000 cm vertical length of this wire is suspended from its top end. How much does it extend under its own weight?   [0·0005 cm]

6.8   A cylindrical rod of uniform dimensions is made from two shorter rods joined end to end by a boundary perpendicular to the rod axis. Density, and velocity of longitudinal stress waves, are ρ_a and c_a, respectively, in the first rod, and ρ_b and c_b in the second. A longitudinal stress wave of initial amplitude A_0 is sent along the first rod towards the boundary. Noting that the displacements of both rods are the same at the boundary, show that the amplitudes of the reflected (A_1) and transmitted (A_2) waves at the boundary are given by

$$A_1/A_0 = (\rho_b c_b - \rho_a c_a)/(\rho_b c_b + \rho_a c_a),$$

$$A_2/A_0 = 2\rho_a c_a/(\rho_b c_b + \rho_a c_a).$$

6.9   Prove that sinusoidal, longitudinal elastic-stress waves in solids are practically adiabatic. Show that thermoelastic damping in such a wave, of wavelength λ and velocity c, in an elastically isotropic solid with a coefficient of thermal diffusion D, is a maximum when $\lambda \simeq D/c$.

6.10   Show, by strain energy considerations, that the thermal strain between neighbouring atoms in a typical crystal at room temperature is of order 0·1. Explain why the thermal strain averaged over larger regions of the crystal is much smaller than this.

6.11   At very low temperatures (i.e. $T \ll \theta$) the integral in Equation (6.79) has the limiting value $\pi^4/15$, so that $C_v = (12\pi^4 Nk/5)(T/\theta)^3$ according to the Debye theory. Deduce the corresponding limiting form of C_v at low temperatures according to Einstein's theory. Why do these differ and why is there no exponential temperature factor in the Debye formula, despite its derivation from the Maxwell–Boltzmann formula? Why do solids in practice follow the Debye formula rather closely, as regards their vibrational specific heat, at low temperatures?

6.12 The thermal conductivity of a pure KCl crystal at $4°K$ is of order 1 cal sec^{-1} deg^{-1} cm^{-1}, and the specific heat is about 10^{-4} cal deg^{-1} cm^{-3}. The conductivity of a single-crystal KCl rod 0·25 cm thick is a maximum at about $7°K$, whereas that of one 0·75 cm thick is a maximum at about 5·5°K. Explain the relation between these facts.

chapter 7

Fluidity and Viscosity

We shall distinguish between a solid and a fluid in terms of the rate of flow produced at small applied stress. This flow is brought about by migrations of individual particles, a process in which, in condensed phases, thermal energy and free volume have to concentrate into the immediate neighbourhood of a particle before that particle can jump. We shall see that this leads to a picture of diffusion, creep and viscous flow, in both crystals and liquids. At low temperatures where quantum restrictions on the motion of a fluid become pronounced, as in liquid helium, viscous resistance to flow can disappear entirely, because there is no way to dissipate the energy of motion. At the other extreme, when the viscosity of a liquid becomes high, often because its molecules join up to form long chains, a glass is formed. Solid particles in a fluid, if large and densely packed, gain little help from thermal energy in sliding over each other, and in this case plastic rather than viscous behaviour is observed. We shall discuss a variety of substances, such as sols, gels, clays, slurries, polymers and rubbers, in terms of these ideas. Finally, we shall solve some simple problems of viscous flow of liquids down tubes, and between rotating cylinders and moving plates, and discuss applications to adhesives, sliding bearings and motion of particles through fluids.

7.1 FLUIDITY AND SOLIDITY

When a substance *flows*, its particles change neighbours. Flow is a process of shear, and the rate of shear $\dot{\gamma}$ is a function,

$$\dot{\gamma} = f(\sigma), \tag{7.1}$$

of the applied shear stress σ. Clearly, $\dot{\gamma} = 0$ when $\sigma = 0$ (in the absence of self-stresses and other sources of spontaneous deformation). For small stresses we can thus expand $f(\sigma)$ as a power series in σ, i.e. $f(\sigma) = \alpha\sigma +$ *higher terms*. The leading term $\alpha\sigma$ satisfies the requirement that $\dot{\gamma}$ has the same sign as σ and is therefore acceptable. Thus, in the limit of small stresses,

$$\dot{\gamma} = \alpha\sigma. \tag{7.2}$$

This is *Newton's law of viscous flow*; α is the *fluidity*, i.e. the reciprocal of the *dynamical viscosity* η (cf. Equation (1.43)).

This is a *linear* law, the strain rate increasing linearly with the stress, and is similar to Hooke's law with η equivalent to the shear modulus μ. The viscous analogue of Young's modulus can be derived from Equation (4.91). The flow usually occurs at constant volume (Poisson's ratio $= 0.5$), and the equivalent elastic relation is then $E = 3\mu$. Hence, the rate of extension $\dot{\varepsilon}$ produced by a uniaxial tension σ is given by

$$\dot{\varepsilon} = \frac{\sigma}{3\eta}. \tag{7.3}$$

Whether a stressed body behaves as a fluid or solid depends on how long the stress is applied. Silicone putty can be poured slowly from a cup or bounced quickly like a rubber ball. Pitch can also be poured slowly or sharply broken with a hammer. Elastic metal springs gradually slacken under prolonged stresses. The distinction between fluids and solids is

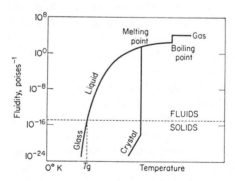

Fig. 7.1. Effect of temperature on fluidity.

drawn conventionally at a (low-stress) viscosity of 10^{15} poises. At this value a 1 in. cube could, for example, support a man's weight for a year without sinking more than 0·1 in.

Figure 7.1 shows how the fluidity of molecularly simple substances usually changes with temperature. The viscosity of a gas is usually about

10^{-4} poises. That of the liquid is about 10^{-3}–10^{-2} at the melting point and rises, slowly at first and then rapidly on cooling. Crystallization sharply increases the viscosity to about 10^{18} and brings a further rapid increase as the crystalline solid cools. If the liquid fails to crystallize, its viscosity still becomes very high at low temperatures and an *amorphous solid* or *glass* is formed. Liquid metals are exceptionally fluid at low temperatures and do not readily form glasses. The ability of a liquid metal to be cast into a mould and to take up fine details of the impression is usually not limited by viscosity but by factors such as speed of crystallization, heat transfer and surface tension. The viscosity of soda-lime–silica glass is shown in Fig. 7.2.

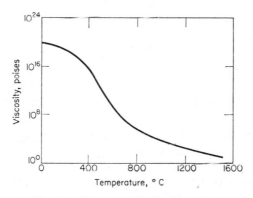

Fig. 7.2.   Viscosity of soda-lime glass.

Gas bubbles escape from glass at a viscosity of about 100 poises. Glass-blowing is done at about 10^7 poises, and stress-relief annealing at about 10^{13} poises.

Maxwell explained the difference between fluids and solids in terms of a *time of stress relaxation* τ. Let the shear stress σ produce elastic shear γ_1 and Newtonian flow $\dot{\gamma}_2$, i.e.

$$\dot{\gamma}_1 = \sigma/\mu \quad \text{and} \quad \dot{\gamma}_2 = \sigma/\eta. \tag{7.4}$$

Let the specimen be strained instantly a small amount γ_0 and then fixed at this strain, as shown in Fig. 7.3. The stress instantly rises to σ_0 ($=\mu\gamma_0$) and then relaxes as elastic strain is replaced by viscous strain. Since $\dot{\gamma}_1 + \dot{\gamma}_2 = 0$, then $\dot{\sigma}/\sigma = -\mu/\eta$, i.e.

$$\sigma = \sigma_0 \, e^{-t/\tau}, \tag{7.5}$$

where

$$\tau = \eta/\mu \tag{7.6}$$

($\tau = 3\eta/E$ for tensile strain). Alternatively, let the specimen be loaded at constant stress σ. Then

$$\dot{\gamma}_2 = \frac{\sigma}{\eta} = \frac{\sigma}{\mu\tau}, \qquad (7.7)$$

i.e. τ is the time to convert the elastic deformation σ/μ to viscous deformation at constant stress. When $t \gg \tau$, the viscous deformation greatly exceeds the elastic deformation and the substance is fluid.

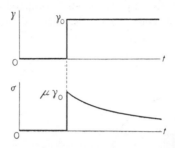

Fig. 7.3.   Stress relaxation due to flow at constant strain.

In a simple liquid of high fluidity τ is the average time the immediate neighbours of a particle remain in place before rearranging themselves. Experiments with ultrasonic waves and scattering of neutrons in such liquids have shown that $\tau \simeq 10^{-11}$ sec, which is sufficient for only 10–100 atomic vibrations. If $\mu \simeq 10^{10}$ dyn cm^{-2}, then $\eta \simeq 0{\cdot}1$ poises. Even at the highest practical rates of shear ($\simeq 10^5$ sec^{-1}), the shear stress is of order $0{\cdot}1$ psi, and only the first term in the power series expansion of $f(\sigma)$ is appreciable; i.e. the fluid is always Newtonian.

When $10^{-4} < \tau < 10^{+4}$ (sec), as in waxes, asphalt and putty, both elastic and viscous strain are readily observable. When $\tau > 10^{+4}$ sec, the substance is a solid. At these larger values of τ additional mechanical effects appear. To produce appreciable rates of flow, the stress must be raised into the range where higher terms in the $f(\sigma)$ power series become important. The flow then becomes *non-linear* or *non-Newtonian*, as shown in Fig. 7.4. In the limit where the rate remains negligible until a critical stress is reached and then becomes very large, non-linear viscosity becomes *ideal plasticity*. Thus, if

$$\dot{\gamma} = A\left(\frac{\sigma}{\sigma_y}\right)^n, \qquad (7.8)$$

$n = 1$ represents linear viscosity, $1 < n < 10$ includes most examples of non-linear viscosity and $n > 10$ approximates to ideal plasticity. We shall discuss the *Bingham flow* in Fig. 7.4 later.

ᵀA large τ also enables the structure and properties of the fluid to be changed by the processes of flow. So long as τ is very small, the statistical distribution of particles is hardly altered by the processes of flow, because the particles quickly readjust their relative positions; but in highly viscous substances the changes in relative positions brought about by the flow persist for long times, because the particles move sluggishly, so that the structure and properties are changed. Thus, *thixotropic liquids*, e.g. oil paints, become more fluid when stirred.

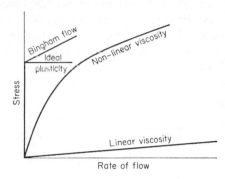

Fig. 7.4. Types of flow

A feature of linear viscosity is that it enables materials such as hot glass to be stretched uniformly in tension, without *necks* (cf. Fig. 6.15). This allows glass fibres to be made by simply pulling their ends apart, and is useful in glass-working generally. Let A be the area of a particular cross-section of a tensile bar. Then $\sigma = F/A$, where F is the tensile load, and $\dot{\varepsilon} = -\dot{A}/A$ for deformation at constant volume. From Equation (7.3),

$$\dot{A} = -\dot{\varepsilon}A = -\frac{\sigma A}{3\eta} = -\frac{F}{3\eta}. \tag{7.9}$$

Since F is constant along the bar, all cross-sections lose area at the same rate, whatever their initial area. An initially narrow section does not neck down faster than elsewhere.

7.2 MOVEMENT OF ATOMS IN SIMPLE SOLIDS AND LIQUIDS

The theory of the transport properties of gases (cf. Section 1.10) does not apply to solids and liquids. The particles no longer move independently. The familiar effect of temperature on the viscosity of stiff liquids such as treacle and engine oil shows that something quite different from gas viscosity is involved. Nor can we automatically assume that the various transport properties—thermal conductivity, diffusion and viscosity—are any longer related. In fact, as discussed in Chapter 6, heat can flow through condensed matter as elastic waves along the network of atomic bonds. We can still expect a relation between viscosity and diffusion, however, since both involve the migration of particles.

Diffusion in crystals is proved by the various heat-treatment processes, such as hardening and tempering of steel, that depend on it. Self-diffusion is proved by experiments with radioactive atoms. The simplest form of diffusion in crystals is that of interstitial solute atoms (cf. Fig. 3.10). In steel, for example, carbon and nitrogen atoms can jump from one interstitial site to the next by squeezing between the iron atoms round those sites. These jumps are the equivalent of the individual mean free paths of gas particles. The iron atoms must be strained apart, temporarily, to let the interstitial atom through, and the extra energy of this strained state forms an energy "barrier" opposing the jump, as shown in Fig. 7.5.

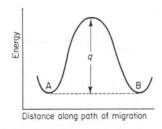

Fig. 7.5. An energy barrier between atomic sites A and B.

Internal friction and other measurements have shown that $q = 0.87$ and 0.81 eV per atom, respectively, for carbon and nitrogen in b.c.c. iron. At room temperature $(kT \simeq \frac{1}{40} \text{ eV})$, $q \gg kT$, and such an atom is unable to jump its barrier during most of its v ($\simeq 10^{13}$) vibrations per second against it. The Maxwell–Boltzmann formula gives the chance that the particle has the necessary thermal *activation energy* q during such a vibration. The number n of successful jumps per second is then given by

$$n = \alpha v \, e^{-q/kT}, \tag{7.10}$$

where α ($\simeq 10$) is an *entropy factor* connected with the changed vibrational and thermodynamical conditions of the barrier atoms in their strained state.

With the above values an interstitial atom jumps about once per second at room temperature. The strong effect of temperature on the jumping rate should be noticed. If $T + \Delta T$ is the temperature at which the rate is 10 times that at T, i.e.

$$e^{-q/[k(T+\Delta T)]} = 10 \, e^{-q/kT}, \tag{7.11}$$

we obtain, by taking logarithms and approximating

$$(T + \Delta T)^{-1} \simeq [1 - (\Delta T/T)]/T,$$

$$\frac{\Delta T}{T} = \frac{2 \cdot 3kT}{q}. \tag{7.12}$$

Thus, if $q = 34kT$ (T = room temperature), a 20°C increase of temperature speeds the jumping tenfold.

Such jumps produce a random walk similar to that described in Chapter 1. We can adapt Equation (1.47), interpreting l as the site-to-site distance and $\bar{c}$ (the particle velocity) as nl, to define the coefficient of interstitial diffusion as $D \simeq \frac{1}{3}nl^2$. A more precise analysis gives

$$D = \tfrac{1}{6}na^2 = D_0 \, e^{-q/kT}, \tag{7.13}$$

where

$$D_0 = \tfrac{1}{6}\alpha v a^2 \tag{7.14}$$

and a is the edge length of the cubic crystal cell. For carbon in b.c.c. iron, $D_0 = 0 \cdot 02$ cm^2 sec^{-1} and $D \simeq 10^{-17}$ cm^2 sec^{-1} at room temperature. It follows from Equation (1.51) that such an atom migrates about 10^{-5} cm in 10^7 sec ($\simeq 4$ months) at room temperature.

Self-diffusion and diffusion of substitutionally dissolved atoms occur mostly by the migration of *vacancies*; Fig. 3.11(b) shows how the atoms along the track of such a vacancy are displaced. To make such a jump, an atom must break away from the binding forces of its own site and squeeze between neighbours to enter the vacant site, as shown in Fig. 7.6. In the

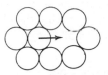

Fig. 7.6.   Jump of an atom into a vacancy.

notation of Section 3.3, the activation energy for this is E_M per mole of jumps. Measured values are given in Table 3.2. Vacancies in gold, for example, are about as mobile as interstitial carbon atoms in b.c.c. iron. A lattice atom is much less mobile, since only rarely is there a vacancy next to it (cf. Equation (3.2)). Combining the Maxwell–Boltzmann factors for E_F and E_M, where E_F is the energy to form a vacancy as discussed in Section 3.3, we obtain for the *coefficient of self-diffusion* in the crystal

$$D = D_0 \, e^{-E_F/RT} \, e^{-E_M/RT} = D_0 \, e^{-E_D/RT}, \tag{7.15}$$

where E_D is the activation energy for self-diffusion. For metals, $D_0 \simeq 10^{-1}$ to 10^{+2} cm^2 sec^{-1} and $E_D \simeq 20RT_m$, where T_m is the absolute melting point. Table 3.2 shows that the expected relation $E_D = E_F + E_M$ is roughly satisfied by the experimental data.

Just below the melting point, $D \simeq 10^{-9}$ cm^2 sec^{-1} in metals and other simple crystals. An atom can then migrate 0·1 cm in about four months. When melting occurs, the increase in free volume is equivalent in a crystal to a 10^3–10^4-fold increase in the number of vacancies. In fact, the self-diffusion coefficient in liquid metals is about 10^{-5} cm^2 sec^{-1}, roughly in agreement with this. However, this is a poor picture of the liquid state. We saw in Section 2.6 that this free volume is dispersed randomly among all the particles. In fact, D_0 and E_D are both much smaller in the liquid than in the solid. For a typical liquid metal, $D_0 \simeq 10^{-4}$–10^{-3} cm^2 sec^{-1} and $E_D \simeq 3RT_m$. These values show that the energy barrier is not the main obstacle to diffusion in a simple liquid. There is actually plenty of free volume among the neighbours of any given particle, but it is of little use for particle jumping except when, by chance, it happens to come together and so temporarily make an atomic-sized hole. This is a fluctuation problem, like the Maxwell–Boltzmann problem of Section 1.3, but a fluctuation of free volume rather than thermal energy, and so it concerns D_0 rather than E_D. In fact, the diffusion coefficient of such a liquid can be expressed in the form

$$D = \alpha a^2 v \, e^{-\gamma v_a / v_f}, \tag{7.16}$$

where $\alpha \simeq 0·1$; a and v are similar to the corresponding quantities in the crystal; v_f is the average free volume per particle (cf. Equation (2.13)); v_a is the volume of the temporary vacancy; and $\gamma \simeq 1$. This resembles the Maxwell–Boltzmann formula, γv_a taking the place of the activation energy and v_f that of the average thermal energy.

7.3 SIMPLE VISCOUS PROCESSES

The processes leading to self-diffusion can also produce viscous flow. From this alone we might expect a crystal to be 10^4 times more viscous than its liquid, at the melting point. The long-range lattice order introduces a second, much larger, effect, however, so that the actual difference

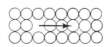

Fig. 7.7. Movement of the vacancy along the central row of sites does not alter the relative positions of the top and bottom rows.

is about 10^{20}. The reason is that in the crystal, but not in the liquid, the atomic sites are defined exactly by the lattice. The migrations of vacancies inside a perfect crystal do not change the shape of that crystal. Figure 7.7 illustrates this. Only where the lattice breaks down, in particular at free surfaces, grain boundaries and dislocations (cf. Fig. 3.16), can the movements of vacancies change the shape of the crystal.

Figure 7.8 shows how such a *vacancy creep* can produce viscous flow by transferring atoms from crystal faces under compression to faces under tension. The atoms move along paths such as those shown by the arrows in the crystal. The vacancies move in the opposite directions, being created

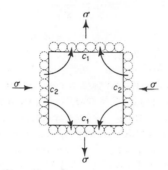

Fig. 7.8. Vacancy creep in a crystal.

at a concentration c_1 at the tensile faces and disappearing at the compressed faces, at a concentration c_2 ($<c_1$). The driving force is the work done by the applied forces when the faces move. If v is the atomic volume, the tensile stress σ exerts a force of about $\sigma v^{2/3}$ on each surface atom and so does work σv each time that atom moves forward by one atomic spacing (about $v^{1/3}$) to create a vacancy in its previous site. The energy of vacancy formation at the tensile faces is thus reduced to $E_F - \sigma v$, and the equilibrium concentration of vacancies near these faces increases to

$$c_1 = c_0 \, e^{\sigma v/kT} \simeq c_0\left(1 + \frac{\sigma v}{kT}\right), \qquad (7.17)$$

where $c_0 = n/N$ (cf. Equation (3.2)). Typically, $\sigma \simeq 10^{-4} \, \mu$, $\mu v \simeq 5 \, \text{eV}$, $kT \simeq 0\cdot1 \, \text{eV}$ (1200°K), so that $\sigma v/kT \simeq 0\cdot005$, which justifies the linear approximation. Similarly, on the compressive faces,

$$c_2 = c_0\left(1 - \frac{\sigma v}{kT}\right). \qquad (7.18)$$

We can now estimate the viscosity. If $2R$ is the length of a crystal face, then, since c_0/v is the equilibrium number of vacancies in unit volume, the vacancy concentration gradient along a typical path from a tensile to a compressive face is of order $(c_1 - c_2)/vR$, i.e. $2c_0\sigma/RkT$. The vacancy diffusion coefficient is D/c_0, since a given vacancy jumps N/n times as often as a given atom. The *flux J* of vacancies, i.e. the net number per unit area and time which enter the crystal through a tensile face or leave through a compressive face, is then of order $(D/c_0)(2c_0\sigma/RkT)$, i.e. $J \simeq 2\sigma D/RkT$. In unit time a face moves a distance Jv, since Jv is the volume of atoms given to or taken from unit area of it. The *vacancy creep rate* is thus $\dot{\varepsilon} = Jv/R$, and the viscosity coefficient is given by

$$\eta \simeq \frac{\sigma}{\dot{\varepsilon}} \simeq \frac{R^2 kT}{\alpha v D}, \tag{7.19}$$

where $\alpha \simeq 2$. A refined version of the calculation gives $\alpha = 4$.

This is the *Nabarro–Herring* equation. If $R = 1$ cm, $kT = 10^{-13}$ erg (i.e. $T \simeq 750°K$), $D = 10^{-10}$ cm^2 sec^{-1} and $v = 10^{-23}$ cm^3, it gives $\eta \simeq 2\cdot5 \times 10^{19}$ poises. The deformation then occurs very slowly, and the crystal is a solid. By contrast, if $R = 0\cdot01$ cm, then $\eta \simeq 2\cdot5 \times 10^{15}$ poises, and the crystal is almost fluid. Vacancy creep is fairly rapid in fine-grained polycrystals and dislocated crystals at high temperatures, because the *internal* sources and sinks of vacancies provided by the lattice imperfections reduce the effective value of R. This is important in *solid-state sintering* (cf. Section 8.9) and *high-temperature creep* (cf. Section 9.8).

Turning now to simple liquids, since there is no longer a lattice holding the external shape fixed, processes such as that of Fig. 7.9, in which the elementary movements distort the atomic "cages", become possible. The

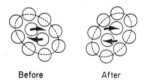

Before   After

Fig. 7.9. Shear due to internal movements in a liquid.

surrounding liquid offers elastic (shear) resistance to such a distortion, but this resistance relaxes as similar movements occur nearby.

We may thus compare viscous flow in the liquid to that of the solid, but with the large R^2 factor removed. In fact, Equation (7.19) indicates the order of magnitude of the liquid viscosity if we replace R by a characteristic atomic dimension, i.e. $R \simeq v^{1/3}$, thus

$$\eta \simeq \frac{kT}{\alpha v^{1/3} D}. \tag{7.20}$$

This is practically the same as the *Stokes–Einstein* formula, which is based on the theory of macroscopic hydrodynamics (cf. Section 7.11). If $kT = 10^{-13}$ erg, $v^{1/3} = 3 \times 10^{-8}$ cm, $D = 10^{-5}$ cm^2 sec^{-1}, then $\eta \simeq 0.1$ poises.

Equation (7.16) shows that the viscosity should be sensitive to the free volume. Highly compressed liquids do, in fact, have greatly increased viscosities. Molecularly simple glasses owe their high viscosities to small free volumes. The liquid–glass transition in such liquids seems to occur when the free volume has practically disappeared.

On the other hand, the constant-volume viscosity of many simple liquids is sensitive to temperature. The explanation of this is not yet clear. It may mean that v_f in Equation (7.16) is sensitive to pressure; or it may mean that an activation energy is involved, as in *Andrade's equation*,

$$\eta = Av^{-1/3}\, e^{q/vkT}, \tag{7.21}$$

which successfully represents the effects of temperature and molecular volume on the viscosities of many simple liquids.

7.4 THERMAL MOTION IN LIQUIDS: SUPERFLUIDITY

The theory of Chapter 6 can be applied to amorphous solids. Both longitudinal and transverse vibrations are possible, with wave velocities given by Equations (6.9) and (6.10), and the "classical" specific heat is

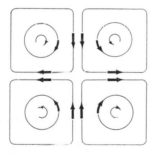

Fig. 7.10. A transverse mode in two dimensions.

3 R per mole. In a liquid, however, the shear modulus is zero, at least for long waves. This reduces the speed of longitudinal waves (cf. Equation (6.11)). More importantly, it turns the transverse motions into *rotations* (cf. Fig. 7.10), since, ideally at least, these no longer experience a restoring force. The classical specific heat of the longitudinal modes remains at R per mole, but that of the transverse modes drops from 2 R (vibrations)

to R (rotations), since the squared potential energy term (cf. Section 1.5) disappears. The classical specific heat of an ideal liquid should then be $2 R$ per mole. In practice, higher values are observed, because some shear resistance remains in the high-frequency transverse modes. In neon and argon, for example, the specific heat of the liquid drops from about $3 R$ at the melting point to $2 R$ at the critical point.

To develop the theory beyond this point is difficult, because in liquids the particles are neither independent nor crystallized and it is hard to find any other simplifying feature that will open up the problem. One approach is to consider behaviour near $0°K$, where the thermal motion is weak. However, only helium is a liquid at $0°K$ and atmospheric pressure. It does not crystallize at pressures below 25 atm. This is because its atoms are light and weakly coherent. Their small mass gives them a high *zero-point energy* ($= \frac{1}{2}h\nu$ where $\nu \simeq \sqrt{\alpha/m}$, α being the spring constant and m the mass), and at atmospheric pressure this zero-point motion enlarges the atomic spacing so much that the crystal lattice is unstable.

Liquid helium is a remarkable substance. When cooled through the "lambda point" ($2.18°K$) it changes to helium II, and below this temperature it becomes increasingly *superfluid*. It can then flow rapidly along narrow channels, e.g. through dense-packed powder beds (channel diameters $\simeq 100$ Å) barely penetrable by ordinary gases or liquids. Its viscosity may, in fact, fall below 10^{-10} poises, a million times smaller than that of ordinary gases.

Landau has developed the theory of liquid helium at low temperatures. There are the usual longitudinal waves with characteristic wavelengths $\lambda = 2l/n$ (cf. Equation (6.57)), where l is the length of the sample and $n = 1$, 2, 3, ... etc. These have frequencies $\nu = c_1/\lambda$ per sec, where c_1 is the velocity of sound in helium. The energy quantum, $h\nu$, acceptable by such a thermal mode is thus hc_1/λ. Since λ can be large, some of these modes are thermally excited even at very low temperatures. This is essentially a repetition of the theory of longitudinal *phonons* in solids. A property of phonons important for superfluidity is the relation

$$\varepsilon = c_1 p \qquad (7.22)$$

which connects the energy ε and momentum p of a phonon. This is quite different from the usual relation $\varepsilon = p^2/2m$ for ordinary particles, but we must remember that the phonon moves through the atoms, not with them, and is a "particle" with zero rest mass. Since $\varepsilon = h\nu$ and $\nu = c_1/\lambda$, Equation (7.22) follows directly from *de Broglie's* quantum-mechanical relation $\lambda = h/p$.

Landau also emphasized that these longitudinal sound waves do not exhaust all the possible modes of motion in the liquid. For the others he postulated *rotons*, which are elementary rotational motions such as micro-scopic *vortex rings* (cf. Fig. 7.10; also Section 12.6). Because the liquid has no long-range elastic shear coherence, the roton equivalents of long-wave, low-frequency phonons do not exist. Only the high-frequency equivalents exist, and the energy quantum for these is large. Hence, at the very lowest temperatures no thermal rotons are formed, only phonons.

The energy–momentum relation for these elementary thermal excita-tions of the liquid is given by curve OABC in Fig. 7.11. The parabolic

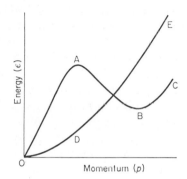

Fig. 7.11. Relation of energy to momentum for elementary motions in a liquid (OABC), compared with that of a free particle (ODE).

part near B is contributed by the rotons and the linear part OA by the phonons.

Consider the liquid flowing with speed v along a tube at 0°K. At this temperature it contains no phonons or other thermal excitations. If the wall friction slows the liquid down slightly, this must occur by the in-jection of elementary excitations into the liquid from the wall. Suppose that one of momentum p and energy ε is injected. Because the liquid is slowed down, p is opposed to v. It can be shown that the liquid thereby loses a kinetic energy pv from its streaming motion. Since the wall of the tube has taken energy from the liquid through the friction, $\varepsilon - pv$ must be negative, i.e.

$$v > \frac{\varepsilon}{p}. \tag{7.23}$$

It is impossible to satisfy this condition if v is small, since there is no point on the OABC curve where $\varepsilon/p \to 0$. If v is below the lowest value of ε/p, the liquid cannot accept the elementary excitation, i.e. it cannot be slowed down and it experiences no frictional resistance from the tube. Super-fluidity occurs, then, because OABC does *not* (unlike an "ordinary" energy–momentum curve such as ODE) approach the origin with zero slope, the reason for this being that the excitations of lowest energy in it are phonons which behave according to Equation (7.22).

7.5 GLASSES

When a liquid is cooled far below its freezing point without crystallizing, it solidifies to a *glass*. The *glass transition temperature* T_g (i.e. at which $\eta \simeq 10^{15}$ poises) is usually 0.25–0.35 $T_b(°K)$, where T_b is the normal boiling point. Some values of $T_g(°C)$ are 1200 (SiO_2), 550 (soda-lime glass), 200 (B_2O_3), 85 (polystyrene), 35 (glucose), 30 (selenium), -70 (rubber), -90 (glycerol) and -180 (alcohol). At temperatures below T_g the glassy state is metastable and endures for long times without *devitrifying* (i.e crystallizing). Most molecularly simple liquids crystallize too quickly to form glasses at ordinary rates of cooling. It is hardly possible to make metal glasses by cooling from the liquid, although this has been achieved in a few eutectic alloys (e.g. Au–Si) of particularly low freezing point by exceptionally rapid cooling. There are of course other ways to make glasses. Metal glasses have been made by condensing vapour directly on to very cold surfaces. Some glasses have been made by electrodeposition. *Silica gel*, a random network of SiO_2 molecules, is made by the decomposition of ethyl silicate or sodium silicate; *silicate cement* depends on this gel-forming reaction for its setting properties.

The *easy glass-formers* are substances with molecules of complex shapes or whose molecules link together strongly in the liquid to form long chains and networks. They have high liquid viscosities (e.g. 10^7 poises for fused SiO_2 at the melting point), because the process of flow involves breaking molecular bonds, with a large activation energy. Many oxides are easy glass-formers, e.g. SiO_2, B_2O_3, GeO_2, P_2O_5, As_2O_5; also silicates, borates and phosphates. Among the elements, S, Se and Te are easy glass-formers. Many organic compounds behave similarly, particularly those which contain hydroxyl groups, e.g. alcohols, and long-chain molecules.

Zachariasen has explained the easy glass-forming oxides. The unit of molecular structure is a small, highly charged, positive ion, e.g. Si or B, surrounded by a polyhedron, e.g. triangle or tetrahedron, of oxygen ions. In the solid or liquid a large number of identical polyhedra are joined together, corner to corner, or edge to edge, or face to face, by sharing oxygen ions in common at their corners. In silica, for example, the tetrahedral grouping of four oxygen ions round one silicon agrees with the SiO_2 composition, because each oxygen is shared by two silicons, i.e. the tetrahedra are joined corner to corner. This principle can be generalized as follows. Let z be the number of unit charges on the positive ion and n the number of positive ions bonded directly to a given oxygen ion. There are $z/2$ oxygen ions per positive ion. The number of bonds emanating from a positive ion is 3 (triangular oxygen polyhedra) or 4 (tetrahedra). The number emanating from $z/2$ oxygen ions is $nz/2$. These numbers are equal. Hence, $n = 6/z$ (triangles) or $8/z$ (tetrahedra). Thus, when z is large, the structure, whether crystalline or amorphous, is an "open" one in which the oxygen ions have a low co-ordination number. When this co-ordination number is 2, the structure forms a rather open network of short molecular chains. In the *cristobalite* structure of crystalline silica, for example, the silicon ions form a diamond structure (cf. Fig. 3.7) and there is an oxygen ion between each pair of silicon neighbours. We can think of this either as an open network, e.g.

$$-O-\underset{\underset{\underset{|}{O}}{\overset{\overset{|}{O}}{Si}}}{}-O-\underset{\underset{\underset{|}{O}}{\overset{\overset{|}{O}}{Si}}}{}-O-$$

or as an array of tetrahedra joined corner to corner. Figure 7.12 shows a *two-dimensional* analogue of this (with oxygen triangles). Similar, but irregular, networks exist in fused silica.

When the structure is such that the co-ordination of nearest neighbours

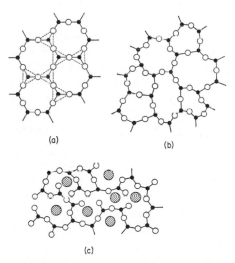

(a)

(b)

(c)

Fig. 7.12. Two-dimensional networks analogous to cristobalite and silica glass: (a) crystalline, showing network of bonds (full lines) and oxygen triangles (broken lines); (b) irregular; (c) effect of sodium ions, as in soda-glass.

is virtually the same in both the crystalline and irregular networks, these alternative forms of network have practically the same energy, and the oxide is an easy glass-former, More precisely, Zachariasen's conditions for a crystalline oxide to be an easy glass-former are as follows:

(1) Each oxygen ion linked to not more than two positive ions.
(2) Small (three or four) number of oxygen ions surrounding a positive ion.
(3) Oxygen polyhedra share corners with each other, but not edges or faces.
(4) Enough polyhedra share at least three corners in common to make a three-dimensional network of bonds.

All the easy glass-forming oxides satisfy these rules. Furthermore, BeF_2, which also satisfies them, is an easy glass-former, whereas CaF_2, which does not, is not.

Viscous network liquids crystallize only slowly, even when seeded with crystal nuclei, because the necessary molecular rearrangements at the liquid–crystal interface are sluggish. Turnbull and Cohen have shown that liquids whose activation energy for viscous flow exceeds about $30 \, RT_m$ per mole (R = gas constant, T_m = absolute melting point) form glass easily. The bond energies in sulphur and selenium chains are much larger than this, and these substances undercool readily to the amorphous state. Silica ($\simeq 30 \, RT_m$) and glycerol ($\simeq 25 \, RT_m$) also undercool readily. Water is a network liquid, structurally similar to fused silica, in which H_2O molecules are held together by "hydrogen bonds", i.e. electrostatic attractions between positively and negatively charged parts of neighbouring molecules. However, its activation energy for fluid flow is only about $10 \, RT_m$, so that pure water is not an easy glass-former. On the other hand, many concentrated water solutions become glasses when quickly cooled (e.g. toffee). Many amorphous hydrogen-bonded substances can be prepared from aqueous solution. For example, $Al(H_2PO_4)_3$ forms a viscous liquid with water; when dried quickly at a low temperature, it sets to a glassy solid.

Ordinary silica-based glasses contain both *network-forming* constituents, such as the oxides of Si, B and P, and *network-modifying* constituents, such as the oxides of Na, K, Ca, Ba and Pb. Examples are *soda-lime* glass ($\simeq Na_2Ca \, Si_5 \, O_{12}$), *light flint* glass for optical use ($\simeq KNaPb \, Si_5 \, O_{12}$) and *borosilicate* glass with a low expansion coefficient ($\simeq Ba \, K_4 \, B_2 \, Si_{10} \, O_{26}$). The addition of, say, 25 % Na_2O to SiO_2 lowers the fusion temperature to about 800°C, which enables a glass to be made by mixing the constituents (e.g. silica sand and Na_2CO_3) at a temperature well below the melting point of SiO_2. Figure 7.12(c) shows a two-dimensional analogue of the structure of soda-glass. The sodium ions fill interstitial sites in the network, as is shown by the *ionic* electrical conductivity of this type of glass, caused by the diffusional drifting of charged interstitial sodium ions in applied electrical fields. The oxygen ions introduced with the sodium go into the silicon–oxygen network. There are then more oxygen bonds than silicon bonds, so that some of the links in the network are broken at oxygen ions, as shown in Fig. 7.12(c). Thus, if we regard silica glass as a *densely cross-linked polymer* (cf. Section 7.7), the sodium addition has the effect of softening the material by reducing the number of cross-links. This is the opposite of the action of *sulphur* in *vulcanized rubber*, which is added to stiffen the material by increasing the number of cross-links between the long-chain rubber molecules. The reduced cross-linking in soda-glass strikingly reduces the viscosity. For example, at 1400°C, 20% Na_2O in SiO_2 reduces the viscosity from 10^{12} to 10^2 poises.

7.6 VISCOUS SOLUTIONS, SUSPENSIONS, GRAVELS AND PASTES

A *sol* is a solution or suspension of *colloidal* particles (sizes between 10 Å and 5000 Å) in a solvent. *Lyophilic* sols contain solutes whose molecules are large enough to be classed as colloidal; *lyophobic* sols contain substances which are normally insoluble but whose particles are small enough to be prevented by their Brownian motion (cf. Section 1.6) from settling. Examples of the former are water solutions of gelatine, starch and dextrin; rubber cement (rubber in benzene); gums (water-soluble resins in water); shellac (animal resin in alcohol); collodion (celluloid in alcohol and ether); varnishes and oil paints. Concentrated solutions of these are highly viscous and readily "set" to *gels* or jellies on resting at constant temperature. Gels are weak solids formed from continuous networks of material with an interstitial filling of liquid. The density of cross-links in them is comparatively low, so that their elastic constants are extremely small (e.g. Young's modulus $\simeq$ 3·5 psi for 10% gelatine in water). Lyophilic sols have non-Newtonian viscosities and are usually *thixotropic*; when allowed to rest for long times, their solute molecules gradually assemble into more densely linked networks and become gels. When disturbed, as when a paint is stirred after standing, these cross-links are broken up, and they become more fluid.

Lyophobic sols are mostly formed from chemical reactions in solution, which produce fine particles of the dispersed constituent. Examples are the formation of a water sol of metallic gold by the reduction of gold ions in aqueous solution; and the precipitation of flocculent hydroxides such as ferrous hydroxide in water. Lyophobic sols are often stabilized by electrical charges which exist on the colloidal particles and which prevent the particles from coagulating. The existence of these charges is shown by *cataphoresis*, the migration of such particles in electrical fields. Coagulation and settling out can usually be produced by adding a salt to the solution, but the reverse process (*peptization*) is more difficult.

If a dilute suspension of isolated particles is formed in a Newtonian fluid, the resulting sol is also Newtonian, but with an increased viscosity. This can be explained by the *perturbed pattern* of fluid flow round such a particle. As it flows past the particle, each element of fluid moves sideways, round the particle, and then back again. These extra movements dissipate energy by the viscous friction within the fluid, but do not contribute to the bulk flow of the fluid. Their contribution to the total viscous energy loss in the fluid can be calculated, and, for dispersed spheres, this leads to *Einstein's relation* for the viscosity,

$$\eta = \eta_0(1 + \tfrac{5}{2}c), \qquad (7.24)$$

where η_0 is the viscosity of the pure fluid and c the volume fraction of dispersed constituent.

Actual viscosities of dilute colloidal suspensions are often much larger than this, usually because the particles are not spherical. Consider, for example, two spheres joined by a long, thin, rigid rod. In an arbitrary field of flow the fluid velocity near one sphere is usually quite different from that near the other. Each sphere is then, by its link to the other, dragged through the liquid, and this dissipates energy. This dragging velocity can be large, since it depends on the difference of fluid velocities at opposite ends of the dumb-bell. For randomly oriented dumb-bells, Equation (7.24) is replaced by

$$\eta = \eta_0(1 + \tfrac{3}{2}p^2 c), \qquad (7.25)$$

where p is the ratio of dumb-bell length to sphere diameter. Similarly large effects, mainly proportional to p^2, are produced by rod-shaped particles. The fluid flow tends of course to rotate the particles into preferred directions, but the orientations of colloidal particles are continually randomized by the Brownian motion, and so the frictional resistance of these very small particles is maintained. There is thus a difference between the behaviour of microscopic and macroscopic rod-shaped particles.

The above formulae break down when the concentration of particles increases to the level at which the perturbed regions of fluid flow round neighbouring particles begin to overlap. Terms in c^2 begin then to appear in the formulae for the viscosity. At the other extreme we meet substances such as *pastes*, *clay* (water-containing aluminium silicate), *sand*, *gravel* and *pebble beds*. These mostly consist of solid grains embedded in a fluid of low viscosity, such as water or air. The volume fraction of grains is high, e.g. $c \simeq 0.65$ for a sand or gravel bed; $c \simeq 0.9$ for sand and gravel mixtures in which fine sand grains fill the spaces between coarse gravel grains (the maximum for equal close-packed spheres, as in the f.c.c. lattice, is $c = 0.74$). Some chemical cohesive forces are exerted between the grains in clay, but in sands and gravels the only important forces are gravity and contact pressures between the grains.

The mechanical properties of such mixtures are sensitive to the volume fraction of grains. Dry clay becomes "plastic" when a moderate amount of water is added to it. The best properties for modelling are reached at about 30% water; there is then a continuous film of water between the grains, about 3000 Å at its thickest points. When more water is added, the grains begin to float apart and the surface of the clay becomes wet and sticky. At still higher water contents the mixture becomes a fluid *slurry* which can be poured into moulds to make *slip castings*. Wet sand with rounded grains behaves as a stiff paste when $c > 0.65$ and as a Newtonian liquid when $c < 0.6$. This explains the dry patches which appear round footsteps on moist sand at the seashore; the sand expands in order to flow under the footprint and water is sucked in from the surface to fill the enlarged pores between its grains. The existence of stable *sand-hills* on beds of dry cohesionless grains shows that the density of packing under gravity is sufficient to give such beds "solid" rather than "fluid" properties. Gravity can be overcome either by immersing the bed in a liquid of the same density as that of the grains or by forcing air or other gas upwards through vertical nozzles beneath the bed. When the change of pressure upwards through the bed equals the weight of the grains, the bed expands slightly, as the contact pressures between the grains are relieved, and then becomes a fluid of low viscosity. Such a *fluidized bed* is an excellent heat-transfer medium, since it combines high fluidity of the gas with high heat capacity of the solid.

Pastes and modelling clays resemble *plastic solids* rather than viscous fluids. Their rate of flow increases sharply at a certain *yield stress* σ_y. Below this stress the substance is almost rigid and preserves its shape against gravity for long periods, but at this stress it can deform rapidly. This property, vital to the moulding of clay pottery, is described approximately by the *Bingham equation*

$$\dot{\gamma} = \frac{\sigma - \sigma_y}{\eta}, \tag{7.26}$$

where $\sigma \,(\geqslant \sigma_y)$ is the applied stress and η is a coefficient of viscosity that is independent of σ. Typically, $\sigma_y \simeq 0.03$ psi and $\eta \simeq 0.2$ poises. The low value of η is evidently due to the fluidity of the water between the grains. The form of the Bingham equation, in contrast to a "quasi-viscous" law such as Equation (7.8) can be understood broadly from the mechanics of flow in a paste. The applied force must be strong enough to slide the grains over each other against the resistance of their contact pressures. In general, the grains are too big for their motion to be affected appreciably by thermal fluctuations. This sliding is therefore essentially a purely mechanical process, which can occur as quickly as required once the critical stress is exceeded. The rate of flow above this stress then depends, not on the mechanical properties of the grains, but on the viscosity of the liquid films between them.

7.7 POLYMERS

To form a long-chain molecule, each unit or *monomer* of the chain must be at least divalent, since it has to link itself to neighbours on either side. The unit, usually repeated many times (e.g. 100–10,000) along the chain, can be a single atom, as in sulphur and selenium, or a molecule, as in asbestos, silicones and organic polymers. Simple *linear polymers* have the form

$$X—A—A— \ldots \text{etc.} \ldots —A—A—Y,$$

where A is the *repeated structural unit* and X and Y are monovalent *terminal units*. An example is *polyethylene*, a long-chain paraffin,

$$
\begin{array}{ccccccc}
& H & H & & & H & H \\
& | & | & & & | & | \\
H— & C— & C— & \cdots\cdots & — & C— & C—H, \\
& | & | & & & | & | \\
& H & H & & & H & H
\end{array}
$$

in which $A = CH_2$ and $X = Y = H$. The carbon atoms are shown as a straight line, but in reality form a zigzag sequence.

The units in the chain need not be all identical. Innumerable variations are possible. In *copolymers* two different units A and B occur randomly along the chain, e.g.

$$—A—B—B—A—B—A—A—A—B—A—$$

and in *block copolymers* they form an ordered sequence. Many different polymers are based on the polyethylene chain, with the hydrogen atoms at least partly replaced by other substances. Examples of repeating units are:

Polyvinyl chloride (vinylite):

$$
\begin{array}{cc}
H & H \\
| & | \\
—C— & C— \\
| & | \\
H & Cl
\end{array}
$$

Polytetrafluorethylene (PTFE, Teflon):

$$\begin{array}{cc} F & F \\ | & | \\ -C-C- \\ | & | \\ F & F \end{array}$$

Polystyrene:

$$\begin{array}{cc} H & H \\ | & | \\ -C-C- \\ | & | \\ H & C_6H_5 \end{array}$$

Polymethlymethacrylate (Perspex, Lucite):

$$\begin{array}{cc} H & CH_3 \\ | & | \\ -C-C- \\ | & | \\ H & COOCH_3 \end{array}$$

Polyisobutylene (butyl rubber):

$$\begin{array}{cc} H & CH_3 \\ | & | \\ -C-C- \\ | & | \\ H & CH_3 \end{array}$$

In natural rubber, i.e. *polyisoprene*:

$$\begin{array}{c} CH_3 \\ | \\ C=CH \qquad CH_2 \qquad\qquad CH_2 \\ \diagup\diagdown \qquad \diagup \qquad\quad C=CH \quad \diagdown \\ CH_2 \qquad CH_2 \qquad\qquad | \\ CH_3 \end{array}$$

some of the carbon atoms in the chain use two of their four valencies to form *double bonds*. The exceptional deformability of the rubber molecule is due to the spiral structure of the carbon chain. In *gutta-percha*, another form of polyisoprene, the carbon chain is a zigzag, not a spiral, and the molecule is much less deformable. In recent years ways have been found of synthesizing *isotactic polymers*, i.e. of controlling the relative orientations of the monomer units along the chain.

Some polymers contain other atoms besides carbon in the main chain, e.g. *cellulose*:

$$-O-C_6H_{10}O_4-$$

superpolyamide (nylon):

$$-\overset{\displaystyle O}{\overset{\displaystyle \|}{C}}-\underset{\displaystyle H}{N}-(CH_2)_6-\overset{\displaystyle H}{N}-\overset{}{\underset{\displaystyle O}{\underset{\displaystyle \|}{C}}}-(CH_2)_{10}-$$

and inorganic polymers, such as *asbestos*:

$$chrysotile, \simeq 3MgO.2SiO_2.2H_2O$$

and *methyl silicone oil:*

Structural units of higher valency in the main chain provide *side chains* by which the main chain can be linked to others, so forming a *network polymer*. Examples are *phenol formaldehyde resin* (*Bakelite*), with three links per unit:

$$-\overset{\displaystyle OH}{\underset{\displaystyle CH_2}{C_6H_2}}-CH_2-$$

and *urea formaldehyde*, with four links per unit:

$$\begin{array}{c} -N-CH- \\ | \\ CO \\ | \\ -N-CH- \end{array}$$

These densely cross-linked polymers are described as *thermosetting*, because, once formed, they do not soften on heating but remain linked together in a hard, rigid form until temperatures are reached at which they disintegrate chemically. Dissolved in organic solvents or combined with drying oils they form varnishes, lacquers and adhesives; when aged in air, and especially when *cured* by heating, they then polymerize and become hard enamels.

Polymers can exist in various mechanical states: fluid, rubbery, gel, glassy and crystalline. Linear polymers without cross-links, such as rubber *latex*, are viscous fluids. They consist of long, convoluted molecules tangled together like a mass of earthworms, and they flow by the sliding of molecular segments over one another. This is a thermally activated process, and the viscosity varies rapidly with temperature. At low viscosities (< 1000 poises) the flow is Newtonian, but not at high viscosities. Such materials are *thermoplastic* polymers. Above their glass temperatures (about 100°C for polymethylmethacrylate; about −80°C for unvulcanized rubber) they are soft, deformable and, if not cross-linked, fluid; below this temperature they are typical glasses. They can also, above their glass temperatures, deform by straightening or crumpling their molecular chains. This process, which is opposed by the randomizing action of thermal motion (cf. Section 4.4) is responsible for the *high elasticity* of rubber. If the molecular chains are long enough, transient rubber elasticity can be observed even when there are no cross-links, as in silicone putty. It becomes non-transient when the molecules are cross-linked sufficiently to make the material a *gel*, consisting of a skeleton network embedded in a viscous fluid. By increasing the density of cross-links, as when rubber is *vulcanized* to various degrees by the addition of sulphur, a continuous range of properties can be produced, from soft elastic to motor tyre rubber and vulcanite.

The rate of chain straightening or crumpling depends on viscosity, because it involves the sliding of segments of chains past one another. The strain develops with time in the manner shown in Fig. 7.13. The total

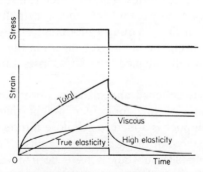

Fig. 7.13. Deformation of a thermoplastic polymer.

deformation is contributed by "true" elasticity (straining of nearest-neighbour bonds), high elasticity and viscous flow, the latter being suppressed in the more highly cross-linked materials. On unloading, the true elasticity is recovered instantly but the high elasticity recovers more gradually because of the viscous drag on the chains, giving *delayed elasticity* or *elastic after-effect*.

At room temperature unvulcanized rubber has a fairly low viscosity (≃ 1000 poises), and its high elastic deformation occurs quickly. The viscous drag on the sliding chains shows itself mainly in the high internal friction which accompanies the deformation of rubber (cf. Fig. 6.23). At low temperatures the chains slide so sluggishly that the rubber becomes a glass. The corresponding temperatures in polymethylmethacrylate are higher, so that this material is glassy at room temperature and rubbery at above 140°C. It can be fabricated easily by stretching or blowing to shape, against a former, while hot; and then cooled down, while still held in this shape, to temperatures where the resistance to sliding prevents the strained molecular network from recovering its initial shape.

Some linear polymers *crystallize*; particularly the more fluid ones with short chain lengths, no cross-links and weak van der Waals forces between the chains. Polymers of simple hydrocarbon chains (e.g. polyethylene) less than about 100 units long crystallize almost completely on cooling. Those with longer chains (100–10,000 units) crystallize partly. In the crystalline regions, about 100 Å across, the chains are parallel, whereas in the non-crystalline regions they are tangled. The chains form crystallites by folding back and forth in straight lengths, about 100 Å long, pleated like a Chinese cracker. A single molecular chain may extend through both crystalline and non-crystalline regions.

A form of crystallization sometimes occurs during deformation, by alignment of the molecular chains along an axis of extension. When the chains are roughly aligned in this way, the chemical forces of crystallization align the molecules more completely and so help the applied forces to stretch the material further. Young's modulus (of high elasticity) thus becomes very small during this change, as shown in Fig. 6.23. Afterwards, when the chains are fully aligned, the stress rises rapidly with further strain.

Glassy polymers are softened and made less brittle by the addition of *plasticizers*, substances of low molecular weight that lower the glass temperature; e.g. camphor is added to cellulose nitrate to make *celluloid*. Brittleness is then reduced at the expense of rigidity. To combine rigidity with toughness, composite substances have been developed in which a bundle, mat or felt of strong fibres, e.g. of silicate glass, cellulose or ceramic oxide, is embedded in a matrix of relatively soft resin or polymer (cf. Section 11.2). Fibrous polymers, such as textile fibres and wood, are strong and tough for the reasons discussed in Section 11.2. *Cellulose* is a natural fibre with a highly oriented crystalline structure, a Young's modulus of 15×10^6 psi, and a breaking strength of up to 15×10^4 psi. Having a density of only 1·5, it is thus competitive with metal wire for tensile thread. *Nylon* (annealed after cold drawing) and several other textile fibres show *yield drops* (cf. Fig. 6.13) and *Lüders bands*, similar to those in mild steel, when stretched. This appears to be caused by an unwinding of the pleated chains in some of the crystallites and recrystallization of them along the axis of extension.

7.8 SIMPLE PROBLEMS OF VISCOUS FLOW

In the remaining sections of this chapter we shall analyse some simple modes of flow in Newtonian fluids. The similarity between Newton's law and Hooke's law suggests that, under some conditions at least, the analysis of viscous flow could be based on elasticity, with strain rate replacing strain. This in fact is so, if the rate of flow is everywhere so small that the inertial forces, which produce *turbulence*, are negligible compared with the viscous forces. We shall assume this in all the following problems and leave the more general aspects of fluid mechanics to Chapter 12. We shall also assume that the fluid is *incompressible* (which is usually a good approximation under the small forces that produce viscous flow in most fluids) and that there is no *slip*, i.e. relative motion, at the interface between the fluid and the walls of its container.

Consider in Fig. 7.14 flow along a tube of length l and radius r_0 under

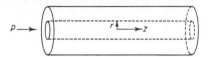

Fig. 7.14. Viscous flow along a tube.

a pressure p applied at one end. Provided the velocity is below a certain value, which depends on the viscosity η and on the radius of the tube, each particle of fluid moves uniformly in a *laminar flow* along a line parallel to the axis (z) of the tube. Let w be the velocity of a particle. Then, w depends on r only, and the shearing stress in the axial direction is given by

$$\sigma = \eta \frac{dw}{dr}. \tag{7.27}$$

Consider the mechanical equilibrium of the cylinder of fluid with radius r, shown in Fig. 7.14. The applied force $\pi r^2 p$ on its end-face is balanced against the viscous force $2\pi r l \sigma$ on its cylindrical surface, i.e.

$$\sigma = -\frac{rp}{2l}. \tag{7.28}$$

Hence, eliminating σ between these equations,

$$dw = -\frac{p}{2l\eta} r \, dr, \tag{7.29}$$

which integrates to give

$$w = \frac{p}{4l\eta}(r_0^2 - r^2).$$ (7.30)

The *velocity profile* from the axis to the wall is thus *parabolic*; w is zero at the wall and increases as the square of distance to its maximum value $pr_0^2/4l\eta$ at the axis. The volume V of fluid discharged from the tube in unit time is given by

$$V = \int_0^{r_0} 2\pi r w \, dr = \frac{\pi p r_0^4}{8l\eta}.$$ (7.31)

This is *Poiseuille's relation*, by which viscosity can be measured from the flow of fluids through capillary tubes.

As a second example, we consider the motion of a fluid between coaxial cylinders, as in Fig. 7.15. Let the outer cylinder, of radius r_1, rotate slowly about its axis with angular velocity ω, i.e. linear velocity ωr_1, and let the inner cylinder, of radius r_0, be at rest. The particles of fluid then move in circular paths round the axis of rotation. In a system of polar co-ordinates (cf. Section 5.8) let u_r, u_θ and u_z be the radial, tangential and

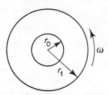

Fig. 7.15. Viscous flow between rotating coaxial cylinders.

axial components of the fluid velocity. Then, $u_r = u_z = 0$, and, from Equation (5.52), the rate of shear is given by

$$\dot{\gamma} = \frac{du_\theta}{dr} - \frac{u_\theta}{r} = \frac{\sigma}{\eta},$$ (7.32)

where σ is the tangential shear stress. The couple exerted by σ on a cylinder of radius r and length l is $2\pi r^2 l\sigma$. This must be the same for all values of r and, hence, equal to M, the couple exerted on the fixed cylinder by the fluid, i.e.

$$\sigma = \frac{M}{2\pi r^2 l}.$$ (7.33)

Hence, eliminating σ,

$$\frac{du_\theta}{dr} - \frac{u_\theta}{r} = \frac{M}{2\pi r^2 l\eta},$$ (7.34)

with $u_\theta = 0$ at $r = r_0$, and $u_\theta = \omega r_1$ at $r = r_1$. The general solution of this equation is

$$u_\theta = Ar - \frac{M}{4\pi r l \eta}. \qquad (7.35)$$

The boundary conditions give

$$M = \frac{4\pi l \eta \omega r_0^2 r_1^2}{(r_1^2 - r_0^2)} \quad \text{and} \quad A = \frac{\omega r_1^2}{(r_1^2 - r_0^2)}.$$

Hence, the velocity profile is given by

$$u_\theta = \frac{\omega r_1^2}{(r_1^2 - r_0^2)} \frac{(r^2 - r_0^2)}{r}. \qquad (7.36)$$

This is the basis of the *Couette method* for determining viscosity from the couple exerted on such cylinders.

7.9 THE EQUATIONS OF SLOW VISCOUS FLOW

We shall now derive the general equations governing slow Newtonian flow, continuing with the same general assumptions as before. The restriction to slow motion, with negligible inertial forces, means that, although the fluid is moving, terms representing accelerations (cf. Chapter 6) do not appear in the equations. This is helpful, because such terms make the equations *non-linear* (cf. Chapter 12) and difficult to solve. Their omission allows us to set up the flow equations in a form similar to the equations of static elasticity.

Referring to Section 4.6, we now regard the $u_i (-u_1, u_2, u_3)$ as the *components of velocity* at the point $x_i (=x_1, x_2, x_3)$, i.e. they are the time derivatives of the displacements. The *components of strain rate* $\dot{\varepsilon}_{ij}$ (with $\dot{\gamma}_{ij} = 2\dot{\varepsilon}_{ij}$ when $i \neq j$) and *rotation rate* $\dot{\omega}_{ij}$ are then defined as in Equation (4.24). In one respect the analysis of fluid flow is easier than that of elastic deformation. In elasticity *finite* deformation raises mathematical difficulties, and, as in Section 4.6, it is usual to avoid these by limiting the analysis to infinitesimal deformation. In fluid flow the deformation is usually both finite and large, but the corresponding mathematical difficulties no longer appear. In elastic deformation we have to compare the deformed and undeformed states, and, if these differ finitely, the problems of finite strain are unavoidable. In fluid flow, however, we deal with *velocities* and thus have to compare states of deformation at successive times t_1 and t_2; there is no difficulty about making t_1 and t_2 as close as we please and so making the difference in states infinitesimal. Ultimately, this difference between elasticity and fluid flow rests on the fact that the mechanical resistance of the fluid (at least in simple fluids) is unchanged by deformation that has gone before, whereas that of the elastic solid increases with the total strain. This same simplifying feature is exploited in the theory of ideal plasticity (cf. Section 10.4).

The *equation of continuity* is an important relation between the velocity coefficients. It expresses the law of conservation of matter; what flows into a given region must either flow out again or accumulate there. For *continuous, incompressible* fluids it reduces to the condition that the outflow must equal the inflow, i.e. that the *dilatation* (cf. Equation (4.43)) is constant,

$$\dot{\Theta} = \dot{\varepsilon}_{11} + \dot{\varepsilon}_{22} + \dot{\varepsilon}_{33} = \frac{\partial u_1}{\partial x_1} + \frac{\partial u_2}{\partial x_2} + \frac{\partial u_3}{\partial x_3} = \frac{\partial u_j}{\partial x_j} = 0. \qquad (7.37)$$

Equations (4.69) define the stress equilibrium in the fluid when the inertial forces are negligible. Substituting $\varepsilon \to \dot{\varepsilon}$, $v \to 0.5$, $E \to 3\eta$ (cf. Equation (7.3)) in Equations (5.9) and defining the hydrostatic pressure p as in Equation (4.63), we can replace Hooke's law by

$$\sigma_{11} = 2\eta\dot{\varepsilon}_{11} - p, \qquad (7.38)$$

$$\sigma_{12} = \eta\dot{\gamma}_{12}, \qquad (7.39)$$

with similar relations for the other components. Using these and also Equations (4.24), the first of Equations (4.69) can be changed to

$$0 = F_1\rho + \frac{\partial \sigma_{11}}{\partial x_1} + \frac{\partial \sigma_{12}}{\partial x_2} + \frac{\partial \sigma_{13}}{\partial x_3}$$

$$= F_1\rho - \frac{\partial p}{\partial x_1} + \eta\left(2\frac{\partial^2 u_1}{\partial x_1^2} + \frac{\partial^2 u_1}{\partial x_2^2} + \frac{\partial^2 u_2}{\partial x_1 \partial x_2} + \frac{\partial^2 u_1}{\partial x_3^2} + \frac{\partial^2 u_3}{\partial x_1 \partial x_3}\right)$$

$$= F_1\rho - \frac{\partial p}{\partial x_1} + \eta\nabla^2 u_1 + \eta\frac{\partial}{\partial x_1}\left(\frac{\partial u_1}{\partial x_1} + \frac{\partial u_2}{\partial x_2} + \frac{\partial u_3}{\partial x_3}\right), \qquad (7.40)$$

where

$$\nabla^2 u_1 = \frac{\partial^2 u_1}{\partial x_1^2} + \frac{\partial^2 u_1}{\partial x_2^2} + \frac{\partial^2 u_1}{\partial x_3^2}. \qquad (7.41)$$

The final term in Equation (7.40) vanishes, because of Equation (7.37). Generalizing, the equations of stress equilibrium become

$$\eta\nabla^2 u_i - \frac{\partial p}{\partial x_i} + F_i\rho = 0. \qquad (7.42)$$

7.10 FLOW BETWEEN PARALLEL PLANES: VISCOUS ADHESION AND LUBRICATION

Consider two large, fixed, parallel planes, with fluid between, very closely spaced with co-ordinates $x_3 = 0$ and $x_3 = h$, respectively. Suppose that this fluid undergoes a *plane flow* in which $u_3 = 0$ and the variation of u_1 and u_2 in the x_3 direction greatly exceeds their variation in the x_1 and x_2 directions; since the velocity is zero at $x_3 = 0$ and $x_3 = h$, and is a maximum at $x_3 = \frac{1}{2}h$, this is reasonable. In the limit of small h we may thus make the approximations

$$\nabla^2 u_1 \simeq \frac{\partial^2 u_1}{\partial x_3{}^2}, \qquad \nabla^2 u_2 \simeq \frac{\partial^2 u_2}{\partial x_3{}^2}, \qquad \nabla^2 u_3 \simeq 0, \tag{7.43}$$

so that, in the absence of body forces F_i, Equations (7.42) reduce to

$$\eta \frac{\partial^2 u_1}{\partial x_3{}^2} = \frac{\partial p}{\partial x_1} \quad \text{and} \quad \eta \frac{\partial^2 u_2}{\partial x_3{}^2} = \frac{\partial p}{\partial x_2}. \tag{7.44}$$

Since p is independent of x_3 (cf. Equations (7.42) and (7.43)), these equations integrate to

$$u_1 = \frac{1}{\eta} \frac{x_3(x_3 - h)}{2} \frac{\partial p}{\partial x_1} \quad \text{and} \quad u_2 = \frac{1}{\eta} \frac{x_3(x_3 - h)}{2} \frac{\partial p}{\partial x_2}. \tag{7.45}$$

The $x_3(x_3 - h)$ term gives a parabolic velocity profile, as in Poiseuille flow. The volume V_1 of fluid flowing through a plane of unit width perpendicular to the X_1 axis is given by

$$V_1 = \int_0^h u_1 \, dx_3 = \frac{1}{2\eta} \frac{\partial p}{\partial x_1} \int_0^h (x_3{}^2 - x_3 h) \, dx_3 = -\frac{h^3}{12\eta} \frac{\partial p}{\partial x_1}, \tag{7.46}$$

i.e. the rate of flow per unit width is equal to the (negative) pressure gradient in the same direction multiplied by $h^3/12\eta$.

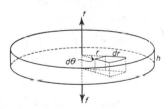

Fig. 7.16. Radial flow between separating disks.

We now consider, as in Fig. 7.16, two circular disks of radius R and spacing h, which are being pulled apart at a speed dh/dt by a force f.

Suppose that the fluid between them is in contact with a reservoir at a pressure p_0, so that additional fluid is sucked in uniformly round the circumference to fill the increasing volume between the plates. This flow is clearly radial. Consider the radial volume element shown in Fig. 7.16, of height h and sides dr and $r\,d\theta$. Its volume is increasing at a rate $(dh/dt)r\,dr\,d\theta$, and fluid is flowing into it at the same rate. From Equation (7.46), the rate of radial flow V through a face of area $hr\,d\theta$ at a radius r is equal to $(h^3 r\,d\theta/12\eta)(\partial p/\partial r)$. The net rate of flow into the element is the change in V across the radial thickness dr of the volume element, i.e. $(\partial V/\partial r)\,dr$. Equating this to the rate of volume increase, we obtain

$$\frac{12\eta}{h^3}\left(\frac{dh}{dt}\right)r = \frac{\partial}{\partial r}\left(r\,\frac{\partial p}{\partial r}\right), \tag{7.47}$$

which integrates to

$$\frac{6\eta}{h^3}\left(\frac{dh}{dt}\right)r^2 = r\,\frac{\partial p}{\partial r}, \tag{7.48}$$

the integration constant being zero, since $V = 0$ (and, hence, $\partial p/\partial r = 0$) at $r = 0$. Integrating again,

$$p = p_0 - \frac{3\eta}{h^3}\left(\frac{dh}{dt}\right)(R^2 - r^2), \tag{7.49}$$

the integration constant being such that $p = p_0$ at $r = R$.

The force f pulling the disks apart at this rate is equal and opposite to the pressure difference $p - p_0$ integrated over the disk, i.e.

$$f = \int_0^R 2\pi r\,\frac{3\eta}{h^3}\left(\frac{dh}{dt}\right)(R^2 - r^2)\,dr$$

$$= \frac{3\pi\eta R^4}{2h^3}\left(\frac{dh}{dt}\right). \tag{7.50}$$

By integrating with respect to time at constant f, we can then find the time t required for the spacing between the disks to increase from h_1 to h_2; i.e.,

$$t = \frac{3\pi\eta R^4}{4f}\left(\frac{1}{h_1^2} - \frac{1}{h_2^2}\right). \tag{7.51}$$

For example, if $R = 0\cdot5$ in. $(=1\cdot27$ cm$)$, $\eta = 0\cdot15$ poises (as in a light mineral oil), $h_1 = 10^{-5}$ cm and $f = 9500$ kg $(=21{,}000$ lb$)$, then $h_2 \simeq \infty$ in 10 sec. One type of viscometer is based on this principle, the time for a ball to fall from a cup being measured.

When h_1 is small, the force required to pull the disks apart quickly is large. We see this in the strong adhesion that exists between accurately ground and oiled metal surfaces that have been "wrung" together. The same principle is used in *viscous adhesives* such as "Scotch tape". Joints made in this way can fail either viscously, by the process just described, or by fracture, usually along the interface or through the adhesive itself. There is no fixed strength for viscous failure, since this depends on the time of loading; the viscous adhesion is measured by the parameter ft, which increases as the viscosity, as the square of the area, and inversely as the thickness of the adhesive layer. Since the viscous strength decreases as the time of loading increases, then, if the joint survives the initial loading without immediately fracturing, its eventual failure will usually be by viscous flow.

A well-made joint of large surface area has a large viscous strength and usually fails, particularly under impact loads, by fracture. The mode of fracture depends on the rate of loading and the viscosity of the adhesive. If the adhesive is glassy, *brittle* fracture usually occurs. If the adhesive is fluid, fracture often occurs by the nucleation and growth of *spheroidal cavities* at weak spots along the joint, such as entrapped gas bubbles, places where the adhesive has failed to bond to the joined surface, and "non-wetting" foreign particles in the adhesive. These cavities expand by the viscous flow of fluid round them, as shown in Fig. 7.17. This is

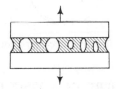

Fig. 7.17.   Failure of a viscous joint by formation and growth of cavities.

clearly a much faster mode of viscous failure than that represented by Equation (7.51), since long-range flow is not required, only local flow round cavities.

Viscous flow between plates is important for *sliding bearings* in machinery. These can be designed so that, as one solid surface slides over the other, they never touch but remain separated by a thin film of oil or other fluid lubricant. The theory of the effect is due to *Reynolds*. The sliding surfaces are slightly inclined, as in Fig. 7.18, and oil is dragged along the wedge-shaped region between them from the wide to the narrow end. The rise in pressure due to the flow of oil down this narrowing channel forces the surfaces apart against the force of the normal load carried by the bearing. The maximum pressure is reached at a point just behind the centre of the slider, and it is common practice in bearings to apply the normal load through a pivot behind the centre. The slider then automatically settles in a stable inclined position when in motion.

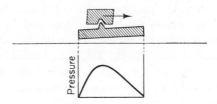

Fig. 7.18. Distribution of pressure in a sliding bearing.

The same principle governs the self-alignment of a journal bearing, as shown in Fig. 7.19. Again, the lubricating fluid is dragged into the wide end of the wedge-shaped region and swept towards the narrow end, where its rise of pressure tends to push the journal back on to the axis of its bearing.

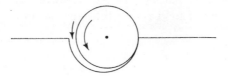

Fig. 7.19. Fluid flow in an eccentric journal bearing.

7.11 MOTION OF A SPHERE IN A VISCOUS FLUID

The important formula

$$f = 6\pi a v \eta \tag{7.52}$$

for the force f exerted by a fluid of viscosity η on a sphere of radius a moving slowly through it with velocity v, was derived by *Stokes*. Its general form, but not the numerical factor, can be obtained from a simple dimensional argument. We suppose intuitively that f depends on the radius and velocity of the sphere and the density ρ and viscosity η of the fluid. Let M, L and T represent the dimensions *mass, length* and *time*. Our variables then have the following dimensions: $f = force = mass \times acceleration = MLT^{-2}$; $a = radius = L$; $v = velocity = LT^{-1}$; $\rho = density = ML^{-3}$; $\eta = viscosity = force\ per\ unit\ area\ divided\ by\ velocity\ gradient = ML^{-1}T^{-1}$. Suppose that the relation is

$$f = ka^p v^q \rho^r \eta^s, \tag{7.53}$$

where k is a number. In terms of dimensions, this is

$$MLT^{-2} = L^p . L^q T^{-q} . M^r L^{-3r} . M^s L^{-s} T^{-s}. \qquad (7.54)$$

Hence, separately equating the powers of M, L and T, we obtain $p = q$, $r = q - 1$, and $s = 2 - q$, so that

$$f = k \left(\frac{av\rho}{\eta} \right)^q \frac{\eta^2}{\rho}. \qquad (7.55)$$

We suppose that, when v is small, f is proportional to v. Hence, putting $q = 1$,

$$f = kav\eta, \qquad (7.56)$$

which is Stokes' law, when $k = 6\pi$.

Experiment has confirmed Stokes' law. It has also shown that at higher velocities the resistance is proportional to v^2. Putting $q = 2$, we obtain in this case the surprising result

$$f = ka^2 v^2 \rho, \qquad (7.57)$$

i.e. f is determined by the *density* of the fluid, but *not its viscosity*! This means that at high velocities the energy is mainly spent in producing turbulent motion in the fluid, not in overcoming viscous forces (cf. Chapter 12).

Stokes' law is important for discussing the suspension and settling of particles in fluids of different densities. The rate of fall of drops of density ρ_1 in a fluid of density ρ_2 under gravity is given by

$$v = \frac{f}{6\pi a\eta} = \frac{2}{9} \frac{(\rho_1 - \rho_2)}{\eta} ga^2, \qquad (7.58)$$

where g is the gravitational constant. For water droplets in air $v = 1\cdot3 \times 10^6 \, a^2$, where a is in cm. If $a \simeq 10^{-3}$ cm, as in mists and clouds, $v \simeq 1$ cm sec^{-1}.

The *Stokes Einstein* formula

$$\eta = \frac{kT}{6\pi aD} \qquad (7.59)$$

(cf. Equation (7.20)) is derived from the consideration that small particles do not settle but are held in suspension by their Brownian motion. Let n be the number of particles per unit volume. In a field of force f each particle has a potential energy fx, where x is its co-ordinate in the field direction. In thermal equilibrium n obeys the Maxwell–Boltzmann distribution, and hence (cf. Section 1.4 and Equation (1.20)),

$$\frac{dn}{dx} = -\frac{nf}{kT}. \qquad (7.60)$$

The rate of flow against f, due to diffusion along this concentration gradient, is $-D(dn/dx)$, where D is the diffusion coefficient of the particles. In equilibrium this is balanced against the drift flow $nf/6\pi a\eta$, given by Stokes' law. Equation (7.59) then follows directly. Combining Equations (7.52) and (7.59), we see that particles with a diffusion coefficient D drift with a velocity

$$v = \frac{Df}{kT} \qquad (7.61)$$

in the direction of the force f, in addition to undergoing random thermal motion. This important formula, due to Einstein, can be derived by more general methods than that above, not involving Stokes' law. It expresses the fact that, when a weak force is applied to a particle that is making otherwise randomly directed thermal jumps, slightly more jumps go with the force than against it, so that there is a net drift systematically in the direction of the force, superposed on the otherwise random thermal motion.

FURTHER READING

FRENKEL, J., *Kinetic Theory of Liquids*, Oxford University Press, London, 1946.

TURNBULL, D., *Trans. AIME*, **221**, 422 (1961).

EWALD, P. P., PÖSCHL, TH. and PRANDTL, L., *The Physics of Solids and Fluids*, Blackie, London, 1936.

HOUWINK, R., *Elasticity, Plasticity, and the Structure of Matter*, Cambridge University Press, London, 1953.

EIRICH, F. R., *Rheology*, Vol. 1, Academic Press, New York, 1956.

VON HIPPEL, A. R., *Molecular Science and Molecular Engineering*, Wiley, New York, 1959.

NORTON, F. H., *Elements of Ceramics*, Addison-Wesley, New York, 1952.

LANDAU, L. D. and LIFSHITZ, E. M., *Statistical Physics*, Pergamon Press, Oxford, 1959.

ELEY, D. D., *Adhesion*, Oxford University Press, London, 1961.

CHALMERS, B., *Physical Metallurgy*, Wiley, New York, 1959.

TRELOAR, L. R. G., *The Physics of Rubber Elasticity*, Oxford University Press, London, 1949.

BUECHE, F., *Physical Properties of Polymers*, Interscience, New York, 1962.

PROBLEMS

7.1 Derive Equation (7.3) without making use of the analogy between linear elasticity and viscosity.

7.2 Discuss the fluidity of crystals and the solidity of liquids, giving examples.

7.3 Measured values of the diffusion of hydrogen in nickel are as follows:

T (°C):   162·5   237   355   496

D (cm^2 sec^{-1}): 9×10^{-8} $4·6 \times 10^{-7}$ $3·1 \times 10^{-6}$ $1·34 \times 10^{-5}$

Determine the activation energy.   [10,000 cal mole^{-1}]

7.4 A polyvinyl chloride mixture has an average molecular weight of 31,200 g. How many repeating units are there in an average molecule? [500]

7.5 An applied strain of 0·4 produces an immediate stress of 1200 psi in a piece of rubber, but after 42 days the stress is only 600 psi. What is the relaxation time? What is the stress after 90 days? [61 days, 275 psi]

7.6 Two parallel circular disks, of radius R, are mounted coaxially at a distance d apart, where $d \ll R$. The lower disk is rotated with angular velocity ω and the upper disk is stationary. Between the disks is a layer of liquid of viscosity η. Assuming that there is no slip between the liquid and the disks, show that the torque exerted on a disk is $\pi\eta\omega R^4/2d$.

7.7 Liquid of viscosity η is flowing under a pressure difference p down a long narrow tube of length l and radius a along the axis of which is a fixed thin wire of radius b. Show that the volume discharged in unit time is given by

$$V = \frac{\pi p}{8\eta l} \left\{ a^4 - b^4 - \frac{(a^2 - b^2)^2}{\ln (a/b)} \right\}.$$

7.8 A perfect gas of viscosity η passes down a long narrow tube of length l and radius a by Poiseuille flow. The pressure is p_1 at the inlet and p_2 at the outlet end of the tube. Show that the volume V_1 of gas entering the tube per second is given by

$$V_1 = \frac{\pi a^4}{16\eta l} \frac{p_1{}^2 - p_2{}^2}{p_1}.$$

7.9 If the material in the cantilever beam of Fig. 5.8 has a Newtonian viscosity η ($>10^{15}$ poises), show from first principles that the loaded end deflects an amount

$$y = \frac{F}{3\eta I} \frac{l^3}{3} t$$

in time t by viscous flow. At ordinary temperatures the viscosity of soda-glass is of order 10^{20} poises. It is commonly believed that if laboratory glass tubes a few feet long are stored horizontally at room temperature by support at their ends they will develop a pronounced permanent bend over a period of a few months. Is this reasonable?

7.10 Falling freely under gravity through air of density 0.0012 g cm^{-3} and viscosity 0.000183 g cm^{-1} sec^{-1}, an oil drop of density 0.92 g cm^{-3} reaches a steady velocity of 0.086 cm sec^{-1}. When an electric field of 300 V cm^{-1} is applied vertically, the steady downward velocity of the drop becomes 0.081 cm sec^{-1}. On the basis of Stokes' law, calculate the number of electronic units of charge on the drop.   [10]

chapter 8

Surfaces

Because of their unique positions, particles at the surfaces or interfaces of liquids and solids have various special properties, which we shall review in this chapter. Surface energy and surface tension largely determine the breaking strengths of elastic solids, the shapes of liquid surfaces, the wetting of substances, the action of adhesives, and many capillarity and interfacial effects. We shall see that the equation of equilibrium of a film, under surface tension, can be used to solve problems of elasticity theory. We shall also discuss the condensation of particles on to surfaces and the effect of the curvature of a surface on the equilibrium between one phase and another, which are important in connexion with adsorption, the growth and shape of crystals, rates of phase changes, and many other aspects of physico-chemical behaviour. Finally, we shall consider questions of grain boundaries in polycrystals, including solid-state sintering and the mechanical properties of such boundaries.

8.1 ENERGY AND TENSION OF A SURFACE

A particle in the free surface of a liquid or solid has no neighbours on one side. Since bond energies are negative, its energy is *higher* than that of an internal particle by approximately its missing share of bond energy. This *surface energy*, E_s per unit area, can be roughly estimated from the heat of vaporization, E_v per unit volume. The vaporization of unit volume of the substance into n constituent particles increases the surface energy by naE_s, where a is the surface area of a particle. Since this is also E_v, then $E_s \simeq E_v/na$. We take $n^{-1} = \frac{4}{3}\pi r^3$ and $a \simeq 4\pi r^2$, where r is the particle radius, and so obtain

$$E_s \simeq \tfrac{1}{3}rE_v. \tag{8.1}$$

231

For water, $E_v \simeq 500$ cal cm$^{-3} \simeq 2 \times 10^{10}$ erg cm^{-3} and $E_s \simeq 100$ erg cm^{-2}, which is consistent with the value $r \simeq 1.5 \times 10^{-8}$ cm.

A liquid can minimize its surface energy by adopting a shape of minimum surface area, as if enclosed in a tensile skin. This *surface tension* is familiar from the behaviour of soap films and from the tendency of small liquid drops to spheroidize. The surface tension γ of a soap film can be

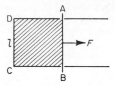

Fig. 8.1.   Force exerted by a soap film on a frame.

measured on a wire frame ABCD, as in Fig. 8.1, from the force F required to hold the movable wire AB (length l) in place. Thus,

$$F = 2\gamma l, \tag{8.2}$$

the factor 2 taking account of the tensions on both sides of the film. For water, $\gamma = 73$ dyn cm^{-1} at 18°C.

We do not regard surface tension as a tensile force between particles in the skin. Stretching in this case is the making of *new surface* by bringing internal particles up to the surface, and the tension opposing it is due to the increase in the numbers of these particles. Unlike elastic tension, the surface tension of a liquid film does not change as the film is stretched (unless the film is stretched so thin that particles on one side feel forces from those on the other side).

Because γ is constant, the energy required to make 1 cm^2 of new surface against the surface tension γ dyn cm^{-1} is equal to γ erg cm^{-2}. However, γ is not identical with E_s, because the energy from the stretching·force is supplied as *work*. By the definition of free energy (cf. Section 2.3), γ is the *free energy* per unit area of surface. In this sense even a solid has a surface tension. To relate γ to E_s, we regard the surface as a heat engine and subject it to a thermodynamical cycle:

(1) Increase its area by δA, stretching it at constant temperature T. Its energy increases by $E_s \delta A$, and the work done on it is $\gamma \delta A$, so that the *heat* absorbed by it is $(E_s - \gamma)\delta A$.

(2) Cool it to $T - \delta T$ at constant area (no work done).

(3) Allow it to contract δA at constant temperature $T - \delta T$. It does work equal to $\left(\gamma - \dfrac{d\gamma}{dT}\delta T\right) \delta A$.

(4) Heat it to T at constant area (no work done).

From the general thermodynamical theory of reversible cycles,

$$\frac{\delta T}{T} = \frac{\text{Net work done by system during cycle}}{\text{Heat absorbed at higher temperature}}$$

$$= \frac{\left(\gamma - \dfrac{d\gamma}{dT}\,\delta T\right)\delta A - \gamma\delta A}{(E_s - \gamma)\delta A} = \frac{\delta T}{(\gamma - E_s)}\frac{d\gamma}{dT} \tag{8.3}$$

and hence,

$$\gamma = E_s + T\frac{d\gamma}{dT}. \tag{8.4}$$

Only at $0°K$ are γ and E_s necessarily equal. In general $d\gamma/dT$ is negative and $\gamma < E_s$ at higher temperatures.

We can write Equation (8.4) as

$$\gamma = E_s - TS_s, \tag{8.5}$$

where S_s is the *surface entropy* per unit area

$$S_s = -\frac{d\gamma}{dT}. \tag{8.6}$$

Particles in the surface, being less bonded than those inside, vibrate less quickly. Hence, by the argument of Section 6.7, the entropy per particle is greater in the surface than inside. This additional entropy possessed by a surface particle is the surface entropy. It can be shown that

$$S_s \simeq n_s k \ln\left(\frac{v}{v_s}\right), \tag{8.7}$$

where n_s ($\simeq 10^{15}$ cm^{-2}) is the number of particles in unit area of surface, k is Boltzmann's constant, and v and v_s are the respective average vibrational frequencies of internal and surface particles. Typically, $S_s \simeq 0·3$ erg deg^{-1} cm^{-2}. We define

$$T_s = E_s/S_s \tag{8.8}$$

as the temperature at which $\gamma = 0$. Usually, T_s is well above the melting point of the substance, so that γ is not greatly different from E_s for solids. We might expect T_s to be the liquid–vapour–gas critical temperature (cf. Section 2.1), but in practice there is only rough agreement, because of effects such as the influence of the saturated vapour on the free energy of the liquid–vapour interface. Usually the surface tension vanishes at a few degrees below the critical point.

When a crystal is allowed to deform as a fluid, its surface tension can be measured. This is possible at high temperatures, through vacancy creep (cf. Section 7.3). For example, gold foil shrinks when heated for long times near its melting point. Experiments similar to that of Fig. 8.1 are then possible, in which the surface tension is measured from the applied force at which a foil or wire neither shrinks nor stretches. There are practical difficulties due to effects of contamination, evaporation and grain boundaries, but the surface tensions of several substances below their melting points have been successfully measured in this way. Some values of γ are given in Table 8.1.

TABLE 8.1
Surface Tensions of Substances, dyn cm^{-1}

	γ		γ
Benzene (in air at 20°C)	28·88	Mercury (in own vapour at	
Oleic acid (in air at 20°C)	32·5	20°C)	480
Water (in air at 20°C)	72·75	Gold (in air at 1130°C)	1100
Water (in air at 100°C)	58·8	Gold (solid, in helium at	
Soap solution (in air at 20°C)	≃25	1030°C)	1400
Oxygen (in argon at −183°C)	13	Copper (solid, in helium at	
Sodium chloride (in nitrogen		1000°C)	1680
at 908°C)	106	Iron (solid, in helium at	
		1450°C)	2300

8.2 IDEAL TENSILE STRENGTH OF A SOLID

Fracture creates free surface. To estimate the *ideal tensile strength* of an elastic solid, suppose in Fig. 8.2 that we pull a cylindrical bar of unit cross-sectional area by a tensile stress σ. When σ reaches the ideal strength σ_t, the atomic bonds joining two neighbouring atomic planes such as

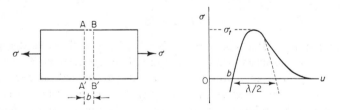

Fig. 8.2. Tensile fracture of a solid bar.

AA' and BB' break. We approximate the atomic stress–strain curve of these bonds by a sine curve of wavelength λ, as shown. This is reasonable provided we make its initial slope agree with Young's modulus E of the material and make the area under it equal to 2γ. Thus,

$$\sigma = \sigma_t \sin\left(\frac{2\pi u}{\lambda}\right), \tag{8.9}$$

where u is the displacement from the equilibrium spacing b. For infinitesimal strains this reduces to

$$\sigma = \sigma_t \frac{2\pi u}{\lambda}, \tag{8.10}$$

and hence, from Hooke's law ($\sigma = Eu/b$),

$$\sigma_t = \frac{\lambda}{2\pi} \frac{E}{b}. \tag{8.11}$$

The surface energy is given by

$$\gamma \simeq \frac{1}{2} \int_0^{\lambda/2} \sigma_t \sin\left(\frac{2\pi u}{\lambda}\right) du = \frac{\lambda \sigma_t}{2\pi}, \tag{8.12}$$

and hence,

$$\sigma_t \simeq \sqrt{\frac{E\gamma}{b}}. \tag{8.13}$$

For iron we take $\gamma \simeq 2000$ erg cm^{-2}, $E \simeq 2 \times 10^{12}$ dyn cm^{-2} and $b \simeq 2 \cdot 5 \times 10^{-8}$ cm, to obtain $\sigma_t \sim 4 \times 10^{11}$ dyn cm$^{-2} \simeq 6 \times 10^6$ psi $\simeq E/5$. For the fracture of a sodium chloride crystal on its cube plane we take $\gamma \simeq 150$ erg cm^{-2}, $E \simeq 5 \times 10^{11}$ dyn cm^{-2} and $b \simeq 2 \cdot 8 \times 10^{-8}$ cm, to obtain $\sigma_t \simeq 5 \times 10^{10}$ dyn cm$^{-2} \simeq 0 \cdot 7 \times 10^6$ psi $\simeq E/10$. Broadly,

$$\sigma_t \simeq \frac{E}{10} \tag{8.14}$$

for most solids. This high value indicates the high strength of atomic bonds (cf. Section 2.5).

These high strengths are rarely reached in practice, because brittle solids break prematurely at cracks and other stress concentrators (cf. Chapter 11) and ductile ones yield by the movements of crystal dislocations (cf. Chapter 9). Ordinary window glass, for example, breaks at about 10^4 psi, but its ideal strength approaches 10^6 psi. It was first shown by *Griffith* that very fine fibres ($\simeq 10^{-4}$ cm diameter), such as freshly drawn threads of glass and silica, can have almost ideal strengths. This property is exploited in quartz fibres for galvanometers; also in *fibreglass*, in which large numbers of fine threads are bonded together by resin to form materials of great strength. The strengths ($\simeq E/70$) of hard-drawn metal wires, e.g. tungsten and steel, are due to work-hardening (cf. Section 9.6) and to the elongation of the crystal grains into a fibrous texture. Higher strengths, approaching $E/10$, are obtained from "whiskers", i.e. single crystals in the form of fine fibres. Many organic fibres, both natural and synthetic, are strong; especially flax, the cellulose molecules of which are particularly well aligned along the fibre axis. Cellulose fibres in cotton and wood are weaker, because they are less well aligned. Woods such as spruce, with well-aligned cellulose molecules, are strong along the fibre axis, but can be split easily between fibres. Table 8.2 gives some representative tensile strengths.

TABLE 8.2
Representative Tensile Strengths of Fibrous Materials in 10^6 psi

Graphite whiskers	up to 3·5	Drawn copper wire	0·08
Silica fibres	up to 3·5	Nylon thread	0·08
Alumina whiskers	up to 2·2	Cotton (cellulose)	0·06
Iron whiskers	up to 1·9	Drawn aluminium wire	0·06
Drawn tungsten wire	up to 0·55	Cat gut	0·06
Ausformed steel wire	up to 0·45	Synthetic fibre (polyester)	0·035
Piano wire	0·35	Spider thread	0·027
Asbestos fibres	0·22	Hemp rope	0·015
Fibreglass	0·2	Hardwood, along grain	0·015
Flax (cellulose)	0·16		

The relation between the ideal strength and Young's modulus is a result of the characteristic range of action of interatomic forces. When the strain in an atomic bond exceeds about 0·2 (cf. Fig. 2.2), the bond begins to weaken and break. Since the elastic work of straining the bond to fracture becomes the surface energy of the separated atoms, we expect a correlation between Young's modulus, atomic spacing and surface energy; in fact,

$$\gamma \simeq \frac{Eb}{20} \simeq \frac{\mu b}{8}. \tag{8.15}$$

For copper this gives $\gamma \simeq 1300$ erg cm^{-2}.

8.3 INTERFACES

The "free" surfaces discussed above are generally liquid–vapour and solid–vapour interfaces. There are many other kinds of interfaces, e.g. solid–liquid, liquid–liquid, and solid–solid, all with surface energies and tensions. In crystalline solids there are also various kinds of grain boundaries, twin boundaries and multiphase interfaces (cf. Section 3.6). Equation (5.88), although deduced for small-angle grain boundaries only, in fact also roughly gives the energy of ordinary large-angle boundaries. We take $\theta \simeq 0{\cdot}5$ and $(A - \ln \theta) \simeq 1$ to deduce the energy of such a boundary as about $\mu b/25$ per unit area, i.e. about one-third of the surface energy (Equation (8.15)). For copper $\mu b/25 \simeq 400$ erg cm^{-2}, the measured large-angle boundary energy being 550 erg cm^{-2}. Stacking faults and coherent twin boundaries have lower energies, because the atomic structure is less disordered there; their measured values range from 100–300 erg cm^{-2} (e.g. Ni, Al, Pb) to below 30 erg cm^{-2} (Au, alpha brass, austenitic stainless steel, aluminium bronze). A large-angle boundary resembles a monolayer of liquid between crystals. We thus expect the interfacial energy between a crystal and its own melt to be about half that of a large-angle boundary, i.e. about $\mu b/50$. Again, this agrees roughly with values deduced from experiment.

The free energies of interfaces can be determined from the equilibrium forms of *triple junctions*, where three interfaces meet. Figure 8.3 shows

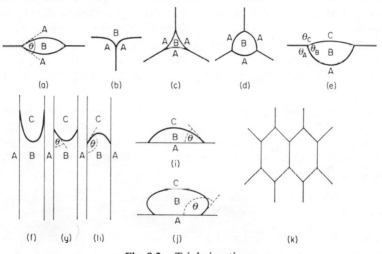

Fig. 8.3. Triple junctions.

examples. Diagram (a) represents an inclusion B in, for example, a grain boundary between two regions of A. If A and B are sufficiently fluid, B will adjust its shape to minimize the total free energy. In polycrystals this is possible at high temperatures. If the two regions of A are identical (e.g. A is liquid, or B is an inclusion inside a single crystal of A), so that the free energy γ_{AA} of the AA interface is zero, the equilibrium shape of B is circular (spherical in three dimensions), unless the surface energy is crystallographically anisotropic (cf. Section 8.8). When $\gamma_{AA} > 0$, B is lens-shaped. Since minimum energy implies equilibrium of forces, the interfacial angle θ is determined from the balance of surface tension forces acting on the junction. Resolving these forces along the AA boundary,

$$\gamma_{AA} = 2\gamma_{AB} \cos \frac{\theta}{2}, \qquad (8.16)$$

so that γ_{AA} can be determined from γ_{AB} and θ, and vice versa. Diagram (b) shows how this enables grain boundary (AA) energies to be found from the angles of grooves formed at high temperatures where grain boundaries meet the free surface (AB).

When γ_{AB} is small, in diagram (a), B spreads along, i.e. *wets*, the AA interface. When $\theta = 0$, there is *complete wetting* and the two regions of A become completely separated by a film of B. This is important in many polycrystalline materials. Liquid gallium disintegrates aluminium by spreading between its grains. Many other polycrystalline metals are severely weakened by contact with liquid metals. Tungsten carbide grains in cutting tools are bonded by wetting films of cobalt between them; oxide grains in fire-bricks are similarly bonded by films of silicate glass. Copper is embrittled by a trace of bismuth which spreads along its grain boundaries; the addition of lead cures this by dissolving in the bismuth and raising the bismuth–copper interfacial energy. *Hot shortness* in steels is mainly caused by the melting of sulphides, which wet and spread along grain boundaries.

Diagrams (c) and (d) show typical wetting and non-wetting forms of inclusions at triple junctions. Diagram (e) shows an inclusion B between two substances A and C, as when a drop of liquid floats on another. Equilibrium in this case requires the three surface tensions to form a triangle of forces; thus,

$$\frac{\gamma_{AB}}{\sin \theta_C} = \frac{\gamma_{BC}}{\sin \theta_A} = \frac{\gamma_{CA}}{\sin \theta_B}, \qquad (8.17)$$

where θ_B, for example, is the interfacial angle of B at the junction. Complete wetting, i.e. $\theta_B \simeq 0$, $\theta_A \simeq \theta_C \simeq \pi$, occurs when $\gamma_{CA} > (\gamma_{AB} + \gamma_{BC})$. This condition is often satisfied. Even so, lenses of insoluble liquids, such as globules of fat, are often seen floating on water. These are in equilibrium, not with clean water but with a monomolecular layer, e.g. of fat, on the surface. We must distinguish between *primary spreading*, in which a monomolecular layer spreads over the surface and lowers its energy, and *secondary spreading*, in which the entire spreading substance covers the surface evenly. The calming of sea by oil is an effect of surface tension.

The oil floats as a thin layer on the surface and damps out wind-driven ripples before these can grow into big waves.

Diagrams (f), (g) and (h) show the forms of meniscus on a liquid (B) in a capillary tube (A), for complete (f), partial (g) and poor (h) wetting. Diagrams (i) and (j) similarly represent drops of liquid on a solid surface. In all these the surface of A is fixed by the rigidity of the solid, and so cannot adjust itself to the surface tensions. The equilibrium of tensions is then confined to directions parallel to this surface; thus,

$$\gamma_{AC} = \gamma_{AB} + \gamma_{BC} \cos \theta. \tag{8.18}$$

If $\gamma_{AC} > (\gamma_{AB} + \gamma_{BC})$, the contact angle θ is zero and complete wetting occurs, a film of B spreading along the AC interface. The action of a *detergent* solution for removing oil from wool depends on this; for efficient detergency the contact angle of the solution, which spreads along the oil–wool interface, should be zero. A molten metal of low melting point will usually spread over the clean surface of a metal of high melting point, since a free surface of high energy is then replaced by one of low energy together with a metal–metal interface of low energy. The action of many *solders* and *brazes* depends on this.

The *adhesion* of a wetting liquid to a solid is measured by the work W per unit area to break the interface. Since a solid–liquid (AB) surface is replaced by a free solid (AC) and free liquid (BC) surface,

$$W = \gamma_{AC} + \gamma_{BC} - \gamma_{AB} = \gamma_{BC}(1 + \cos \theta). \tag{8.19}$$

When the liquid completely wets the solid,

$$W = 2\gamma_{BC}. \tag{8.20}$$

From Equations (8.13), (8.15) and (8.20), the tensile strength of the interface is approximately

$$\sigma_t \simeq \sqrt{\frac{E\gamma}{b}} \simeq \frac{\gamma\sqrt{20}}{b} \simeq \frac{2W}{b}. \tag{8.21}$$

For a typical organic adhesive $\gamma \simeq 30$ dyn cm^{-1}, and thus $\sigma_t \simeq 4.5 \times 10^9$ dyn cm$^{-2} \sim 65,000$ psi. This is for a normal force, of course; the liquid layer has little resistance to sliding along the interface.

Equation (8.20) shows that W for a wetting liquid is simply the work required to make two liquid surfaces, without regard to the solid. Adhesive joints, in fact, usually break in the adhesive, not along the interface. Even interfacial separations usually expose, not bare solid, but a surface covered with a monolayer of the liquid substance. The observed value of γ_{AB} is reduced by the presence of this layer.

The height h to which a wetting liquid of density ρ rises against gravity g in a vertical tube of radius r (or falls below the level of the surface outside, if the liquid is non-wetting) is easily found. The meniscus is pulled by a circumferential force $2\pi r(\gamma_{AC} - \gamma_{AB})$, i.e. $2\pi r\gamma_{BC} \cos \theta$. This is opposed by the weight of liquid in the length h, which is $\pi r^2 h\rho g$. Hence,

$$h = \frac{2\gamma_{BC} \cos \theta}{r\rho g}. \qquad (8.22)$$

Usually, h is measured from the surface level of liquid outside the tube to the bottom of the meniscus, and it is necessary to take account also of the weight of liquid in the meniscus; for example, by replacing h by $h + r/3$, if the meniscus is hemispherical. Improved forms of Equation (8.22) have been developed which take account of the shape and weight of the meniscus. These are useful as the basis of the simple capillary rise method for determining surface tensions.

Figure 8.3(k) shows a two-dimensional *foam* structure, e.g. a soap froth, or grain boundaries in a polycrystal. From Equation (8.17), if $\gamma_{AB} = \gamma_{BC} = \gamma_{CA}$, then $\theta_A = \theta_B = \theta_C = 120°$. This condition can be satisfied in two dimensions by hexagonal straight-sided cells. Polygonal cells of other types cannot generally coexist in equilibrium. Those with more than six sides tend to grow at the expense of those with less. This disequilibrium exists more generally in three-dimensional foams and is the cause of the cell growth observed in soap froths when rested after vigorous shaking. It is also a main cause of *grain growth* in annealed polycrystals. In polycrystals and multiphase solids there are complications due to variations of interfacial energy with the structure and orientation of each of the several types of interfaces possible in them, but long annealing at high temperatures nevertheless produces grain structures strikingly similar to the cell structures of soap froths. Interfacial energies largely determine the geometry of microstructures in metals, alloys and minerals.

8.4 PRESSURE BENEATH A SURFACE

Although surface tension is not really caused by a "tensile skin", this does give a simple interpretation of the pressure difference across a curved

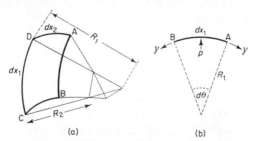

(a)   (b)

Fig. 8.4. Equilibrium of forces on a curved film.

surface. In Fig. 8.4(a) let ABCD be a small element of the surface, with principal radii of curvature R_1 and R_2, and sides dx_1 and dx_2. In diagram (b) we consider the surface tensions acting on AD and BC. The total force from these tensions, resolved radially, is $\gamma\, dx_2\, d\theta$, i.e. $\gamma\, dx_1\, dx_2/R_1$. Similarly, that from the tensions on AB and CD is $\gamma\, dx_1\, dx_2/R_2$. These balance the force $p\, dx_1\, dx_2$ on the element caused by the pressure difference p. Hence,

$$p = \gamma\left(\frac{1}{R_1} + \frac{1}{R_2}\right). \tag{8.23}$$

For a spherical bubble or droplet of radius R, this becomes simply

$$p = \frac{2\gamma}{R}. \tag{8.24}$$

For *synclastic* surfaces, such as the sphere and ellipsoid, both centres of curvature lie on the same side. For *anticlastic* surfaces, i.e. saddle shapes, the centres lie on opposite sides, and the positive sign in Equation (8.23) is then replaced by a negative. In all these formulae, γ is of course replaced by 2γ for two-sided films. The condition $R_1 = -R_2$ enables curved surfaces to exist even when there is no pressure difference, as, for example, when a soap film is stretched across a circle of wire which is then twisted to make a skew surface.

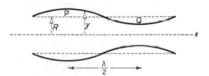

Fig. 8.5. Longitudinal section of a nearly cylindrical figure of revolution with axis x.

In Fig. 8.5 we consider the *stability* of a cylinder, initially of radius R, which is slightly distorted to the shape

$$y = R + A\cos\left(\frac{2\pi x}{\lambda}\right), \tag{8.25}$$

where λ is the wavelength and $A \ll R$. The cylinder resists this distortion if the pressure required to balance the surface tension in the bulges exceeds that in the necks. In Equation (8.23) we take $R_1 = y$ and, since dy/dx is small,

$$\frac{1}{R_2} = \frac{d^2y}{dx^2}. \tag{8.26}$$

Hence,

$$\frac{1}{R_1} + \frac{1}{R_2} = \frac{1}{R + A \cos\left(\dfrac{2\pi x}{\lambda}\right)} \pm \frac{4\pi^2 A}{\lambda^2} \cos\left(\frac{2\pi x}{\lambda}\right), \qquad (8.27)$$

where the positive sign holds for the bulges and the negative for the necks. At the maxima and minima, P and Q, we have $\cos(2\pi x/\lambda) = +1$ and -1, respectively. The pressure at P thus exceeds that at Q by

$$\gamma\left(\frac{1}{R + A} + \frac{4\pi^2 A}{\lambda^2} - \frac{1}{R - A} + \frac{4\pi^2 A}{\lambda^2}\right) \simeq \frac{2A\gamma}{R^2}\left(\frac{4\pi^2 R^2}{\lambda^2} - 1\right). \quad (8.28)$$

When $2\pi R > \lambda$, this is positive and the cylinder is stable. The cylinder becomes unstable when its length exceeds its circumference, and it can then break up into separate drops at intervals $2\pi R$ along its length.

A viscous liquid such as molten glass deforms too slowly to show this effect obviously, and it can be drawn into long uniform fibres; but when a fluid jet such as water emerges from an orifice, the viscous forces are negligible. The cylinder of liquid first undulates and then breaks up into oscillating spheroidal drops, as shown in Fig. 8.6. The small drops form

Fig. 8.6. Oscillation and break-up of a liquid jet.

from the narrow necks between the large ones. The instability of a liquid cylinder is also shown by the way that water gathers itself up into separate beads along a glass fibre or on the threads of a spider's web.

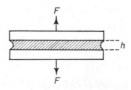

Fig. 8.7. Adhesion due to surface tension.

Surface tension can contribute, in addition to viscosity (cf. Section 7.10), to the effectiveness of a liquid adhesive. Figure 8.7 shows two flat plates held together by a thin film of liquid. If the radius of curvature ($\simeq \frac{1}{2}h$) at the free surface of the liquid is small compared with the radius of the plates, the pressure in the liquid is approximately $2\gamma/h$ below that outside. This pressure difference acts uniformly over the whole area A of a plate. The adhesive force F of the joint due to this effect is given by

$$F = 2A\gamma/h. \tag{8.29}$$

Thus, if $\gamma = 30$ dyn cm^{-1}, $A = 5$ cm^2 and $h = 10^{-5}$ cm, then $F \simeq 3 \times 10^7$ dyn $\simeq 66$ lb. This partly explains the adhesion of slip gauges and other flat surfaces when wrung together.

The pressure under a surface also accounts for the familiar attraction or repulsion of small bodies floating on a liquid. Figure 8.8 shows how this force is related to the wetting of the bodies. We consider first the balance of horizontal forces acting on a meniscus ABC of liquid. These forces are: (a) the surface tension γ applied at A by the liquid surface

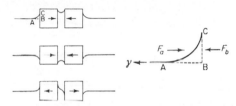

Fig. 8.8. Forces between bodies floating on a liquid.

which continues the line BA beyond A; (b) the force F_b exerted by the body on BC; and (c) the force F_a of atmospheric pressure on AC. We note that F_a is the same as would be exerted by the atmosphere on the body at BC if the actual liquid surface AC were replaced by the level surface AB. For equilibrium, $\gamma + F_b - F_a = 0$. Hence, the liquid exerts a force $\gamma - F_a (= -F_b)$ on the body at BC, but this is the same as if the liquid surface AC were replaced by the level surface AB, since then there would be a tension γ applied at B and a pressure F_a applied along BC. For calculating forces, we may thus replace actual liquid surfaces by the level surfaces shown in the diagrams.

If both bodies are wet, the liquid rises between them and is held there at a pressure *below* atmospheric. The atmospheric pressure on their outer faces thus pushes the bodies together. If neither body is wet, the liquid between them falls, and the hydrostatic pressure (proportional to depth below surface) at a given point between the bodies is smaller than that at the corresponding point outside. Again the bodies are pushed together.

If one is wet and the other not, we can again apply the level-surface argument, through the point of inflexion in the meniscus between the bodies, but in this case, since the mean slope of this meniscus is not horizontal, only the *horizontal component* of the surface tension pulls the bodies together. The bodies are thus pulled apart by the greater horizontal force from the surface tensions on their outer faces.

8.5 PRANDTL'S SOAP-FILM ANALOGY

It is convenient to digress here to discuss a useful analogy by which problems of elastic torsion can be solved with soap films, due to *Prandtl*. A film is stretched over a frame with the same outline as that of the transverse cross-section of the twisted bar, and a uniform pressure p is applied to one side, as shown in Fig. 8.9. When the deflexion ϕ of any

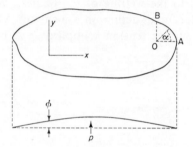

Fig. 8.9. Deflexion of a soap film on a frame by a pressure p.

point x, y, in the film is everywhere small, Equation (8.23) becomes

$$\frac{\partial^2 \phi}{\partial x^2} + \frac{\partial^2 \phi}{\partial y^2} = C, \tag{8.30}$$

where $C = p/\gamma$. This is mathematically identical to the *warping equation* which governs the warping out of a plane of transverse cross-sections in non-circular twisted bars (Fig. 8.10).

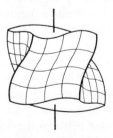

Fig. 8.10. Torsion of a prismatic bar.

Let x, y and z be co-ordinates in the bar, with z along the axis of twisting, and let θ be the angle of twist per unit length. In a round bar, transverse cross-sections of which remain plane, the displacements parallel to these co-ordinate axes are, respectively, $u = -yz\theta$, $v = xz\theta$, $w = 0$. They give the simple deformation described in Section 5.3. When the cross-section is non-circular, as in Fig. 8.9, the displacements are

$$u\ (= -yz\theta), \qquad v\ (= xz\theta), \qquad w, \tag{8.31}$$

where the *warping displacement* w depends on x, y and θ. From Equations (4.24), the strains are

$$\varepsilon_{xx} = \varepsilon_{yy} = \varepsilon_{zz} = \varepsilon_{xy} = 0,$$

$$2\varepsilon_{zx} = \frac{\partial w}{\partial x} - y\theta, \tag{8.32}$$

$$2\varepsilon_{zy} = \frac{\partial w}{\partial y} + x\theta.$$

Eliminating w between these relations, we obtain

$$\frac{\partial \varepsilon_{zy}}{\partial x} - \frac{\partial \varepsilon_{zx}}{\partial y} = \theta. \tag{8.33}$$

As regards stresses, we have $\sigma_{xx} = \sigma_{yy} = \sigma_{zz} = \sigma_{xy} = 0$. If body forces are absent, the equations of stress equilibrium (Equations (4.69)) reduce to

$$\frac{\partial \sigma_{zx}}{\partial x} + \frac{\partial \sigma_{zy}}{\partial y} = 0, \tag{8.34}$$

which is satisfied by stresses derived from a *stress function* ϕ, where $\sigma_{zx} = \partial \phi / \partial y$ and $\sigma_{zy} = -\partial \phi / \partial x$. Multiplying by 2μ in Equation (8.33) to convert the strains to stresses and then substituting ϕ for the stresses, we obtain Equation (8.30) again, with C multiplied by $-2\mu\theta/(p/\gamma)$.

The sides of the bar are free from forces. Consider the equilibrium of a small element of the bar, AOB, at the surface. This element is equivalent to that in Fig. 4.22, if we make ABC parallel to X_3, i.e. to the z axis. We then have $\alpha_3 = \pi/2$, $l_3 = 0$, and $\alpha_1 + \alpha_2 = \pi/2$. Let $\alpha_1 = \alpha$, as shown in Fig. 8.9. Then, $l_1 = \cos\alpha$ and $l_2 = \sin\alpha$. Since the surface forces on AB are zero, we have, from Equations (4.55),

$$p_z = 0 = \sigma_{xz} \cos\alpha + \sigma_{yz} \sin\alpha = \frac{\partial \phi}{\partial y} \cos\alpha - \frac{\partial \phi}{\partial x} \sin\alpha. \tag{8.35}$$

This, however, is the amount by which ϕ changes as we move a unit distance along the surface of the bar from A towards B. Hence, ϕ is constant along the boundary. For solid bars we can take $\phi = 0$ on the boundary, as in Fig. 8.9. By measuring the gradients $\partial\phi/\partial x$ and $\partial\phi/\partial y$ at points on the soap film, we can thus determine the stresses in the bar.

8.6 ADSORPTION AND ADHESION

The adsorption from vapour or solution of foreign atoms or molecules on to a surface can range from a complete monolayer or multilayer of particles to a mere slight abnormality of concentration. Many substances are *surface-active*; they tend to collect on a surface but not to dissolve in the medium beneath. Often this is because their particles form weak bonds with the medium, sufficient to anchor them to its surface but too weak to force the underlying solvent particles apart in order to make room for themselves inside the solvent. *Positive* adsorption occurs when the surface tension is lowered by the adsorbed substance (e.g. soap in water). If the surface tension is increased, the adsorption is *negative* and the substance avoids the surface (e.g. inorganic salts in water).

A quantitative relation between adsorption and surface tension is provided by the *Gibbs theorem*. We shall prove this for adsorption from a vapour, but the argument also applies to adsorption from solution, if the vapour pressure p is replaced by the osmotic pressure of the solution. We consider adsorption of the vapour on the wall of its closed container and take this system round a closed reversible cycle, thus:

(1) Deform the container at constant volume to increase its surface area by δA, so changing the free energy by $\gamma \delta A$.

(2) Allow the container to increase its volume by δV at constant area, so changing the free energy by $-\left(p + \dfrac{\partial p}{\partial A}\delta A\right)\delta V$.

(3) Reduce the area by δA at constant volume, so changing the free energy by $-\left(\gamma + \dfrac{\partial \gamma}{\partial V}\delta V\right)\delta A$.

(4) Compress the container by δV at constant area, so changing the free energy by $p\delta V$.

The sum of these four free energy changes is zero. Hence,

$$\left(\frac{\partial p}{\partial A}\right)_{V=\text{constant}} = -\left(\frac{\partial \gamma}{\partial V}\right)_{A=\text{constant}} \tag{8.36}$$

Let n_1 be the number of vapour particles per unit volume before adsorption and n the number after supplying c particles per unit area of surface.

Then, $n = n_1 - Ac/V$. Since $\partial\gamma/\partial V = (d\gamma/dn)(dn/dV) = -(n/V)(d\gamma/dn)$ and $\partial p/\partial A = (dp/dn)(dn/dA) = -(c/V)(dp/dn)$, Equation (8.36) becomes

$$c\frac{dp}{dn} = -n\frac{d\gamma}{dn}. \tag{8.37}$$

We use the perfect gas law $p = nkT$ and obtain

$$c = -\frac{n}{kT}\frac{d\gamma}{dn}, \tag{8.38}$$

which is Gibbs' formula for the excess c adsorbed in the surface.

The *Langmuir isotherm* is important for the further understanding of adsorption. Suppose that the surface in equilibrium with the above vapour offers s ($\simeq 10^{15}$ cm^{-2}) adsorption sites per unit area. The fraction θ of surface covered is thus c/s. The rate at which particles, of mass m, strike the surface from the vapour is given by the number per unit volume in the vapour, p/kT, multiplied by the mean speed towards the surface ($\simeq\sqrt{kT/m}$). The rate of adsorption is thus given by $\alpha p(1 - \theta)(mkT)^{-1/2}$, where α ($\simeq 0.25$) takes account of numerical factors including the *condensation coefficient*, i.e. the tendency of a particle to be adsorbed rather than bounce back, and where $(1 - \theta)$ measures the proportion of surface available for adsorption. In equilibrium this is equal to the *rate of desorption*, which we write as $cv \exp(-\psi/kT)$, where v is the vibrational frequency of an adsorbed particle, since the detachment of such a particle from the surface is obviously a thermally activated process opposed by an *energy of adsorption* ψ. Hence,

$$\frac{\theta}{1 - \theta} = p\kappa \quad \text{and} \quad \theta = \frac{p\kappa}{1 + p\kappa}, \tag{8.39}$$

where

$$\kappa = \frac{\alpha\, e^{\psi/kT}}{sv(mkT)^{1/2}}.$$

Equation (8.39) is the Langmuir isotherm. We see that θ ranges from about $p\kappa$ to about unity as $p\kappa$ increases. When $\theta \ll 1$, the adsorbed particles are too far apart to interact with one another and ψ is then constant. At higher concentrations they interact, and ψ may then increase or decrease with θ. When a nearly complete monolayer is formed, a second layer may begin to form, with a new ψ determined by the forces from the first layer. Multiple layers form when the vapour is nearly saturated. They represent incipient liquefaction or crystallization of the vapour.

The analysis leading to Equation (8.39) can also be used to describe evaporation and condensation of a pure substance in equilibrium with its own vapour. Clearly, the *maximum* rate of evaporation under equilibrium conditions can never exceed the rate at which particles of vapour strike the surface. Pure liquid mercury has been shown to evaporate at the maximum rate, whereas contaminated mercury is some 1000 times slower.

Strong adsorption is possible when $\psi \gg kT$. Typical values of ψ are 0·06 eV for argon on KCl and 0·3 eV for ammonia on charcoal, whereas $kT \simeq 0.025$ eV at room temperature. At low temperatures ψ/kT becomes large, and thus charcoal at the temperature of liquid air adsorbs gases to extremely low pressures and so is useful for producing high vacua. A vapour particle which lands on a smooth close-packed crystal face is only weakly adsorbed, since it can bond only through a small area of its surface, as shown at A in Fig. 8.11. It can bond strongly at a ledge such as B (which may be linked to a screw dislocation; cf. Section 3.4) and more

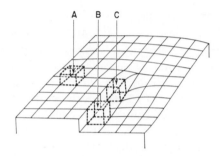

Fig. 8.11.   Adsorption sites on a crystal surface.

strongly still in a corner, such as C, or crevice where contact can be made over much of its surface area. The value of ψ thus depends on the nature of the site, and particles tend to be adsorbed first in corners and crevices.

A chemically saturated molecule such as H_2 is often adsorbed by *van der Waals* bonding, with an adsorption energy usually below 0·2 eV. This weak *physical adsorption* is easily overcome by thermal energy; for example, hydrogen freely desorbs from nickel at $-200°C$. At higher temperatures, however, e.g. above $-100°C$ for hydrogen on nickel, adsorption becomes strong again and the energy of adsorption rises to values comparable with those of chemical bonding (a few eV). This is *chemisorption*. The saturated molecule dissociates on the surface and its constituent atoms separately bond themselves, by chemical forces, to the surface. Figure 8.12 shows how such a molecule interacts with the surface. Curve 1 with a shallow minimum at A represents physical adsorption, and curve 2 with a deep minimum at B represents chemisorption. When the physically adsorbed molecule gains enough energy to bring it to C, it can then, at constant energy, dissociate and transfer to curve 2, so becoming chemisorbed as atoms.

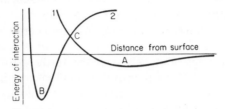

Fig. 8.12. Interaction of a molecule with a surface: 1, physical adsorption; 2, chemisorption.

Adsorption can occur by various kinds of interactions. Chemically active molecules can bond themselves directly and strongly to chemically suitable surfaces. A good example is provided by the *fatty acids* and their *soaps*. The molecules of these, e.g. stearic acid, $CH_3(CH_2)_{16}COOH$, and sodium stearate, $CH_3(CH_2)_{16}COONa$, consist of homopolar hydrocarbon chains ("non-polar") which end in a COOH group (fatty acid) or a similar group in which the end hydrogen atom is replaced by a metal atom such as sodium (soap). This "polar" hydrogen or metal atom is ionically bonded to the oxygen and carries a positive electrostatic charge. It can therefore attach itself to the electronegative parts of other molecules, such as water and ionic compounds. The polar end of a short fatty-acid molecule can pull its insoluble hydrocarbon chain into water, so giving some solubility, but cannot do so when the hydrocarbon chain is long. The molecule then sits on the surface as in Fig. 8.13.

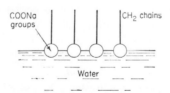

Fig. 8.13. Sodium soap molecules on water.

Since the metallic bond is exerted through free electrons and is chemically unspecific, all *clean* metals adhere to one another, which makes soldering and brazing possible. In practice, chemical fluxes are usually needed to remove oxide films, which otherwise prevent the metals from meeting. Inside a crystalline solid, solute atoms are sometimes pulled into grain boundaries and dislocations, because they can reduce their elastic energies of misfit there. In the distorted crystal structure that exists at such places there are sites in which an oversized or undersized solute atom can fit more easily than in a normal crystal site. This *elastic energy of adsorption* is usually less than about 0·1 eV but is sometimes significantly large, e.g. about 0·5 eV for carbon and nitrogen in b.c.c. iron.

We have already mentioned some practical applications of adsorption. There are many others. *Wetting agents* are added to paints, dyes and sprays, to make them spread evenly over the surfaces to which they are applied. *Frothing agents* (e.g. pine oil, saponins) are used to stabilize foam bubbles; if part of a film stretches unduly, its surface tension increases because the concentration of adsorbed substance is reduced there, and this opposes further stretching. *Metal catalysts*, e.g. platinum used in the manufacture of sulphuric acid and ammonia, act by enabling chemical reactions to take place between particles chemisorbed on their surfaces. *Hydrophobic sols* such as metal sols are stabilized by repulsive electrostatic forces between their colloidal particles, due to charged ions adsorbed on them. *Froth flotation* is an important process for separating powdered minerals such as metal sulphides. The mineral particles are agitated in water with a frothing agent. A *collecting agent* (e.g. a *xanthate*) is added; the polar ends of its molecules attach themselves to the ionic mineral surface and the non-polar hydrophobic ends attach themselves to air bubbles which carry the mineral particles up into the froth, where they can be separated. By adding various *activators* (e.g. Cu salts for ZnS) and *depressants* (usually alkalis and cyanides) which attach themselves to the mineral surface and encourage or discourage the collectors, it is possible in a sequence of operations to separate the various constituents in mixed minerals almost completely.

Hydrodynamic lubricants must adhere well to the surfaces they separate, if "dry spots" are to be avoided in bearings under load. Traces of fatty acids such as stearic acid are added to lubricating oils to improve their adhesion. When the lubricated surfaces are brought to rest, hydrodynamic lubrication is no longer possible (cf. Section 7.10), and the separation of the surfaces then depends on *boundary lubrication*, i.e. the maintenance of a monolayer, e.g. of fatty acid, in the bearing interface.

Adhesives must be strongly attracted chemically to their surfaces of application, to form strong bonds. Thus, vulcanized rubber bonds well to copper and its alloys because of the affinity of its sulphur for copper. Chemically active metals are usually covered with oxide films, the ionic surfaces of which attract other polar substances; hence the value as metal cements of the phenolic and epoxy resins which, through their polar hydroxyl groups, bond strongly to ionic surfaces. Shear strengths of up to about 7000 psi can be obtained in well-made joints. Most strong adhesives are solids, having solidified by cooling (solder, thermoplastic adhesives), by evaporation of a solvent (rubber adhesive, cellulose solutions), or by chemical change (synthetic resins). They usually "set" rigidly, and are liable to fail by brittle fracture at stresses determined by the factors discussed in Chapter 11. It is important in these, as in metal–glass seals, to equalize the coefficients of thermal expansion, so as to minimize thermal stresses. Organic adhesives in metal joints are usually much weaker than the metals joined. To counteract this, the area of joint is made large, so spreading the load thinly over the adhesive, and the adhesive is made elastically pliant, e.g. by including some rubber in it, to spread the load uniformly and so reduce the effectiveness of stress concentrators, such as cracks and "dry spots" in the joint.

8.7 STABILITY OF SMALL BUBBLES AND DROPLETS

Many physico-chemical effects depend on the fact that small bodies have more surface area, relative to volume, than large ones. The standard vapour pressures refer to equilibrium with large samples of liquid or solid. Small droplets have higher vapour pressures because of their higher energies. A single uniform vapour at a given temperature cannot be in equilibrium simultaneously with droplets of different sizes. It is *undersaturated* to the small ones, which tend to evaporate, and *supersaturated* to the large ones, which tend to grow. A given drop may grow at first, until all the smaller ones have gone, and then shrink as larger ones capture its vapour.

Suppose that a spherical drop of radius r is slightly increased in size by the addition of one atom or molecule of volume v. Its surface area is increased by $2v/r$, and the work done against the surface tension γ is then $2\gamma v/r$. This work must come from the free energy of the vapour, so that, if the drop is in equilibrium with the vapour, the vapour pressure p must exceed its standard value p_0 by a certain amount. To compress a gas ($pV = kT$ per particle) from p_0 to p at temperature T, the work required is

$$\int_V^{V_0} p \, dV = \int_V^{V_0} \frac{kT \, dV}{V} = kT \ln (V_0/V) = kT \ln (p/p_0) \qquad (8.40)$$

per particle. The free energy per particle is increased by this amount. Hence, for equilibrium

$$p/p_0 = \exp (2\gamma v/rkT). \qquad (8.41)$$

This is *Kelvin's equation* for the vapour pressure of a droplet. For a simple substance near its melting point, $2\gamma v/kT \simeq 10^{-7}$ cm, so that a droplet of radius 10^{-6} cm requires about 10% supersaturation. The equilibrium is unstable, since, if the droplet begins to grow, the vapour pressure causes it to grow further; and vice versa if it shrinks.

This and similar equations for other states of matter have many applications. The lifting of roads and pavements in cold weather by ice beneath, i.e. "frost heaving", is far too large an effect to be explained by the expansion on freezing. What appears to happen is that a large ice crystal, growing in a cavity in water-logged soil, captures freezing water from a large region round it because this is energetically preferable to forming fine ice crystals of high surface area uniformly through the pores of the soil. Soils of mixed grain sizes are particularly vulnerable, since the large pores enable large ice crystals to form and the fine pores hold a lot of water to feed them.

Most *phase changes* in matter (cf. Section 2.4) start by the formation of small nuclei of the new phase in the old, which then grow; e.g. vapour bubbles in a boiling liquid, droplets in a condensing vapour, crystallites in a freezing liquid, small precipitates in supersaturated solids, liquids or vapours. The extra surface energy of a small nucleus opposes such a change, however, and the old phase must become supersaturated (or supercooled or superheated as the case may be) before stable nuclei can form.

The work to make a liquid droplet of radius r in a vapour at pressure p and temperature T is $4\pi r^2 \gamma - \frac{4}{3}\pi r^3 w$. The first term is the surface energy of the droplet; the second, with

$$w = \frac{kT}{v} \ln \left(\frac{p}{p_0}\right), \tag{8.42}$$

is the free energy of condensation. This work varies with the droplet radius as in Fig. 8.14, and reaches a maximum, A, at the critical radius ($r = 2\gamma/w$) which satisfies Equation (8.41). Eliminating r, we obtain

$$A = \frac{16\pi\gamma^3}{3w^2}. \tag{8.43}$$

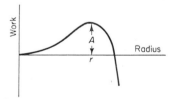

Fig. 8.14. Work of droplet formation in a supersaturated vapour.

Until the critical radius is reached, the droplet prefers to evaporate rather than grow. The work A is thus a barrier which must be overcome by thermal activation when making a nucleus. The *rate of nucleation I*, per unit volume per second, is then given by

$$I = \eta\, e^{-A/kT}, \qquad (8.44)$$

where η depends on the rate of arrival of vapour particles at the droplet.

It follows that the vapour pressure critically affects the rate of nucleation. Below a certain p/p_0, nuclei hardly form; just above, they form abundantly. Figure 8.15 shows typical results for the condensation of water in a cloud chamber.

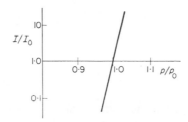

Fig. 8.15. Rate of nucleation of water drops from vapour (after M. Volmer and H. Flood, *Z. Phys. Chem.*, **170** A, 273 (1934)).

Nucleation is greatly helped by the presence of foreign particles wetted by the new phase. The surface energy and work of nucleation are then reduced, sometimes to zero. Such foreign grains act as *seeds* on which nuclei form almost as soon as the saturation point is reached. This is called *heterogeneous* nucleation, to distinguish it from the homogeneous type described above. Most phase changes under ordinary conditions are heterogeneously nucleated, e.g. by specks of dust or by points on the walls of containers. The industrial process of metal-casting would be impracticable without the seeded nucleation of crystals in a melt.

8.8 GROWTH AND FORM OF SINGLE CRYSTALS

Different faces of a crystal generally have different surface energies. Lamellar crystals such as graphite are extremely anisotropic in this respect, and often grow in special polycrystalline forms (*spherulitic* graphite) to avoid exposing high-energy faces. Even cubic crystals are anisotropic. Calculations have shown that the energy of the cube face in crystals of the rock-salt type is about one-third that of high-energy faces. The calculated surface energy of diamond is 5650 erg cm^{-2} on close-packed (octahedral) faces and 9820 erg cm^{-2} on cube faces. Close-packed faces have low energies because close-packed planes are joined together by fewer bonds than other planes. A face which differs slightly in orientation from a close-packed plane can be regarded as a stepped terrace of close-packed planes, with its energy raised by the energies of these steps. The surface energy γ thus increases whichever way the orientation ϕ deviates from the close-packed plane. Similar but weaker *cusps* in the γ,ϕ relation occur at other crystallographically simple faces. The γ,ϕ relation is best shown by a *polar diagram*, as in Fig. 8.16, in which the energy of any surface plane is measured by the length of the *radius vector*, perpendicular to the plane, from the centre to the boundary.

Fig. 8.16.   Schematic polar diagram of surface energy.

What shape minimizes the surface energy of a crystal? The sphere satisfies this condition only when γ is isotropic. When γ is anisotropic, the shape can be found from a *Wulff construction*. A plane is drawn perpendicularly through the end of each radius vector in the polar diagram. The body formed by all points which can be reached from the origin without crossing any of these planes has the shape of minimum surface energy (e.g. the broken lines in Fig. 8.16).

Although crystals often form in this shape (e.g. cubic ionic crystals), they usually do so for other reasons; e.g. because high-energy faces tend to grow quickly and thereby disappear, leaving the crystal bounded by slowly growing (usually close-packed) faces. Unless the crystal is smaller than about 10^{-3} cm, the degree of supersaturation required for a reasonable rate of crystal growth is usually great enough to allow non-equilibrium faces, and the shape is then mainly determined by the *kinetics* of growth.

A crystal growing from *dilute* vapour or solution has a special nucleation difficulty. Even when it has become large, there remains the problem of nucleating new layers on its faces, particularly the slow-growing close-packed faces. Sites such as A in Fig. 8.11, on smooth close-packed faces, provide poor lodgement for condensing particles. Usually, the particles on such faces are quickly thrown off again, before they have time to migrate far along the surface and join others to form the nucleus of a new layer. This is a difficulty peculiar to growth from dilute vapours and solutions, and does not apply to a crystal grown from a melt which wets its surface.

The difficulty disappears when there are steps on the surface which provide good lodgement sites, such as B and C in Fig. 8.11. What is particularly required is a *self-perpetuating growth step*, so that, however much the face grows, it is never without good lodgement sites. Such steps are, in fact, provided at the ends of screw dislocations, as shown in Fig. 8.11. Such a dislocation turns the crystal into a spiral "staircase", with the surface step as the topmost "stair", and growth can then occur by adding more stairs. Since only a few condensed particles are required to produce a complete rotation of those parts of the step near the dislocation, the step in fact winds up into a *growth spiral*, as shown in Fig. 8.17. Many crystals have been observed to grow in this manner. When a crystal is evaporated or dissolved, this process is reversed, and *evaporation pits* or *etch pits* may then be formed at the dislocations. This is important both as a means of detecting dislocations in crystals and also as a mechanism of *chemical corrosion* on crystalline solids. Etch-pitting is often

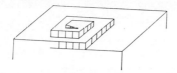

Fig. 8.17.   A growth spiral on a crystal.

strongly accentuated by the localization of impurities in dislocations, which encourage chemical attack there.

Traces of impurity adsorbed on crystal faces can "poison" the growth process by attaching themselves to the growth steps, so blocking the flow of matter and thereby considerably altering the shape of the growing crystal. This is one reason why a particular type of crystal often appears in a variety of shapes. This blocking effect leads to an *instability* in crystal growth, as shown in Fig. 8.18. If a growth step such as A chances

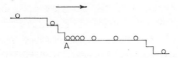

Fig. 8.18.   Effect of adsorbed impurities on crystal growth.

to move slowly, more time is allowed for impurities to collect on the surface ahead of it, and these slow it down still more. Conversely, if a step happens to move quickly, less time is allowed for the impurities to gather and it can then run on still faster. There is thus a tendency for the growth steps to "bunch" closely behind a slow one, so forming steep-sided growth "hills" or deep etch pits.

The growth of some *whisker* crystals, long filamentary crystals about 10^{-4} cm thick, is believed to be due to bunching. Some crystalline materials, such as cellulose and asbestos, grow as fibres simply because their long-chain molecules are shaped that way. Whiskers can, however, be grown from practically all crystals, even cubic ones, under certain conditions. Figure 8.19 shows how the poisoning of a growth step can cause a whisker crystal to grow with a screw dislocation along its axis. A similar growth, without the dislocation, could occur from a heavily supersaturated vapour or solution.

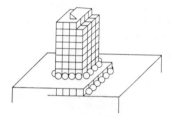

Fig. 8.19.   Growth of a whisker due to poisoning of a growth step.

The shapes of growing crystals, particularly from melts and highly supersaturated solutions, are often governed by *diffusion*, i.e. by the transport of solute or heat to or from the centres of growth. The fastest growth then occurs at places where the surface meets fresh liquid, and this causes the crystal to send out long *dendritic* arms (cf. Fig. 3.18) into distant, untapped regions of the liquid.

8.9   PROPERTIES OF GRAIN BOUNDARIES IN POLYCRYSTALS

The disordered structure of grain boundaries gives rise to many effects in polycrystals, including some of great importance to engineering materials. They can be understood most easily by picturing a boundary as a monolayer of liquid between the adjoining crystals. This can act, for example, as a *source* or *sink* of lattice vacancies. The emission of vacancies from the boundary is equivalent to removing atoms from inside the crystals, placing these in the liquid layer, and then immediately crystallizing them on to the adjoining crystal faces. *Vacancy creep* is thus strongly promoted by the presence of grain boundaries.

Solid-state sintering depends, at least partly, on vacancy creep at grain boundaries. Many polycrystals of refractory materials, e.g. alumina, tungsten, metallic carbides, are made by heating pressed powders at high temperatures. To obtain the highest mechanical strength, the last traces of porosity must be eliminated, but this is difficult because the last 10% or so of shrinkage involves closing isolated spheroidal pores. Vacancy creep can do this, under the driving force $2\gamma/r$ of the surface tension, provided the pore is near a vacancy sink such as a grain boundary, as shown in Fig. 8.20(a).

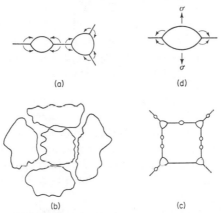

(a) (d)

(b) (c)

Fig. 8.20. Pores on grain boundaries: (a) sintering by vacancy creep (the directions of atomic migration are indicated by arrows); (b) and (c) formation of boundaries and pores from sintered powder; (d) negative sintering under a tensile stress.

Diagrams (b) and (c) show that the formation of a sintered compact leads naturally to an association of boundaries and pores. This enables sintering to occur fairly rapidly up to about 95% full density. When the pores become very small, however, they can no longer anchor the grain boundaries against their tendency to migrate under the forces of grain growth. Pores are then left stranded inside the grains, and, being a long way from boundaries, they sinter very slowly. To promote total sintering, small grains of foreign substances are added. These act as additional anchors to the grain boundaries and prevent grain growth, so enabling dense, strong, fine-grained solids to be made.

Negative sintering, i.e. growth of pores, can occur when the driving forces are reversed. For example, if the pores contain gas at a pressure p ($>2\gamma/r$), the net pressure $p - (2\gamma/r)$ forces them to act as vacancy *sinks*. Another example is shown in Fig. 8.20(d). The vacancy concentration c produced near the grain boundary by the normal tensile stress σ is, from Equation (7.17), $c_0 \exp(\sigma v/kT)$. That produced near a pore of radius r is, from Equation (8.41) (with p proportional to c), $c_0 \exp(2\gamma v/rkT)$. Hence, vacancies flow from the boundaries to the pores when

$$\sigma \geqslant \frac{2\gamma}{r}. \tag{8.45}$$

The pores can then grow, coalesce and fracture a grain boundary as shown in Fig. 8.21. This is similar to the multiply nucleated viscous

Fig. 8.21. Fracture of a transverse boundary, under tension, by the growth and coalescence of pores.

fracture discussed in Section 7.10. If $\gamma = 1000 \text{ erg cm}^{-2}$ and $r = 10^{-4} \text{ cm}$, the critical stress is only $2 \times 10^7 \text{ dyn cm}^{-2}$ ($\simeq 300 \text{ psi}$). Creep-resistant alloys, as used, for example, in gas turbine blades, after prolonged stressing at high temperatures sometimes break by a process of *creep cavitation* which strongly resembles that of Fig. 8.21. In a few cases non-wetting foreign particles in the grain boundaries have been identified as the nuclei of these cavities.

The disordered structure enables atoms to diffuse relatively quickly along grain boundaries; e.g. 10^5 times faster than self-diffusion within the grains, in silver at 500°C. The activation energy for grain boundary self-diffusion in silver has been measured as 0·88 eV, i.e. $9kT_m$ ($T_m = $ absolute melting point). At the melting point an atom in the boundary makes about 10^{10} jumps sec^{-1}, which compares with that in the liquid at this temperature. The high atomic mobility in and near grain boundaries leads to enhanced rates of solid-state reactions in these regions, in some cases with harmful results in engineering materials. The fine intermetallic precipitates responsible for the strength of precipitation-hardened aluminium alloys, for example, become coarse in zones along the grain boundaries. These "over-aged" grain boundary zones are mechanically weak and also susceptible to corrosive attack. Stainless steels of the 18 % Cr, 8 % Ni type owe their corrosion resistance to the chromium in solid solutions. Heating at 500–800°C causes the chromium near grain boundaries to precipitate as chromium carbide, thus creating zones along the boundaries which are *denuded* of chromium in solid solution and are thereby vulnerable to corrosive attack. In this condition a stainless steel can quickly crumble to powder, falling apart along its grain boundaries, if exposed to moist air. This effect is prevented by adding carbide-forming elements (e.g. Ti, Nb) to the steel. These capture the carbon and release the chromium for solid solution.

Referring to Equation (7.20), we may write

$$\eta \simeq \eta_0 \, e^{9T_m/T}, \qquad\qquad (8.46)$$

where $\eta_0 \simeq 0.1$ poises, as a rough estimate of the viscosity of the grain boundary monolayer at a temperature T, near the melting point. This gives the glass temperature of the boundary at $T_g \simeq 0.25T_m$, but the formula is certainly not valid at temperatures as low as this. Since the activation energies of viscous liquids increase at low temperatures, we expect the glass temperature to be somewhat higher. The fluidity of a grain boundary has several effects. Grain boundaries in very pure metals have been observed to migrate quickly, up to about 1 mm min^{-1} in lead and tin, in grain growth at temperatures near the melting point, which suggests that the migration occurs by melting on one adjoining crystal face, and simultaneous crystallization on the other, at the high rates typical of the crystallization of simple liquids. Extremely pure aluminium can *recrystallize* (i.e. replace highly dislocated, cold-worked grains by relatively undislocated grains in a process of grain boundary migration) at $-70°C$, so that some mobility persists even at $0.215T_m$. In commercial metals and alloys the grain boundaries become mobile only at higher temperatures ($\simeq 0.4T_m$), because they are pinned by foreign inclusions and adsorbed solute atoms.

The fluidity of a boundary also allows one crystal to *slide* past another, along their common boundary, under a suitably directed applied shear stress. This grain boundary sliding can be detected at very low applied stresses by internal friction measurements (cf. Section 6.5). If the shear stress σ produces a relative velocity v between the sliding crystals, the viscosity η of the boundary, of thickness d, is given by

$$\eta = \sigma d / v. \tag{8.47}$$

The values of v measured by internal friction give viscosities which extrapolate approximately to that of the liquid at the melting point, assuming $d \simeq 5$ Å. Again, this type of grain boundary movement is greatly slowed down by impurities. It is also restricted by the interlocking of the crystals at triple junctions and at geometric irregularities along the boundaries. We see from Fig. 8.22 that the apparent speeds of *large* grain boundary

Fig. 8.22. Obstruction of a sliding grain boundary AB at a triple junction B. The boundary can continue sliding, if the displacement at the junction is accommodated by plastic deformation in the grains, as in BC.

movements, such as may be observed in alloys after long periods at high temperatures, must actually be determined by the rates at which plastic deformation within the grains allows the boundaries to continue sliding.

The *strengths* of grain boundaries are extremely important for engineering materials. The ideal tensile strength of a transverse boundary in a metal is expected to be very high, approaching the values estimated in Section 8.2, and this is supported by the fact that, when a ductile metal such as copper is stretched at low temperatures, it withstands a stress of order $E/100$ before it breaks ($E = $ Young's modulus), and the fracture passes through the grains, not along the boundaries. In ionic and other non-metallic polycrystals the boundaries are expected to be weaker because of the difficulty of satisfying the conditions for chemical bonding there. At high temperatures the tensile strength of a cavitated boundary is extremely low, for the reasons discussed above (cf. Fig. 8.21).

Except at low temperatures the shear strength of a boundary is low. A sliding boundary acts in effect as a *shear crack*, and, if the stress concentrated at its ends is not dispersed by processes such as that of Fig. 8.22, it may produce a *grain boundary crack*, as shown in Fig. 8.23. This is a

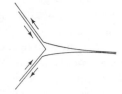

Fig. 8.23.   Fracture produced by sliding at grain boundaries.

common form of high-temperature fracture in creep-resistant materials. It can be overcome by introducing foreign inclusions (wetting) into the boundaries to peg the sliding surfaces together; also by forming a fibrous grain structure, as in Fig. 8.24, to reduce the stress acting on the boundaries.

Fig. 8.24.   Avoidance of high stresses across boundaries in a fibrous polycrystal.

The strengths of boundaries are affected in various ways by alloy elements and impurities. Both the mode of freezing (cf. Section 3.5) and surface adsorption encourage the segregation of foreign substances into grain boundaries, where they may produce weak grain boundary layers (e.g. antimony in copper, which is adsorbed at the grain boundaries and is believed to weaken and embrittle the metal by reducing the energy of the free surfaces produced by fracture along the boundaries). Such segregates may also lead to *hot-shortness*, i.e. weakness at high temperatures due to melting of fusible constituents (e.g. sulphides in iron) in the boundaries, and to various forms of severe and localized chemical corrosion, particularly in the presence of tensile stress (*stress–corrosion cracking*). These difficulties have to be overcome mainly by careful metallurgical treatments, involving the elimination of harmful substances, or the addition of alloys which counteract them, or heat treatments which distribute them in harmless forms. A longitudinal fibrous grain structure, as in Fig. 8.24, is also helpful for reducing applied stresses on the boundaries.

FURTHER READING

NEWMAN, F. H. and SEARLE, V. H. L., *The General Properties of Matter*, Arnold, London, 1961.

ADAM, N. K., *The Physics and Chemistry of Surfaces*, Oxford University Press, London, 1941.

DAVIES, J. T. and RIDEAL, E. K., *Interfacial Phenomena*, Academic Press, New York, 1961.

BURDON, R. S., *Surface Tension and the Spreading of Liquids*, Cambridge University Press, London, 1949.

AMERICAN SOCIETY FOR METALS, *Metal Interfaces*, A.S.M., Cleveland, 1951.

GOMER, R. and SMITH, C. S., *Structure and Properties of Solid Surfaces*, Chicago University Press, 1953.

DOREMUS, R. H., ROBERTS, B. W. and TURNBULL, D., *Growth and Perfection of Crystals*, Wiley, New York, 1958.

BUCKLEY, H. E., *Crystal Growth*, Wiley, New York, 1951.

MCLEAN, D., *Grain Boundaries in Metals*, Oxford University Press, London, 1957.

PROBLEMS

8.1 A sphere of water, 0·1 cm radius, is broken up into a million equal droplets. What is the minimum amount of work required to do this?
[900 erg]

8.2 A conical glass tube is held vertically with its lower end, of diameter 0·3 cm, just touching the surface of water, of surface tension 80 dyn cm^{-1}. The tube is 20 cm long and its diameter at the upper end is 0·1 cm. If the water completely wets the glass, how high will it rise up the tube?
[0·6 cm]

8.3 A gold wire of radius 0·008 cm is held at a temperature just below its melting point, where creep is rapid, and is strained in tension by a small suspended weight. Under a load of 0·075 g the wire maintains a constant length. Under smaller loads it shortens; under larger ones it lengthens. What is its surface tension?   [1500 dyn cm^{-1}]

8.4 Two soap bubbles of radii r_1 and r_2 are joined by a common boundary. Show that the radius of curvature of this boundary is $r_1 r_2/(r_1 - r_2)$.

8.5   If n molecules of an adsorbed substance are free to move in all directions on a surface and occupy a small fraction of the surface area A, show that they may behave as a two-dimensional "gas" for which the perfect gas law is $FA = nkT$, where F is the force exerted by them per unit length of surface.

8.6   Helium atoms are injected into a metal crystal by a particle accelerator. Helium is insoluble in this metal, but the atoms remain in forced solid solution until the metal is heated to $800°K$, when they diffuse together and form small spherical gas "bubbles" which expand by capturing lattice vacancies. If the pressure in the bubbles comes to equilibrium with the force due to the surface tension of the surrounding metal and the helium obeys the perfect gas laws, calculate the number of gas atoms in bubbles of diameters 10^{-6} and 10^{-4} cm, assuming that the surface tension is 1000 dyn cm^{-1}. If the maximum rate of vacancy transport through the metal is 600 times that of helium transport, and the volume per vacancy is 2.2×10^{-23} cm^3, at what size does a bubble first exert an outward force on its surroundings?

$$[2 \times 10^4; 2 \times 10^8; 5 \times 10^{-4} \text{ cm}]$$

8.7   A single soap bubble of radius R is formed from the coalescence of two soap bubbles of radii r_1 and r_2, under an external pressure p. Show that the surface tension of the film is given by

$$\frac{p(R^3 - r_1{}^3 - r_2{}^3)}{4(r_1{}^2 + r_2{}^2 - R^2)} .$$

8.8   Two equal soap bubbles are formed on the ends of a cylindrical tube. The radius of one of them is slightly altered by an external disturbance. Deduce the effects of this if the bubbles are initially (a) smaller, (b) larger, than one hemisphere.

8.9   Deduce a formula for the equilibrium radius r of an axial hole along the centre of a screw dislocation of Burgers vector $\mathbf{b}$ in a crystal of shear modulus μ and surface energy γ. $[r = \mu b^2/8\pi^2\gamma]$
       Do you expect unit dislocations in an f.c.c. metal to be hollow?

8.10 Show that the melting point T of a small spherical crystal of radius r, in equilibrium with its liquid, is given by

$$T_m\left(1 - \frac{2\gamma V}{rL_m}\right),$$

where T_m is the melting point of a massive crystal, γ is the crystal–liquid interfacial energy, V is the volume per mole and L_m is the latent heat of melting per mole. Assuming that γ, V and L_m are independent of temperature, deduce the free energy of the liquid, relative to the crystal, from the entropy of melting, which in turn is deduced from L_m and T_m.

8.11 Explain why, in a stable froth, the number of films which meet at an edge is always three, and the number of edges at a corner always four.

8.12 A glass sphere of radius R rests on a flat glass surface. Between them is a drop of water which is spread out, by its contact with them, into the shape of a thin concave-planar lens of diameter d, where $d \ll R$. The contact angle is zero and the tension of the free surface of the water is γ. Show that the force F of adhesion between the sphere and the plate is independent of the volume of water and is given by $F = 4\pi R\gamma$.

8.13 The following are approximate values of interfacial energies in erg cm^{-2} for polycrystalline copper in contact with air and with molten lead: $\gamma_{CA} = 1800$, $\gamma_B = 600$, $\gamma_{CL} = 400$, where C = copper, A = air, B = grain boundary, and L = lead. Show that, so far as interfacial energies are concerned, there is little to choose between intercrystalline and transcrystalline paths for a crack, when copper is broken in air; but that, when the copper is in contact with molten lead, an intercrystalline path is energetically preferred, even if it is nearly four times longer than a transcrystalline path.

chapter **9**

Plastic Crystals

*We shall now discuss the remarkable ability of dislocations
to glide rapidly through crystals, under shear stress, so
producing plastic deformation. In soft metals the crystal
lattice offers almost no resistance to this glide motion. The
strengths of most commercial metals and alloys are due to
various obstacles placed in the paths of the glide dislocations;
for example, foreign particles, lattice defects (including other
dislocations) and grain boundaries. In typical non-metallic
crystals the dislocations cannot move quickly except at high
temperatures or stresses. Such materials are hard and brittle
at low temperatures. Plastic deformation itself increases the
number of dislocations, but only under limited circumstances
does this produce softening; more commonly it makes the
crystal harder, because the dislocations mutually obstruct one
another (work-hardening). At high temperatures various
additional processes occur. These soften the material and
allow it to creep under constant stress.*

9.1 SLIP AND TWINNING

The plastic properties of ductile solids are remarkable. Despite the great
forces holding its atoms in place, a suitably oriented crystal of a pure
metal such as cadmium can be stretched plastically at room temperature
to about 20 times its initial length, quickly if desired (e.g. $\dot{\gamma} \simeq 1000 \text{ sec}^{-1}$),
by a stress so small ($\simeq 10^{-5}E$, where E is Young's modulus) that such a
crystal can hardly support its own weight. Moreover, this severe deformation leaves the crystal structure almost as perfect as before. The
deformation is of course permanent, in the sense that, unlike elastic
deformation, it does not reverse on unloading but leaves the specimen with
a *"permanent set"*.

The crystallinity of the structure is the prime cause of this behaviour, for it enables whole slabs of crystal to *glide* past one another, as in the *slip* process of Fig. 9.1. Each slip is a displacement, by a whole number of atomic spacings, in a certain *glide direction*, generally the crystal direction of closest atomic packing. The crystal structure is preserved in the *glide surfaces* between the slabs, and the *slip steps* on the outer surface provide the only obvious indications of these surfaces. The heights of slip steps vary widely from one to another. Steps some thousands of atomic spacings

Fig. 9.1.   Plastic deformation by slip.

high, visible to the naked eye, are often formed. These large *slip lines* (or *slip bands*, when composed of clusters of smaller steps) are separated by layers, some thousands of atoms thick, in which slip steps are either absent or finely distributed.

Glide surfaces often follow certain crystal planes, called *slip planes*. These are mainly close-packed planes, such as octahedral planes in f.c.c. metals; and basal planes in graphite and c.p. hex. metals of large axial ratio. In crystals such as b.c.c. iron there is no fixed slip plane. Instead *pencil glide* occurs, giving corrugated and forked glide surfaces as if a bundle of parallel hexagonal pencils were being deformed by sliding the pencils along their axes. True (or *conservative*) glide is always a *tangential* movement along the glide surface, which preserves the atomic spacing perpendicular to that surface. The glide surface, whether plane or not, is therefore always parallel at all points to the glide direction. If this were not so, neighbouring slabs would move *into* or *away from* each other as they glided.

Engineering metals and alloys owe their unique combination of strength and ductility to their slip properties. Practically all crystalline solids can slip, however, even diamond and sapphire, when the temperature is high or when there is a large hydrostatic pressure acting to prevent them fracturing. Some geological processes are brought about by the plastic deformation of normally brittle minerals under high pressures at great depths below the earth's surface.

Plastic deformation can also occur by *twinning*. Here the atoms slide, layer by layer (sometimes also rearranging their relative positions locally), to bring each deformed slab into mirror-image lattice orientation relative to the undeformed material. Figure 9.2 shows a simple example. The atomic displacements in twinning are *imperfect* (cf. Section 3.4), whereas those in slip are *perfect*. Twinning is less common than slip, in soft metals, because its critical stress is usually higher, i.e. generally above about $E/500$. When b.c.c iron is strained quickly at room temperature, e.g. by a hammer blow, or slowly below about 100°K, long thin deformation twins called *Neumann bands* are often formed. Copper can be twinned at temperatures

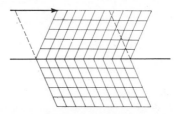

Fig. 9.2.   A simple type of deformation twinning.

in the range of 4°K, but the most obvious examples of deformation twinning are provided by non-cubic metals such as zinc and bismuth, and by non-metallic crystals such as calcite. The creaking sound, or "cry", heard when polycrystalline tin is bent is the noise of deformation twins suddenly bursting into existence. Twins form in an abrupt manner, giving jerks on the stress–strain curve, probably because a higher stress is required to create the first monolayer of twin (i.e. a stacking fault) than to add other twinned layers to it afterwards.

9.2   GLIDE DISLOCATIONS

Since conditions can never be *exactly* uniform over the whole of a large glide surface, some areas must begin to slip before others. They can of course do this, because the material is elastically flexible. Slip therefore begins on some small area of the surface and then spreads outwards from it over the rest of the surface. The important feature is the *slip-front*, i.e the line of demarcation between the slipped and unslipped areas, which sweeps across the glide surface as the slipped area grows. This line is by definition a *dislocation line*, since, if it were allowed to cut through one arm of a closed Burgers circuit (cf. Fig. 3.14), such as ABC or DEF in Fig. 9.3, that circuit would be distorted in the manner characteristic of a dislocation. It follows from Section 3.4 that the Burgers vector **b** of this dislocation PQRS is equal to the discontinuity of slip displacement at the slip-front. Clearly, a dislocation line must end at two points P and S on the outer surface, as in Fig. 9.3, or form a closed loop, or be joined at its ends to other dislocation lines. In other words, the Burgers vector of a single dislocation line is *constant* all along the line. Parts of the line may be *edge* (e.g. PQ), *mixed* (e.g. QR) or *screw* (e.g. RS) dislocation, according to the path followed. The boundary of a *slipped* region must be a *perfect* dislocation, with a Burgers vector equal to a lattice vector. If PQRS

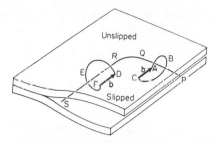

Fig. 9.3. The Burgers circuits ABC and DEF show that the boundary PQRS between slipped and unslipped areas of a glide surface is a dislocation line.

were, however, the boundary of a stacking fault, as in deformation twinning, the dislocation would be *imperfect* or *partial*. Depending on the length of **b**, the dislocation may be a *unit* or *multiple* dislocation; if a multiple, it generally consists of a bundle of parallel unit dislocation lines lying in a flat *ribbon* of glide surface along PQRS.

The existence of slip-fronts is proved by observations of slip lines which run only part way across crystals; also, by the fact that plastic crystals can be bent, like thick telephone directories, to sharp curvatures, as shown in Fig. 9.4. Since opposite ends of the layers slip in opposite

Fig. 9.4. Plastic bending, with edge dislocations in the glide surfaces.

directions, there must be edge dislocations between the layers. Because of the elastic forces between them (cf. Section 5.11), these dislocations tend to align themselves into vertical "walls" or *small-angle boundaries*. Such walls of glide dislocations, often seen in plastically bent crystals, are called *bend planes* or *kink bands*. High-resolution electron microscopy has in recent years enabled individual dislocation lines to be seen directly, even in some cases while gliding.

   The glide motion of a dislocation is a remarkable property of a periodic crystal structure. Figure 9.5 is a sectional view of a simple type of edge dislocation. This is a *wide* dislocation, i.e. the transition from the slipped to the unslipped region is spread over several atomic distances. The distance w, defined as in the diagram, measures this width. As the centre of the dislocation moves from A to B, each atom in the transition region

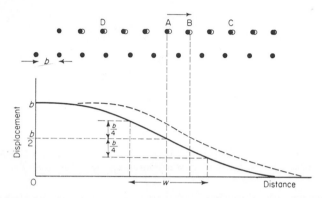

Fig. 9.5. Displacement of atoms, from full to open circles, caused by the glide of an edge dislocation from A to B. For simplicity of illustration, the lower row is supposed fixed.

moves forward by a distance of order b^2/w. Thus, an atom ahead of the dislocation, such as that at C, is pushed a little further out of its initial site. It resists this and so exerts a force against the dislocation. But this force is countered by an equal and opposite force on the dislocation from an atom behind it, such as that at D, which is trying to complete its slip movement. At symmetrical positions such as A and B the dislocation is thus in *neutral* equilibrium, since the atomic forces acting on it from either side precisely cancel. As the dislocation moves from A to B, this conclusion no longer remains exactly true, because atoms such as C and D no longer remain equidistant, but it remains approximately true if the dislocation is wide, because the atoms almost symmetrically placed on either side of it ensure a near-equilibrium of atomic forces for *all* positions of the dislocation. *The atomic forces of the crystal thus offer practically no resistance to the motion of a wide dislocation.* This is the basis of the remarkable plasticity of soft crystals. The glide of such a dislocation is an almost *purely mechanical process*, like the motion of an elastic wave packet, and is quite different from viscous flow or vacancy creep or, indeed, dislocation *climb* (cf. Section 3.4). Dislocations can, in fact, glide at speeds (e.g. 10^5 cm sec^{-1}) almost up to those of elastic waves in crystals.

The rate of plastic shear due to dislocations is easily deduced. In Fig. 9.6 we show *positive* and *negative* edge dislocations (i.e. half-planes above and below the slip plane, respectively) gliding oppositely to produce shear in the direction of the applied shear stress σ. The *density* ρ of dislocations is the number in unit area of the section. In a time L_1/v one dislocation of speed v and Burgers vector of length b completely crosses the crystal and

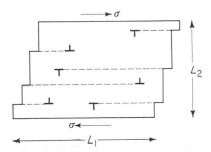

Fig. 9.6. Shear produced by dislocations.

produces a shear strain b/L_2. The total rate of plastic shear is thus given by

$$\dot{\gamma} = \frac{b}{L_2}\frac{v}{L_1}\rho L_1 L_2 = b\rho v. \tag{9.1}$$

The value of this can be very large, e.g., if $b = 3 \times 10^{-8}$ cm, $v = 10^5$ cm sec^{-1} and $\rho = 10^7$ cm^{-2} (a common value for unworked metals), then $\dot{\gamma} = 30{,}000$ sec^{-1}, which could double the length of a tensile bar in less than 10^{-4} sec. Only explosions, high-speed impacts and certain metal-working processes demand such high rates as this. At more conventional rates, e.g. 1 sec^{-1}, only a few dislocations need move quickly at any instant. Equation (9.1) also applies to less simple distributions of dislocation lines, including both edges and screws, if ρ is defined as the total length of mobile dislocation lines in unit volume.

The dislocations move because their movement enables the applied stress to do work. The *applied force F* per unit length on a dislocation line is defined as follows. In Fig. 9.7 let a dislocation line of length L_3 move under the stress σ. In crossing the slip plane completely, it moves the top face of the crystal by a distance b and so causes the force $\sigma L_1 L_3$ on this face to do the work $\sigma b L_1 L_3$. The applied force FL_3 on the dislocation line has moved a distance L_1 and so done the work $FL_1 L_3$. We define F by making these two amounts of work equal, i.e.

$$F = \sigma b. \tag{9.2}$$

Most of this work is converted to heat in the glide surface. Clearly, components of applied stress other than *that acting on the glide surface in the glide direction* do no work when the dislocation moves, and so do not

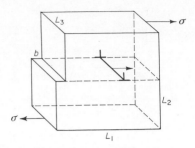

Fig. 9.7. A dislocation moved by the applied stress.

influence its motion (unless so strong as to alter the elastic properties of the crystal). This is the *law of resolved shear stress*, proved experimentally by subjecting crystals to variously oriented stresses and observing that glide always occurs when a critical resolved shear stress is applied to the active *glide system* (i.e. glide surface and direction), irrespective of the other stress components.

9.3 HARD AND SOFT CRYSTALS

The softness of many pure metals contrasts strikingly with the high shear strength expected of a perfect crystal. In Fig. 9.8 we consider the *simultaneous* slip of one sheet of atoms over another. Let a and b be the

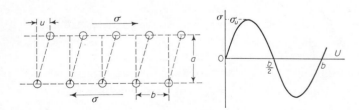

Fig. 9.8. Simultaneous shear.

respective spacings between and along the sheets. The shear stress σ clearly varies periodically with the shear displacement u, with a period b, since (1) $\sigma = 0$ at the lattice positions $u = 0, b, 2b, \ldots$ etc.; (2) $\sigma = 0$ at the mid-way positions $u = \frac{1}{2}b, \frac{3}{2}b, \frac{5}{2}b, \ldots$ etc., by symmetry; and (3) $\sigma > 0$ when

$\frac{1}{2}b > u > 0$, and $\sigma < 0$ when $b > u > \frac{1}{2}b$, since the atoms are attracted to the nearest lattice position. For simplicity, we assume

$$\sigma = \sigma_u \sin\left(\frac{2\pi u}{b}\right) \tag{9.3}$$

with an initial slope that conforms to the shear modulus μ, i.e.

$$\sigma = \sigma_u \frac{2\pi u}{b} = \mu \frac{u}{a} \tag{9.4}$$

at small strains. The *ideal shear strength* is then

$$\sigma_u = \frac{\mu}{2\pi} \frac{b}{a}. \tag{9.5}$$

If $a \simeq b$, then $\sigma_u \simeq \mu/6$. Refined calculations suggest $\mu/10$ or $\mu/30$. The equivalent *ideal tensile yield stress*, about twice this value, is thus given in order of magnitude by Equation (8.14). In essence, σ_u is the stress to *create* a dislocation loop in a glide surface of a *perfect* crystal. Its largeness is confirmed by the great strengths of some whisker crystals and other flawless crystals (cf. Section 8.2). Figure 9.9 shows how the strength disappears as soon as dislocations are made, in a copper whisker.

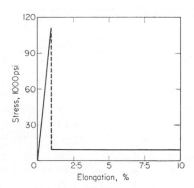

Fig. 9.9. Stress–strain curve of a copper whisker (after S. Brenner, *J. Appl. Phys.*, **27**, 1484 (1956)).

We turn now from σ_u, the *highest* shear strength a crystal can have, to σ_L, the *lowest* strength. Leaving aside processes such as vacancy creep (cf. Section 7.3), σ_L is evidently the stress to glide an existing dislocation in an otherwise perfect crystal. By extending the argument of Fig. 9.5

and evaluating the unbalanced atomic forces on the dislocation, *Peierls, Nabarro* and others derived approximate formulae of the type

$$\sigma_L \simeq \mu \, e^{-2\pi w/b}, \tag{9.6}$$

where w is the width of the dislocation,

$$w \simeq \frac{\mu b}{2\pi(1 - v)\sigma_u}, \tag{9.7}$$

v being Poisson's ratio. The mobility of the dislocation thus increases rapidly as σ_u decreases, which favours close-packed glide systems (i.e. small b/a).

The atomic forces contribute both directly and indirectly to σ_L: directly, by providing directional or non-directional bonds; and indirectly, by providing crystal structures with favourable or unfavourable b/a values. In ionic crystals the glide system is chosen so as to avoid bringing together ions of like charge. In covalent crystals, such as diamond or silicon, the directional bonds tend to produce narrow dislocations, which leads to *intrinsic hardness*, i.e. even when purified and annealed, the crystal remains hard at low temperatures.

The free-electron bond in metals favours intrinsic softness, because it is mainly non-directional and so can tolerate relatively wide dislocated regions, in which the atoms are sheared well out of their normal lattice positions, in the glide surfaces. Close-packed planes in f.c.c. and c.p. hex. metals are particularly favourable for widely *extended* dislocations. As well as having a small b/a factor, they provide the gliding atoms with excellent crystal sites near the half-way mark. This is because, as Fig. 3.9(b) shows, each close-packed layer provides *two* sets of sites for the layer above it. One of these (say B) gives the proper crystal structure, and the other (C) gives a stacking fault. A unit dislocation gliding between the layers slides each atom in the upper layer from one B site to the next. This B → B transition can occur most easily in two separate zigzag stages, B → C → B, taking the atoms through the C sites on the way. The unit dislocation may thus *extend* itself into a flat ribbon in the slip plane. The leading edge of this ribbon is a partial dislocation that produces the B → C slip; the ribbon itself is a stacking fault; and the trailing edge is a second partial dislocation that completes the total B → B slip by producing the second slip C → B. In metals of high stacking-fault energy, such as aluminium and nickel, the ribbon is probably only about two atoms wide. In gold, however, where the fault energy is lower, it may be 30 atoms wide, and in some f.c.c. alloys (certain brasses, bronzes and austenitic steels) it is even wider and can easily be seen in the electron microscope.

Crystals with wide dislocations are intrinsically soft. Their hardness is due mainly to various extraneous obstacles to dislocations, such as foreign atoms, precipitates, lattice defects and grain boundaries, which can be largely removed by purification and annealing. The free-electron bond makes metals soft mainly by producing simple close-packed crystal structures in which there are glide systems with favourable b/a values. When b/a is unfavourable, the free electrons do not seem particularly effective. Thus, f.c.c. metals rarely slip on systems other than the close-packed one, and c.p. hex. metals are hard along glide directions lying out of the basal plane. B.C.C. transition metals, such as iron, tungsten and molybdenum, generally remain hard at low temperatures, although the extent to which this hardness is intrinsic is still questionable. Intermetallic compounds such as $CuAl_2$, with complex crystal structures and unfavourable b/a factors, are hard and brittle despite their metallic bonding. The importance of the b/a factor is also shown by the converse effect, the *softness* of many layer crystals of non-metals and semi-metals on layer planes; e.g. graphite, molybdenum disulphide, talc, soap flakes and metal hydroxides.

Intrinsically hard crystals become soft at high temperatures. When the dislocation is very narrow, i.e. $w \simeq b$, an atom at its centre has to jump a distance of order b as the dislocation moves forward by one spacing, and this is similar to a vacancy jump or to the elementary atomic movement in a flowing liquid. It can occur at low stresses, if there is thermal energy to overcome the activation energy barrier. To express the argument in a general form, we can regard *any* barrier to a dislocation as equivalent to a certain internal stress σ_i which opposes the applied stress σ over a length mb of the dislocation line to a depth nb in the direction of dislocation motion. If $\sigma < \sigma_i$, the net stress $\sigma_i - \sigma$ is equivalent, by Equation (9.2), to a force $(\sigma_i - \sigma)mb^2$ opposing the length of dislocation facing the barrier. This force moves a distance nb as the dislocation crosses the barrier. The work thus done against it is the activation energy

$$Q = mn(\sigma_i - \sigma)b^3 = mn\sigma_i b^3 \left(1 - \frac{\sigma}{\sigma_i}\right) \qquad (9.8)$$

which the thermal motion must overcome. The frequency of thermal fluctuations is given by the usual Maxwell Boltzmann formula, with a dislocation vibration frequency of order v/m, where v is the atomic frequency. Starting at $\sigma = \sigma_i$ at $0°K$, the glide stress σ decreases with increasing temperature, *not necessarily linearly* because m and n may depend on σ/σ_i. Thermal fluctuations for which $Q \leqslant 1$ eV occur fairly often at room temperature. Thus, if $\sigma_i = 10^{-2} \mu$, $\mu b^3 = 5$ eV, $Q = 1$ eV, then $\sigma \simeq \sigma_i(1 - 20/mn)$; i.e. $\sigma \simeq \sigma_i$ when $mn \gg 20$, and $\sigma \simeq 0$ when $mn \simeq 20$. Obstacles to dislocations thus fall into two classes, according to size: *large* ones (which may be weak or strong, according to σ_i), for which the glide stress hardly varies with temperature (or strain rate); and *small* ones which, if strong (i.e. large σ_i), produce a rapidly varying glide stress. The obstacles to the atomic movements at the centre of a narrow dislocation are of this second type, and the yield stresses of intrinsically hard crystals thus depend strongly on temperature and strain rate.

Figure 9.10 shows measured yield stresses at various fractions of the absolute melting point T_m. Silicon and sapphire, for example, are very

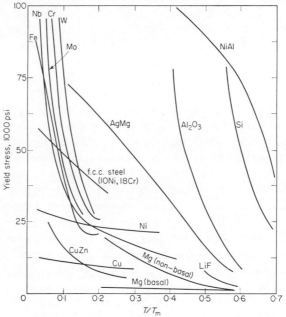

Fig. 9.10. Yield stresses at various temperatures.

hard when $T < 0.5T_m$, whereas copper and nickel hardly change with temperature. Non-basal slip in c.p. hex. metals such as magnesium is hard at $T < 0.4T_m$, and b.c.c. transition metals are hard at $T < 0.15T_m$. The alloys CuZn, AgMg, and NiAl, all with the CsCl structure (cf. Fig. 3.7(f)), harden at various temperatures.

By locating dislocations with etch pits, dislocation speeds have been directly measured in some crystals. Figure 9.11 shows an example. We see that screw dislocations move less quickly than edges and that, at high stresses, speeds approaching that of elastic waves are reached. The speed v varies rapidly with the applied stress σ; in fact, if the empirical formula

$$v = \left(\frac{\sigma}{\sigma_y}\right)^n \tag{9.9}$$

is fitted, where σ_y is the stress for unit speed, then $10 < n < 40$. For *ideal plasticity* (large obstacles) we have $n = \infty$. Intrinsically hard crystals at high temperatures and low stresses most nearly approach the Newtonian limit $n = 1$ (e.g. $n \simeq 2$ for germanium at 935°C).

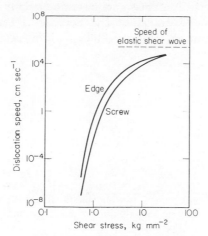

Fig. 9.11. Dislocation speeds in lithium fluoride (after W. G. Johnston and J. J. Gilman, *J. Appl. Phys.*, **30**, 129 (1959)).

9.4 GEOMETRY OF GLIDE

Figure 9.12 shows that both edge and screw dislocations are geometrically capable of gliding along *corrugated* glide surfaces. An edge dislocation with a corrugated line is called a *prismatic dislocation*. By

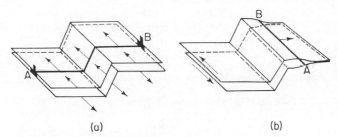

(a)   (b)

Fig. 9.12. Corrugated glide surfaces produced by dislocation lines, AB: (a) prismatic edge dislocation; (b) cross-slipping screw dislocation.

definition, the line of a screw dislocation is always parallel to the Burgers vector, and so *every* surface traced by it is a glide surface. The corrugations in this case are produced by the screw dislocation *cross-slipping* from one slip plane to another, parallel to the Burgers vector.

Cross-slip occurs readily in cubic ionic crystals (e.g. LiF, MgO, NaCl) and in b.c.c. transition metals, all of which have narrow dislocations, and it produces the broad, interlaced slip bands often seen in these materials. In aluminium it occurs fairly readily at room temperature but rarely at low temperatures. In gold, f.c.c. brass and austenitic steel it is sufficiently rare that straight slip lines parallel to the octahedral plane predominate. Figure 9.13 shows that cross-slip is difficult for an extended dislocation,

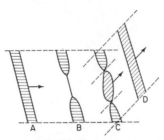

Fig. 9.13.   Stages A (before), B and C (during) and D (after) the transfer of an extended screw dislocation with a stacking fault ribbon (shaded) from one slip plane to another.

because the stacking fault ribbon is confined to a single plane. An activation energy is needed to remove the fault and re-form it on the cross-slip plane. Thus, the process occurs more readily at high temperatures than at low.

An extended dislocation can glide only in the plane of its stacking fault, because the low-energy fault is an atomic configuration specific to this plane. The edge dislocation of Fig. 3.15 is therefore *sessile*, i.e. cannot glide in any direction, because its Burgers vector and its stacking fault require it to glide in different directions. Sessile dislocations are important because their stress fields can block the movements of glide dislocations. Figure 9.14 shows a second kind of sessile dislocation, formed at the intersection of two slip planes. Here the leading partials of two extended glide dislocations A and B coalesce, so forming a sessile

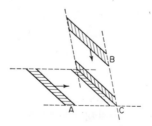

Fig. 9.14.   A sessile dislocation C formed by the coalescence of two extended glide dislocations A and B at the intersection of their slip planes.

group of three partials joined by two stacking faults in different slip planes. Such dislocations have been seen in austenitic steel.

Small steps in a dislocation line are called *jogs*. Figure 9.15 shows two ways in which jogs may form: by a dislocation line cutting through an intersecting screw dislocation, and by a part of a screw dislocation undergoing cross-slip. A jog on a screw dislocation is a short length of mainly or

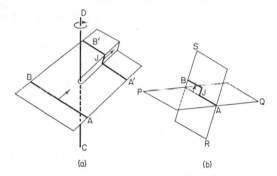

(a)   (b)

Fig. 9.15. Formation of a jog J: (a) by a dislocation cutting through a screw CD as it glides from AB to A′B′; (b) by part of a screw dislocation line AB cross-slipping from the primary slip plane PQ into the plane RS.

wholly edge dislocation, and so can glide only *along* the screw dislocation, since this is the direction of its Burgers vector. As the screw dislocation glides forward, in the direction $P \rightarrow Q$ in Fig. 9.15(b), the jog J must trail behind, since it cannot glide in this direction. Thus, as shown in Fig. 9.16, two parallel edge dislocations of opposite signs, called a *dislocation dipole*

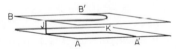

Fig. 9.16. Formation of an edge dislocation dipole JK by the trailing of a jog J behind a gliding screw dislocation AJB $\rightarrow$ A′JB′.

JK, are formed. These are often seen in plastically deformed ionic and covalent crystals, and b.c.c. transition metals. If the two edge dislocations lie only one atom apart, they form a line of *vacancies* or *interstitials*, according as their half-planes point inwards or outwards. This is one way in which lattice point defects may be produced during plastic deformation.

Turning now to more macroscopic aspects of dislocation geometry, we distinguish between *necessary* and *redundant* dislocations. The dislocations in Fig. 9.4 are clearly necessary to the plastically bent state, since opposite ends of the slip planes slip in opposite directions. Whenever glide surfaces are plastically bent or twisted, in fact, dislocations are necessary to accommodate the non-uniform strains. By contrast, dislocations are not *geometrically* necessary in deformations such as the simple shear of Fig. 9.1. Although slip is propagated by dislocations, the overall geometry of the deformation does not require any of these dislocations to remain in the crystal. Any dislocations that do remain are thus geometrically *redundant*. In practice, redundant dislocations almost always do accumulate in plastically worked materials, up to a density of order 10^{12} cm^{-2} in extreme cases, and the deformation is *"turbulent"* in the sense that different regions of the crystal shear and rotate by different amounts, even when not required geometrically to do so.

When a ductile *polycrystal* is plastically deformed, each grain remains joined to its neighbours and so deforms to a shape dictated by them. The polycrystal, in fact, deforms roughly as if it were a block of uniform rubber with the positions of the grain boundaries pencilled in. To a first approximation, the dislocations entering each grain boundary from the grains on either side of it mutually annihilate. The deformation in the grains, particularly near the boundaries, is, however, very turbulent, because of the mutual interference of the dislocations gliding on various intersecting systems. Simultaneous glide on differently oriented systems has to occur to allow the grain to deform to the shapes demanded by its neighbours. Glide on a single family of parallel planes could not produce these shapes. Of the six independent strain coefficients required to produce a general strain, only five are necessary to produce plastic shear without dilatation, since one is determined by the condition $\varepsilon_{11} + \varepsilon_{22} + \varepsilon_{33} = 0$ (cf. Equation (4.43)). Hence, for general "three-dimensional" plasticity the crystal must glide on at least five differently oriented and geometrically independent systems. Such a crystal can always conform in shape to its neighbours.

Their high crystal symmetries provide the f.c.c. and b.c.c. metals with more than enough glide systems for this polycrystalline ductility. These metals are thus well suited to *mechanical working* processes such as rolling, forging, drawing and pressing. They often appear *plastically isotropic*, the principal axes of plastic strain coinciding with those of stress, a fact which misled some people in the nineteenth century into thinking that such metals really were non-crystalline. After heavy working they usually become anisotropic, because the grains are then drawn out into a fibrous or lamellar *texture* with slip directions rotated towards the axis of elongation.

Non-cubic crystals which normally glide only on one or two systems, e.g. basal slip in graphite and hexagonal metals, are often hard and brittle in polycrystalline form, because their grains cannot cohere except by using hard glide systems to achieve the necessary three-dimensional plasticity. One reason for *hot-working* polycrystalline metals such as zinc, magnesium, beryllium and uranium, is to take advantage of the thermal softening of these hard systems at high temperatures.

9.5 THE FORMATION OF GLIDE DISLOCATIONS

The low yield stresses $(10^{-5}E–10^{-2}E)$ of most ductile solids show that glide dislocations must either exist in them initially or be formed at points of high stress concentration. Most crystals contain "grown-in" dislocations, usually in small-angle boundaries and networks, and some of the links in these networks are geometrically capable of acting as unlimited sources of glide dislocations. Figure 9.17 shows one such *Frank–Read*

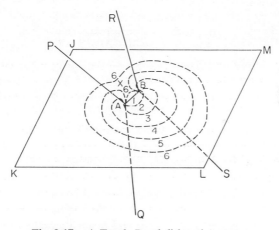

Fig. 9.17. A Frank–Read dislocation source.

source. Two dislocation lines PAQ and RBS intersect the slip plane JKLM at the *nodes* A and B where they are joined by the *source* dislocation line AB. Under a shear stress on the slip plane, AB tries to glide but is held back at its nodes, when PAQ and RBS are immobile, and so passes through the sequence of configurations numbered 1 to 6. Somewhere between configurations 5 and 6, as the opposite arms sweep round behind the nodes, they meet and coalesce in the region X and there form a complete dislocation ring together with a new source dislocation (configuration 6). The ring expands outwards across the plane and the source dislocation moves forward into position 1 to repeat the whole process once more. So long as the Frank–Read source retains its form, this repetition can continue indefinitely, so producing intense slip on its plane.

A minimum stress σ_{FR} is needed to bend the source dislocation into a semicircular shape (configuration 3) before it can expand into the slip plane. Let l be the distance between the nodes. The elastic strain field of the source dislocation cannot spread much beyond a distance of order l from the nodes (cf. the argument leading to Equation (5.87)), and thus the energy per unit length of the source dislocation must be of order

$$U \simeq \frac{\mu b^2}{4\pi} \ln \left(\frac{l}{r_0} \right)$$
(9.10)

in the notation of Section 5.11. By analogy with surface tension and energy, U is equivalent to a *line tension* T along the dislocation line which tries to shorten it. By analogy with Fig. 5.3(b), we balance this tension against the force $\sigma_{FR}bl$ pushing the source dislocation forward, and obtain the minimum stress as

$$\sigma_{FR} = \frac{2T}{bl} \simeq \frac{\mu b}{2\pi l} \ln \left(\frac{l}{r_0} \right) \simeq \frac{\mu b}{l},$$
(9.11)

the simplified form $\mu b/l$ being limited to the fairly common case where $10^2 < (l/r_0) < 10^4$. Thus, stresses of order $10^{-4}\,\mu$ are required to operate Frank–Read sources in crystals where the dislocation spacing is about $10^4 b$.

Active Frank–Read sources have occasionally been seen in crystals, but glide dislocations are often formed in other ways. Grain boundaries sometimes provide glide dislocations; as also do foreign inclusions, particularly when these, through thermal shrinkage, are centres of high thermal stress. Where cross-slip occurs easily, dislocations often multiply by a variant of the Frank–Read mechanism which derives from the configuration of Fig. 9.16. If the jog J is large, the dipole dislocations can glide past each other and expand into their slip planes, as shown in Fig. 9.18. By repeated

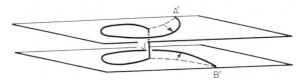

Fig. 9.18. Slip sources operating from the configuration of Fig. 9.16.

cross-slip from one primary slip plane to another, a thick *band* of active planes is built up. Slip bands of this type are commonly observed in ionic crystals and some b.c.c. transition metals.

When an intrinsically hard crystal is strained quickly, the stress rises high at first, because the few glide dislocations initially present cannot produce the applied strain rate except by moving quickly. Thus, if $\rho = 10^3$ cm^{-2} (e.g. well-prepared Si or Ge crystals) and $\dot{\gamma} = 3$ sec^{-1}, a dislocation speed of 10^5 cm sec^{-1} would be necessary. As the dislocations multiply, however, they no longer need move so quickly and their glide stress then falls, giving a *yield drop* such as that of Fig. 9.19(a). Such

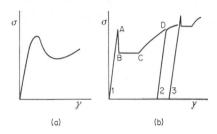

(a)   (b)

Fig. 9.19. Yield drops: (a) due to increase of dislocations in an intrinsically hard crystal; (b) due to impurity-pinning of dislocations.

yield drops are often seen in non-metallic crystals. According to Equation (9.9), they should appear mainly when n is small and should become insignificant in intrinsically soft crystals where the glide stress is insensitive to dislocation speed.

Large, sharp yield drops can occur, even in intrinsically soft crystals, when there are *no* free glide dislocations initially. The yield drop in Fig. 9.8 is evidently due to the *creation* of glide dislocations in the whisker. Almost as effective as removing the dislocations before straining, is to immobilize any which exist initially in the crystal. One way to do this is to *pin* the dislocations by allowing foreign atoms to segregate to them in an *ageing* process. This happens in mild steel, and many of the unique mechanical properties of this metal are due to interstitial carbon and nitrogen atoms which migrate to dislocations because the interstitial sites there have more room for them. When mildly segregated, foreign atoms form "atmospheres" or "adsorbed" lines of impurity along the dislocations; when strongly segregated, they usually precipitate as definite crystalline particles (e.g. carbide or nitride) at points along them.

Figure 9.19(b) shows a typical stress–strain curve obtained from mild steel. The stress drops sharply from the *upper yield point* A to the *lower yield point* B, where the yield elongation BC occurs and a *Lüders band* (cf. Fig. 6.14) runs through the material at roughly constant stress. After this, the "normal" part of the plastic stress–strain curve, CD, begins. If the plastically strained specimen is unloaded and immediately reloaded while in the *overstrained* state, as in curve 2, no yield drop is seen, because the free glide dislocations in it can simply resume their motion; but if the specimen is aged while unloaded, to allow segregating atoms to pin these dislocations, the yield drop reappears on reloading, as in curve 3. In mild steel this *strain ageing* usually takes a few days at room temperature or a few minutes at 100°C, its rate being governed by the interstitial diffusion of carbon and nitrogen in b.c.c. iron. On sheet steel pressings the Lüders bands produce unsightly markings called *stretcher strains*; these can be eliminated by lightly rolling the sheet plastically just before pressing.

The high stress ($\simeq E/50$) needed to pull a fully aged dislocation away from its pinning atoms is rarely detected in practice, because yielding usually starts prematurely in regions of stress concentration. The observed upper yield stress is usually the stress at which an embryonic Lüders band breaks out from the region of stress concentration where it began. Within a single crystal or grain, mobile dislocations can multiply by the cross-slip process of Fig. 9.18, but in a polycrystal they are confined by the grain boundary to their own grain. Figure 9.20 shows a grain boundary blocking

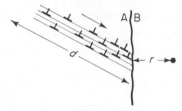

Fig. 9.20. Dislocations in a yielded grain A held up at the boundary of an unyielded grain B.

a *pile-up* of dislocations in a slip band at a Lüders front. We can regard this band as similar to a *shear crack* (cf. Section 5.11) on which the stress has locally dropped from its applied value σ_y to some lower value σ_i which represents the "frictional" resistance encountered by a mobile glide dislocation due to the various obstacles in its way. By analogy with stress concentrations at cracks (cf. Equation (5.92)), the stress at a distance r ahead of the pile-up is of order $(\sigma_y - \sigma_i)(d/r)^{1/2}$. Suppose that the next grain yields when this stress concentration reaches a certain value σ_d. Writing $k_y = \sigma_d r^{1/2}$, we then obtain the *Petch equation*

$$\sigma_y = \sigma_i + k_y d^{-1/2} \tag{9.12}$$

for the *lower yield stress* σ_y, where d is generally the grain size.

This equation, which shows that fine-grained metal is stronger than coarse-grained, has been confirmed on mild steel and other metals that deform similarly. The interpretation of k_y depends on the degree of strain ageing. The next grain can yield either by creating dislocations from the boundary (large σ_d, small r) or by releasing pinned dislocations within itself (smaller σ_d, large r). The preferred process is that with the smaller $\sigma_d r^{\frac{1}{2}}$. If the pinning is weak, as in lightly aged material, the yield spreads by unpinning; if strong, as in most annealed steel, by grain boundary creation. This affects the temperature-dependence of the yield stress, because unpinning from a line of adsorbed atoms is highly sensitive to temperature, whereas grain boundary creation is not. Measured values of k_y, in fact, vary as shown in Fig. 9.21. Heavily aged specimens give the

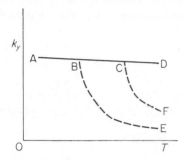

Fig. 9.21. Variation of k_y with temperature.

almost horizontal line AD. Their temperature-dependence of yield stress σ_y is contributed almost entirely by the friction stress σ_i (cf. Section 9.3). Mildly and very lightly aged samples are represented by curves ACF and ABE, respectively, in which the steep high-temperature parts CF and BE represent unpinning and the low-temperature parts AC and AB represent grain boundary creation.

If strain ageing develops into precipitation along the dislocations, as in Fig. 9.22(b), the unpinning process itself becomes insensitive to tempera-

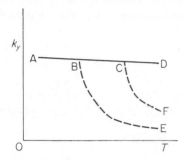

(a)   (b)

Fig. 9.22. (a) Thermally activated unpinning of a dislocation line from adsorbed atoms; (b) bowing of a line between pinning precipitates.

ture. The dislocation can then break free from the precipitates by bowing between them at a stress of about $\mu b/l$ (cf. Equation (9.11)). If *overageing* occurs and the precipitates become coarse, the increase in l reduces the yield stress. This effect is observed in some non-ferrous metals, in which dislocations are pinned by precipitates.

9.6 THE PLASTIC STRESS–STRAIN CURVE

Except for the effect shown in Fig. 9.19(a), plastic deformation and multiplication of glide dislocations, particularly in soft metals such as pure copper, *increase* the glide stress. This *work-hardening* can be very striking. A 1 in. diameter single-crystal copper bar, for example, can easily be bent to a horseshoe shape between the hands, but not straightened out again. Figure 9.23 shows the usual shape of (shear) stress–strain curve obtained from an f.c.c. metal crystal. Stage I (also called *easy glide*) of the plastic range is entered at the initial yield stress. In this stage the

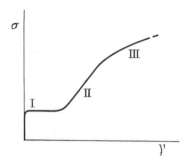

Fig. 9.23.   Typical stress–strain curve of an f.c.c. metal crystal.

crystal glides to a strain usually 0·01–0·1, on a single set of parallel slip planes, and there is little work-hardening, because the glide dislocations mostly escape through the free surface.

Eventually the deformation becomes "turbulent", and strong work-hardening (stage II, also called *linear hardening*) sets in as the dislocations moving on *intersecting* systems mutually entangle one another. The electron microscope shows dense tangles of dislocation lines, i.e. extremely irregular three-dimensional networks, formed at this stage. The dislocations can interact in various ways. One may coalesce with another to form a sessile, as in Fig. 9.14. Screw dislocations may be held up by the jogs formed where they have cut others. The cutting process itself is difficult. Wide stacking faults obstruct dislocations on other planes trying to pass through them. The stress fields of immobilized dislocations obstruct mobile dislocations. The importance of these *intersection* effects is shown by the fact that f.c.c. metal crystals, oriented to glide from the start on two or more intersecting systems, usually start stage II immediately at the initial yield stress, omitting stage I altogether; conversely, in c.p. hex. crystals such as zinc and cadmium, where there is little glide except on the basal plane, stage I continues to very large strains (e.g. 1–10).

At low temperatures and stresses both edge and extended screw dislocations can move only in their primary slip planes. This geometrical constraint helps stabilize the dislocation tangles, since a dislocation cannot escape round obstacles by changing its direction of motion. Stage II hardening at low temperatures thus continues to high stresses (e.g. $10^{-2}\mu$). At higher temperatures the onset of cross-slip at a certain stress causes stage II to give way to stage III (or *parabolic*) hardening. Here the slip lines become forked and interlaced with short cross-slip lines. Work-hardening continues, but at a smaller rate, mainly because screw dislocations can cross-slip from one plane to another and so bypass obstacles and annihilate dislocations of opposite signs in neighbouring planes. Thus, as Fig. 9.24 shows, metals with narrow dislocations tend to have lower work-hardened strengths at room temperature than those with wide dislocations. High work-hardened strengths can be obtained in metals and alloys whose dislocations can be widely extended to form stacking faults over large areas of slip plane. The strong work-hardening abilities of certain brasses, bronzes and austenitic steels are, in fact, exploited industrially. In some austenitic steels work-hardening is further intensified by a partial *strain*

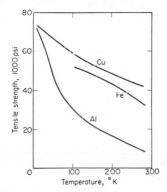

Fig. 9.24. Effect of temperature on the tensile strengths of pure metals.

transformation of the crystal structure to b.c.c. The exceptional resistance to abrasion of *Hadfield's steel* (12 % Mn) seems to be due to this effect.

The limit to which a metal can be hardened by working is determined by the ability of its densely tangled dislocations to *polygonize* themselves into small-angle boundaries, which depends on the temperature of deformation. At temperatures where *climb* (cf. Section 3.4) can occur, even edge dislocations can move out of their glide surfaces. Polygonization can then soften the material by producing relatively undistorted sub-grains. Extreme cold-working at room temperature can give tensile strengths of about $E/100$ in pure metals of high melting point, such as tungsten and molybdenum; about $E/200$ in iron, nickel and copper; and about $E/300$ in aluminium.

If sub-grains of widely different orientations form in the worked crystal, some of their boundaries may become large-angle boundaries which can *migrate* by the process discussed in Section 8.9, so *recrystallizing* the material by sweeping up the dislocations of the highly worked regions ahead of them and producing relatively undistorted crystal behind them. This *primary recrystallization*, which commonly occurs at $0.4–0.5T_m$ (°K) in ordinary metals and alloys, is complete when all the worked material has been eliminated; but it may be followed by *grain growth* or even, at high temperatures, by a second wave of recrystallization, called *secondary recrystallization*, in which the primary recrystallized grains are replaced by another set of grains.

Cold-working is mechanical working done at temperatures too low for recrystallization. If the working temperature is sufficiently high, recrystallization may occur while working is still being done; this is *hot-working*. Under these conditions the material remains soft, because the work-hardened regions are continually replaced by soft, recrystallized grains, and the tensile strength usually remains well below $10^{-3}E$. Pure lead and tin owe their familiar softness to the fact that room temperature is a hot-working temperature for them.

Polycrystals (and single crystals which contain "gritty" inclusions round which the deformation is turbulent) do not usually show a stage I in their stress–strain curves, because intersecting glide and work-hardening begin immediately the plastic range is entered. Polycrystalline metals and alloys are generally examined mechanically by *tensile tests*. The simplest form of tensile specimen is a round bar with enlarged ends, for gripping, which taper smoothly through shoulders to a central uniform *gauge length*, on which the measurements are made. Figure 9.25 shows such a specimen and also a typical stress–strain curve, in which the *nominal strain* or *tensile elongation*,

$$\varepsilon = \frac{l_1 - l_0}{l_0}, \qquad (9.13)$$

where l_0 and l_1 are the initial and final gauge lengths, is plotted against the *nominal stress*

$$\sigma = \frac{F}{A_0}, \qquad (9.14)$$

i.e. the tensile load F divided by the *initial* cross-sectional area A_0. In materials which do not show a yield drop the initial yield stress Y is ill-defined. It is often then measured as a *proof stress*, the stress at which a *permanent set* of, typically, 0·1 % is first produced. Other measures are the *elastic limit*, the lowest stress at which a detectable permanent set is produced, and the *limit of proportionality*, the lowest stress at which deviation from Hooke's law can be detected during loading. From Y to U the rate of work-hardening is sufficiently high to keep the deformation *stable*, i.e. if one part of the gauge length strains more than the rest, it hardens so much that the next increment of strain occurs elsewhere. The point U is

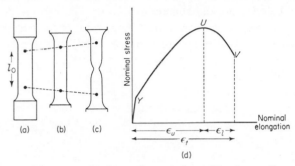

Fig. 9.25. The tensile test: (a) form of specimen; (b) after uniform elongation; (c) after further, non-uniform, elongation; (d) nominal stress–strain curve.

the *tensile strength* or *ultimate tensile stress* of the material. It is technologically important because, multiplied by the initial area, it gives the highest load a tensile member can support without breaking. The material still continues to work-harden beyond U, but at a rate too small to compensate for the loss of cross-sectional area. The result is that the *load* and *nominal* stress supported by the test piece decrease beyond U. The plastic deformation becomes unstable and a *neck* forms in the gauge length, as shown in diagram (c). Because of the fall in load, the gauge length outside the necked region stops deforming; but the necked region itself continues, until fracture occurs at V, because the *true* stress on it continues to rise. The *total* elongation ε_t of the gauge length thus occurs in two stages, *uniform* (ε_u) and *localized* (ε_l), respectively, before and after the point of instability. The localized elongation has little meaning, since it does not give the actual strain in the neck. A measure of the latter is the *reduction of area*, $(A_0 - A_1)/A_0$, where A_0 and A_1 are the initial and final areas of the narrowest section of the neck.

This necking instability occurs in tensile tests. In a *compression* test, usually made by compressing the flat ends of a short cylindrical specimen between parallel compression plates, the cross-sectional area *increases* as the specimen is squashed down and the *nominal* stress–strain curve rises throughout. Very late in the test, when the material has practically exhausted its capacity for work-hardening, a *dynamic instability* may set in through an effect called *adiabatic softening*. If the material begins to slide along an inclined *shear zone*, as in Fig. 9.26, this zone may become so hot, through the conversion of work into heat, that it *softens* and continues to deform even more. The faster the deformation in the zone, the less chance there is for the heat to leak away and the better the opportunity for the effect. Thus, less work is consumed and less distortion produced in punching a hole through a metal plate, if the punch is shot through at high speed than if slowly pressed through.

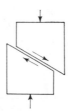

Fig. 9.26. Localized shear in a compression test.

The nominal stress–strain curve gives a distorted record of the plastic properties of a ductile material, because of the gross changes in gauge length and cross-sectional area at large strains. To overcome this, *true stress–strain* curves are often plotted. The *true stress* at any stage in the test is the load F divided by the cross-sectional area A at that stage. If necking has begun, A is measured across the neck. The *true strain* is derived as follows. Consider a stage at which the gauge length alters infinitesimally from l to $l + dl$. The corresponding increment of true strain is then defined as dl/l, i.e. by taking l, not l_0, as the gauge length. The total true strain in a deformation from l_0 to l_1 is then

$$\varepsilon = \int_{l_0}^{l_1} \frac{dl}{l} = \ln\left(\frac{l_1}{l_0}\right). \tag{9.15}$$

This is positive for extension ($l_1 > l_0$) and negative for compression ($l_1 < l_0$). Small nominal and true strains are almost equal, since $\ln(1 + x) \simeq x$ when x is small, but, as Table 9.1 shows, large ones differ greatly.

TABLE 9.1
True and Nominal Strains

True strain:	0·01	0·1	0·2	0·5	1·0	2·0	3·0	4·0
Nominal extension:	0·01	0·105	0·22	0·65	1·72	6·39	19·1	53·6
Nominal compression:	0·01	0·095	0·18	0·39	0·63	0·86	0·95	0·98

True strain provides a more consistent measure of plastic deformation than nominal strain. We expect an elongation from l_0 to l_1 to be physically equivalent to a compression from l_1 to l_0, apart from the change of sign, and in fact these are numerically equal when measured as true strains, but when measured as nominal strains they are $(l_1 - l_0)/l_0$ and $(l_1 - l_0)/l_1$, respectively. When plotted as true stress–strain curves, the tension and compression curves of a ductile metal generally coincide, as shown in Fig. 9.27. This coincidence shows that the hydrostatic component of the

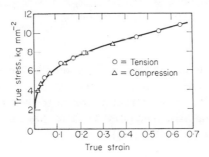

Fig. 9.27. True stress–strain curve of aluminium, determined from tension and compression tests.

stress state has little effect on the plastic properties, as is to be expected from the fact that plastic gliding is essentially a shear at constant volume. True stress–strain curves often follow the empirical relation

$$\sigma = C\varepsilon^n, \tag{9.16}$$

where C is a constant for the material and n is the *strain-hardening exponent*, usually between 0·1 and 0·3.

At the point of tensile instability an increment of strain gives no increase in the load supported by the specimen, i.e. $dF = 0$. Thus, the relation

$$dF = d(\sigma A) = A\,d\sigma + \sigma\,dA = 0 \tag{9.17}$$

defines the point of instability, where σ and A are true stress and cross-sectional area, respectively. Because plastic glide occurs essentially at constant volume, i.e. $dV = 0$, we also have

$$dV = d(lA) = A\,dl + l\,dA = 0. \tag{9.18}$$

Combining these, we obtain

$$\frac{d\sigma}{\sigma} = -\frac{dA}{A} = \frac{dl}{l} = d\varepsilon = \frac{d\varepsilon_n}{1 + \varepsilon_n}, \tag{9.19}$$

$$\left(= \frac{dl}{l_0 + l - l_0} = \frac{\frac{dl}{l_0}}{1 + \frac{l - l_0}{l_0}} = \frac{d\varepsilon_n}{1 + \varepsilon_n} \right)$$

where ε and ε_n are true and nominal strains, respectively. The _tensile strength_ is thus reached when

$$\frac{d\sigma}{d\varepsilon} = \sigma, \quad \text{or} \quad \frac{d\sigma}{d\varepsilon_n} = \frac{\sigma}{1 + \varepsilon_n}. \tag{9.20}$$

The first of these enables us to locate the instability on the true stress–strain curve. For example, when Equation (9.16) is valid, we have

$$C\varepsilon^n = \sigma = \frac{d\sigma}{d\varepsilon} = nC\varepsilon^{n-1}, \tag{9.21}$$

i.e.

$$\varepsilon = n \tag{9.22}$$

at the instability. The second form of Equation (9.20) enables us to find the instability by _Considère's construction_. We plot true stress against nominal strain, as in Fig. 9.28, and construct a tangent to the curve from the point $\varepsilon_n = -1$ on the strain axis. The point P where they touch is the instability and the tensile strength is $\sigma/(1 + \varepsilon_n)$.

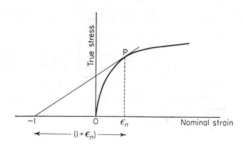

Fig. 9.28. Considère's construction.

If the initial yield stress of the material has been sufficiently raised by previous work-hardening or by other means, or if the test is performed at high temperatures where there is little work-hardening, P can coincide with the initial yield stress. We then have $Y = U$ and $\varepsilon_u = 0$ in Fig. 9.25. The tensile ductility of the material then depends entirely on the localized elongation. If the specimen consists of a long thin wire (length l, diameter d), this appears as a very small strain, even though the material may be _fully ductile_, because the extension in the neck cannot exceed the order of magnitude of the diameter of the wire and this gives an overall plastic elongation only of order d/l before the wire breaks. For example, when $(d/l) < (Y/E)$, the overall plastic strain of the wire is less than the elastic strain, even when there is 100% reduction of area in the neck. Provided U is higher than Y, however, the contribution from ε_u usually makes the plastic elongation orders of magnitude larger than the elastic elongation. The ratio of tensile to yield strength is thus important in engineering design, particularly where long thin wire or sheet is used.

The occurrence of tensile necking also explains why it is not possible to make metal wires by simply pulling their ends apart, as is possible with hot glass (cf. Section 7.1). Wire-drawing is, in fact, done by pulling the metal through a die, as in Fig. 9.29. The plastic resistance of the metal is overcome by the large forces exerted on it by the die, although of course it is

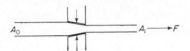

Fig. 9.29. Wire-drawing through a die.

the pulling force F that does the mechanical work. This pulling force can be made small, because it undergoes a large displacement as the metal comes through the die. In drawing a volume V of metal from area A_0 to area A_1, it does the work $F(V/A_1)$. We equate this to the plastic work done inside the metal against its tensile yield strength Y (assumed constant), which is

$$\int_{l_0}^{l_1} YA \, dl = YV \int_{l_0}^{l_1} \frac{dl}{l} = YV \ln\left(\frac{l_1}{l_0}\right) = YV \ln\left(\frac{A_0}{A_1}\right), \qquad (9.23)$$

and so obtain

$$\frac{F}{A_1} = Y \ln\left(\frac{A_0}{A_1}\right). \qquad (9.24)$$

Thus, provided $\ln (A_0/A_1) < 1$, i.e. $A_0 < 2 \cdot 7A_1$, the tensile stress F/A_1 applied to the drawn wire is less than the tensile yield stress. This makes wire-drawing possible. In practice the limiting value of A_0/A_1 is affected by work-hardening of the wire as it passes through the die, and by friction between the metal and the die.

9.7 ALLOYS

There are many different ways in which the mechanical properties of metals are altered by alloying, and one of the tasks of the metallurgist is to exploit them to improve the strength and ductility of metals. Some of them are indirect. Thus, metals such as nickel, chromium, manganese and molybdenum are added to steel in small amounts to improve its quench-hardening and tempering characteristics, and in large amounts to produce austenitic steel; a trace of aluminium is added to steel, and of zirconium to magnesium, to refine their grain size; many alloy additions are made (e.g. manganese in sulphur-bearing steel; niobium and titanium in austenitic steel; phosphorus in copper) to counter the embrittling effects of other elements, particularly at grain boundaries; zinc, tin and other solid-solution alloy elements, added to copper produce brasses and bronzes with great capacity for mechanical working; many alloy additions (e.g. calcium and antimony in lead; arsenic and nickel in copper) raise the recrystallization temperature and so maintain work-hardening to higher temperatures.

Alloy atoms in solid solution can interact with glide dislocations through their elastic strain fields and in other ways. *Random* solid solutions provide only mild hardening, however, because on average a glide dislocation is pulled equally to and fro by the solute atoms on opposite sides of it. Greater hardening is possible in *non-random* solutions. Solute atoms often arrange themselves in special places in crystals to reduce the free energy of the material, e.g. by reducing elastic energy or by increasing the number of strong atomic bonds. If the movement of dislocations disturbs or destroys these special distributions, this free energy has to be returned to the crystal and the applied stress has to be increased to do the necessary extra work. Since these distributions are produced by diffusion processes in the solid solution, the hardening due to them gradually develops as the alloy is rested or *aged* at a suitable temperature. The three main examples of this type of hardening are *segregation to dislocations*, already discussed in Section 9.5; *clustering*, in which the solute atoms gather together in small groups; and *ordering*, in which the solute atoms arrange themselves in a regular pattern through the crystal. The work done by gliding dislocations in breaking up clusters and ordered regions leads to increased glide stress.

The outstanding example of alloy hardening is *martensite*, the hard constituent of quench-hardened carbon steel. The carbon is first taken into solid solution, up to about 4 atomic per cent in a typical high-carbon steel, by heating the metal at temperatures (e.g. 900°C) where it is austenitic. It is then held in solution by *quenching* to room temperature, rapidly in water if a plain carbon steel, more slowly in oil or air if an alloy steel. The crystallographic transformation to the b.c.c. structure occurs during the quench, but, in the short time available, the carbon is forced to remain in *supersaturated* interstitial solution and it distorts the b.c.c. crystal cell into a *body-centred tetragonal* shape, the crystal structure of martensite. The carbon atoms are ordered, in the sense of occupying only those interstitial sites with a common tetragonal axis. In addition to carbon solution hardening, martensite owes some strength to crystallographic effects of the transformation, in particular to finely distributed twin interfaces formed in martensite crystals. Tensile yield strengths of 300,000 psi ($\simeq 10^{-2}E$) or more can be produced in steel by martensitic transformation.

If *hardness*, i.e. resistance to dislocation motion, were the only property that mattered for mechanical strength, we could simply use intermetallic compounds, oxides, carbides and other *intrinsically hard* solids for engineering materials. The trouble of course is that they are *brittle*. Because of this, they are weak against fracture, for the reasons discussed in Chapter 11, even though strong against yielding. The best combination of strength and ductility is obtained from an intrinsically soft metal stiffened by finely dispersed particles of hard substances. Since the ductile metal is the *geometrically continuous* constituent of this microstructure (provided the grain boundaries are not lined with a brittle constituent), its properties determine the fracture behaviour.

Alloys of this type are said to be *precipitation-hardened* or *dispersion-hardened*. Aluminium alloys, for example, are hardened by first quenching from about 500°C, to retain copper and other elements in supersaturated solid solution, and then ageing at about 150°C, a process in which the copper atoms, for example, first cluster together in the aluminium grains and then precipitate as fine particles of an intermetallic compound. Tensile strengths approaching $10^{-2}E$ can be obtained in this way. Quench-hardened steel is usually *tempered* at 100–300°C to precipitate the carbon as fine carbide particles, which reduces brittleness without greatly reducing hardness.

If the clusters or precipitates are dispersed too finely, the glide dislocations can cut right through them. This process is helped by thermal activation, because mn is small (cf. Equation (9.8)); and, in fact, the yield strengths of lightly aged alloys are fairly sensitive to temperature. Greater hardness is gained by increasing the ageing treatment to produce larger precipitates (though fewer in number, many having redissolved to feed the larger ones). There is a limit to this hardening, however, when the spacing between particles becomes large enough for the dislocations to pass between the particles. This process, shown in Fig. 9.30, is closely related to the bowing processes of Figs. 9.17 and 9.22(b). The glide stress, in fact, is given by

$$\sigma \simeq \frac{\mu b}{l}, \qquad (9.25)$$

where l is the spacing between particles; this is *Orowan's formula*. For most alloys the maximum hardening is obtained when $l \simeq 100$ Å. Softening becomes pronounced when *overageing* increases l much beyond this.

We notice that, when a glide dislocation bypasses obstacles, it leaves closed loops of dislocation line round them. This leads to the *Fisher, Hart and Pry (FHP) effect.* As successive dislocation rings form round the particles, they produce a stress in the particles much higher than that on the slip plane as a whole. This effect can be understood by regarding the dislocations simply as agents through which the plastically deforming metal transfers all applied load beyond its own strength on to the particles. The particles, in fact, act almost as rigid *pegs* across the slip plane. Initially they are no more highly stressed than the remainder of the plane,

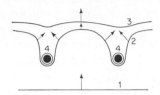

Fig. 9.30. A dislocation line which (1) approaches, (2) bypasses, dispersed obstacles and then, (3) continues its motion, leaving behind closed dislocation loops (4) round the obstacles.

but later, as the plane slips and they do not, all the excess load associated with the rising applied stress is carried by them. The pegged slip plane is like a bolted joint, the faces of which could slide if the bolts were not carrying the load.

The FHP effect thus allows the intrinsic strength of the dispersed particles to contribute directly to the overall strength. These particles, being very fine, usually have ideal strengths ($\simeq E/10$) and so are capable of contributing strongly. This contribution does not appear immediately, however, but only when the alloy is plastically deformed to a strain of 0·01–0·1; it thus gives an abnormally high rate of work-hardening during the first few per cent of plastic strain. An overaged aluminium–copper alloy, for example, typically has an initial yield stress of about 15,000 psi but hardens to about 35,000 psi in the first 5% strain. As a second example, Fig. 9.31 shows the effect of increasing content of iron carbide, Fe_3C, on the stress–strain curve of annealed plain carbon steel. The effect is large, partly because a little carbon produces a lot of carbide, e.g. 0·8 wt. % C converts 10% of the Fe atoms to Fe_3C, and partly because the carbide is mostly distributed in a lamellar *eutectoid* (similar to a *eutectic* but produced from a solid solution instead of a liquid solution) called *pearlite*; 0·8 wt. % C produces 100% pearlite. The high stresses carried by carbide particles in worked steel have been confirmed by X-ray diffraction measurements of the lattice constants.

The limit to the FHP effect is set by the strength of the particles and their immediate surroundings. If the particles are soft or weak, they may yield or break under the pressure of the dislocation loops on them and so allow these loops to pass through them. If the metal nearby is soft, it may flow plastically round the particles, as a fluid flows round obstacles. It is important therefore, in utilizing the effect, to make the particles fine, so that both they and the metal immediately adjacent to them have almost ideal strength. An overall strength of order $10^{-2}E$ can then be obtained with 10% by volume of particles. An example is strong carbon-steel wire

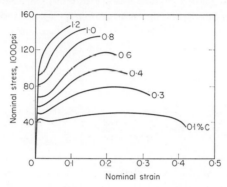

Fig. 9.31. Effect of carbon content (weight per cent) on the stress–strain curve of annealed steel.

($\simeq 0\cdot4$ wt $\%$ C) made by cold-drawing a microstructure in which the carbide is finely dispersed.

An alternative to a fine dispersion is a microstructure in which the hard constituent is arranged in long fibres. The softness of the metal is then less important for the overall strength, since the metal in this case serves merely as a binder or "glue", holding the fibres together (cf. Section 11.2). One is no longer "strengthening" the metal in this case but directly using the high strengths of the fibres.

9.8 CREEP

It might be thought from the plastic stress–strain curve that the same stress always produces the same strain, no matter how long applied. This is never strictly true, and not even approximately true at high temperatures. Plastic glide is a form of *flow*, like fluid flow except that the atomic movements are crystallographically organized, and can occur, as *creep*, at constant stress. Closely related to creep is *stress relaxation*, the slow replacement of elastic strain by plastic strain at constant total strain, which, for example, causes high-temperature bolts to lose their grip.

The slowness of creep, compared with the high strain rates possible from dislocations, shows that it is a thermally activated process; and, in fact, creep rates at a given stress usually double or treble for each 10°C rise in temperature. The amount of creep produced depends on the experimental conditions. At high temperatures and high stresses creep strains exceeding 0·1 are readily produced, but in tests of *creep-resistant alloys* the temperature and stress are usually chosen to give a strain of order 0·01 in 10^3–10^5 hr; the applied stress in this case is usually well below the short-time yield stress. Figure 9.32 shows typical strain–time curves obtained at constant stress (or in less precise tests at constant *load*). The standard curve shows

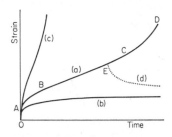

Fig. 9.32. Creep curves: (a) standard type; (b) at low temperatures and stresses; (c) at high temperatures and stresses; (d) creep recovery after unloading at E.

an immediate strain OA, elastic and plastic, when the stress is applied, then a *primary* creep AB, a *secondary* creep BC, and a *tertiary* creep CD, leading to fracture at D. At low temperatures and stresses only primary creep occurs; and at high temperatures and stresses tertiary creep predominates. In tensile tests at constant load the rise of stress due to loss of cross-sectional area contributes to tertiary creep. More generally, however, the acceleration of creep towards the end of the creep life is due to the growth of cracks in the material, particularly along grain boundaries (cf. Section 8.9). At high temperatures, under stresses too low to deform the grains appreciably, a polycrystal may slowly break along its boundaries at a creep strain of only about 0·01, even though very ductile in short-time tests.

Leaving aside tertiary creep, the strain (ε)–time (t) curve often fits one of the two empirical formulae

$$\varepsilon = \varepsilon_0 + \alpha \ln t + \kappa t, \tag{9.26}$$

$$\varepsilon = \varepsilon_0 + \beta t^{1/3} + \kappa t, \tag{9.27}$$

where ε_0 is the immediate strain, α, β and κ are *creep coefficients*, κt is *steady-state* creep and $\alpha \ln t$ and $\beta t^{1/3}$ are two forms of *transient* creep, respectively called *logarithmic* and *Andrade* creep. Primary creep is usually logarithmic at low stresses and of Andrade type at high stresses.

The decrease of creep rate during the primary or transient phase can be interpreted in terms of work-hardening. For example, if work-hardening contributes to σ_i, in Equation (9.8), then Q increases with increasing ε, and the rate at which the dislocations cross their obstacles decreases. It is not easy to explain the precise shape of the transient creep curve, because the relation between σ_i and ε itself depends on strain rate; the elementary processes producing the creep strain may be the same as those producing thermal softening. *Mott* has shown that, if flow at a point in the material depends on a favourable fluctuation in the internal stress at that point, caused by flow occurring nearby, then Andrade creep is to be expected.

The simplest interpretation of steady-state creep is that the properties of the material remain constant during this stage, i.e. the work-hardening is balanced by thermal softening. Let $h = \partial\sigma_i/\partial\varepsilon$ be the *coefficient of work-hardening*, measured from the slope of a stress–strain curve in the absence of thermal softening, and let $r = -\partial\sigma_i/\partial t$ be the *rate of softening*, measured from the fall of yield stress with time when work-hardened material is annealed. Then, for steady-state creep,

$$d\sigma_i = \frac{\partial\sigma_i}{\partial t}\,dt + \frac{\partial\sigma_i}{\partial\varepsilon}\,d\varepsilon = 0, \qquad (9.28)$$

so that

$$\dot{\varepsilon} = \kappa = \frac{r}{h}. \qquad (9.29)$$

A difficulty in the quantitative application of this formula is that r may be affected by the applied stress. A confirmed result is that, if the applied stress is suddenly reduced slightly during steady-state creep, a period of induction follows, during which the creep rate is very small (while the material is softening to the new level of applied stress), before the new steady state is attained.

Two forms of thermal softening, which we may call *glide-softening* and *climb-softening*, can be recognized. Glide-softening can occur at low temperatures and its main cause is cross-slip (cf. Section 9.6), whereas climb-softening is caused by vacancy creep. Glide-softening is a process which occurs while the dislocations are gliding, and therefore mainly shows itself, not in slow creep, but in the shape of the short-time stress strain curve. Climb-softening, which enables edge dislocations to move out of their glide surfaces, to annihilate or polygonize with other dislocations, and to pass round obstacles, is pronounced at temperatures above about $0.5T_m$ (°K). The dislocations climb out of their glide surfaces by absorbing or emitting vacancies, the driving force coming from the local stress fields, e.g. from other dislocations, acting on them. The driving force is generally small enough for this vacancy creep to be Newtonian (cf. Section 7.3), occurring with the activation energy for self-diffusion. Because temperature, not stress, is responsible for vacancy diffusion, climb-softening can occur at stresses well below those for glide. Thus, unlike glide-softening, climb-softening in a stressed material allows large creep strains to develop gradually over long periods of time at stresses well below the short-time yield stress.

Even when controlled by Newtonian vacancy creep, the creep produced by climb-softening is generally non-Newtonian. This is because the number of climbing dislocations, the local stresses acting on them, the distance they have to climb, and the area swept by each of them, all depend on the dislocation density, and this in turn depends on the applied stress. The rate of steady-state creep in practice is fairly well represented by the *Weertman* formula

$$\dot{\varepsilon} = C\sigma^n \, e^{-E_D/RT}, \tag{9.30}$$

where E_D is the energy of self-diffusion (cf. Equation (7.15)), $n \simeq 4$, and the value of C depends on the microstructure. Other formulae have also been developed in which the activation energy for creep depends on the applied stress.

We now consider *creep-resistant* materials. The success of such a material depends of course on many other factors besides creep strength: for example, *cost* limits the large-scale use of many alloys; *oxidation* is a severe problem for molybdenum and graphite; *low-temperature brittleness* makes chromium, beryllium, and many oxides and ceramic materials mechanically unreliable (although of course ceramics are used in *fire-bricks*, to support small, mainly compressive, loads at very high temperatures under conditions where most other materials would be too chemically reactive); *weight* rules out tungsten and other heavy metals in applications where lightness is important; *instability of microstructure* is a problem in high-chromium steels, in which a brittle intermetallic compound called *sigma phase* tends to form along grain boundaries during service, and also in aluminium-bearing steels, in which *graphite* tends to form during service.

Where utmost stability of dimensions is essential, as in precision instruments, an intrinsically hard material such as sapphire at a temperature well below that for dislocation mobility is preferable. Impurity-pinning of dislocations can also suppress traces of creep at low temperatures, and this is important for preventing stress relaxation in steel tension bars used in prestressed concrete. There are two general methods for raising the temperature at which a material shows creep-resistance; by maintaining work-hardening properties to higher temperatures; and by raising the initial yield stress at high temperatures. One way to maintain work-hardening is to stabilize the dislocation tangles by pinning the dislocation lines with precipitates along them. Solid-solution alloys, slightly super-saturated, are used for this. To minimize transient creep, a solute which precipitates quickly at the creep temperature is required; but to ensure long-time stability, a solute which only slowly precipitates is necessary. A complex alloy which contains several solutes with different diffusion coefficients and which allows various precipitates to form at widely different rates may thus be necessary. Low-carbon steels have useful creep strengths up to about 450°C (provided the aluminium content is low), probably due to the pinning of dislocations by nitrogen. The addition of molybdenum ($\simeq 0.5\%$) and vanadium ($\simeq 0.25\%$) enables alloy carbides to precipitate on dislocations, and the creep range is then extended to about 550°C. Such steels are useful in steam turbines. The creep strength of some *austenitic* steels (containing 13–18% Cr for oxidation resistance and 8–14% Ni to stabilize the f.c.c. structure) is improved by the addition of niobium ($\simeq 1\%$), which precipitates on dislocations as niobium carbide. Similar strain-ageing effects appear to be responsible for the effect of small Ce, Th, Mn and Zr additions in improving the creep strength of magnesium, and for the effect of 0.05% Ag in improving the creep strength of copper used in electrical generator windings.

The alternative to stabilized work-hardening and dislocation-pinning is stabilized *dispersion-hardening*. Many creep-resistant alloys owe their strength to fine, uniform dispersions of stable particles in the grains. For the strength to exceed $10^{-3}E$, the particles need to be less than 10^{-5} cm apart. The problem is to maintain such fine microstructures at high temperatures. To prevent *coarsening* by re-solution and precipitation on larger particles, the solubility of the dispersed substance in the metal must be very low. For example, iron carbide is much more soluble than many alloy carbides, in iron, so that quench-hardened plain carbon steels rapidly lose their strength at temperatures above about 450°C, whereas many high-alloy steels retain their strengths to much higher temperatures. Aluminium oxide dispersions are very stable and are capable of giving exceptional high-temperature strength in some metals; they enable sintered aluminium, for example, to maintain good strength at temperatures up to $0.75T_m$ (°K). Thorium oxide acts similarly in nickel.

Ordinary precipitation-hardening alloys, used for low-temperature strength, are not generally suitable for high-temperature strength, because the alloy solubility on which their heat treatment depends leads to rapid coarsening at high temperatures, large particles growing at the expense of small ones through the general effect described in Section 8.7. Some ways of meeting this problem are: (1) to form a fine dispersion by methods other than solution and precipitation—for example, by sintering mixed powders of metal and insoluble particles together; (2) to use precipitates whose solubility is small except very near the melting point (e.g. some alloy carbides in steel); (3) to use precipitates whose crystal structure matches that of the parent metal so well that there is almost no interfacial energy to induce coarsening (e.g. Ni_3Al in nickel-base alloys). The combination of (2) and (3) has produced several creep-resistant alloys, particularly for gas turbines. Austenitic nickel–chromium steels, hardened by various alloy carbides, have good strengths up to about 800°C. For temperatures in the range 800–1000°C nickel-base alloys hardened by intermetallic compounds containing nickel, aluminium and titanium, and cobalt-base alloys hardened by complex alloy carbides have been developed. In all these about 10–20% chromium is added to give oxidation resistance.

FURTHER READING

NABARRO, F. R. N., *The Mathematical Theory of Stationary Dislocations*, *Advan. Phys.* (1952).

READ, W. T., *Dislocations in Crystals*, McGraw-Hill, New York, 1953.

COTTRELL, A. H., *Dislocations and Plastic Flow of Crystals*, Oxford University Press, London, 1953.

FRIEDEL, J., *Les Dislocations*, Gauthier-Villars, Paris, 1956.

SEEGER, A., "Theorie der Gitterfehlstellen", *Handbuch der Physik*, Vol. VII, p. 383, Springer-Verlag, Berlin, 1955.

FISHER, J. C. (ed.), *Dislocations and Mechanical Properties of Crystals*, Wiley, New York, 1957.

VAN BUEREN, H. G., *Imperfections in Crystals*, North-Holland, Amsterdam, 1960.

McLEAN, D., *Mechanical Properties of Metals*, Wiley, New York, 1962.

SMALLMAN, R. E., *Modern Physical Metallurgy*, Butterworths, London, 1962.

PROBLEMS

9.1 A 1 in. cube of copper is sheared at the rate of 1 in. min^{-1}. Estimate in order of magnitude the minimum number of unit dislocations in motion at a given instant.								[10]

9.2 In an aluminium–copper alloy the precipitates are 10^{-5} cm diameter and the average distance between the centres of neighbouring particles is 10^{-4} cm. Estimate the tensile yield stress.					[approximately 2000 psi]

9.3 A Frank–Read dislocation source is operated by an applied shear stress of magnitude 10^{-4} μ. If the limiting speed of a dislocation is 10^5 cm sec^{-1}, show that the source could nucleate a slip band containing 1000 dislocations in about 10^{-4} sec.

9.4 The energy stored in heavily cold-worked copper (i.e. the mechanical work not converted into heat) is about 5×10^8 erg cm^{-3}. Assuming this to belong to dislocations, what is the approximate density of dislocations? [10^{12} cm^{-2}]

9.5 Deduce a formula similar to that of Equation (9.2) for the force tending to make an edge dislocation climb due to a normal stress applied in the direction of its Burgers vector. By considering the "vapour pressure" of lattice vacancies in a crystal, show that a concentration c of vacancies at a temperature T exerts a "chemical stress" equal to $kT \ln (c/c_0)/b^3$ on an edge dislocation, tending to make it climb, where c_0 is the equilibrium concentration of vacancies.

9.6 Equation (9.2) can be used to give the force on a dislocation due to any stress field which applies a shear stress of intensity σ at the position of the dislocation. Show that this enables Equation (5.88) to be derived for the force between two parallel screw dislocations. Use this method to deduce the force in the glide direction exerted on an edge dislocation by a similar edge dislocation of the same sign, in a parallel but different glide plane, and to explain the tendency to form small-angle boundaries, described in Section 5.11.

9.7 The true stress–strain relation of a ductile material can be represented by $\sigma = 27\varepsilon^{1/2}$. What is the ultimate tensile stress? [15·4]

9.8 A frictionless uniaxial compression test of a copper cylinder gave the following results:

True stress:	2	5	8	12	16	19	20	20·7	21
Reduction in height, %:	0	2·5	5	10	20	30	40	50	60

Construct the true stress–strain curve, calculate the nominal strain at which necking begins in uniaxial tension, calculate the ultimate tensile stress, and construct the nominal tensile stress–strain curve as far as the ultimate tensile stress. [0·38; 13·3]

9.9 Under an applied shear stress σ a circular area of a slip plane inside a perfect crystal slips by one atomic spacing, so forming a circular dislocation line of radius r. Taking account of the elastic energy of the dislocation and of the work done by the applied stress, show that the total energy change is given approximately by

$$U = 2\pi r \left(\frac{\mu b^2}{4\pi} \right) \ln \left(\frac{r}{r_0} \right) - \pi r^2 \sigma b$$

in the notation of Section 5.11. Then show that the slipped region can expand by the action of the applied stress alone if its radius exceeds the value

$$r_c = \frac{\mu b}{4\pi\sigma} \left[\ln \left(\frac{r_c}{r_0} \right) + 1 \right].$$

What is the activation energy to produce slip by such a process if $\sigma = \mu/120$, $r_0 = 2b = 5 \times 10^{-8}$ cm, and $\mu b^3 = 5$ eV? [20 eV]

9.10 An edge dislocation in a close-packed plane in an f.c.c. crystal dissociates into two parallel partial dislocation lines, separated by a ribbon of stacking fault, as described in Section 9.3. Show that the equilibrium spacing of these partial dislocations is approximately $\mu a^2/24\pi\gamma$, where μ is the shear modulus, a is the edge length of the f.c.c. cubic cell, and γ is the energy per unit area of the stacking fault. Show that a shear stress of suitable sign and suitable direction acting on this plane can separate the partial dislocations completely if its magnitude exceeds about $5\gamma/a$.

chapter 10

Plasticity

We now consider how the plastic properties of solids determine the overall deformation of bodies under various stresses. At first we consider only frames and beams. These are important in engineering and show some of the principles of plasticity rather simply. Then we go on to general plastic stress–strain relations and to various applications, including pressures exerted by soils, slip-line fields, plastic rupture, notch effects, hardness, adhesion and friction. Finally, we look at a few simple problems of deformation by creep.

10.1 PLASTIC COLLAPSE

When designing engineering structures with ductile materials, it is important to know the *collapse load* at which such a structure first suffers large-scale plastic distortion. For a plain tensile bar the collapse load is simply YA (Y = uniaxial yield stress, A = cross-sectional area). In more complex structures, however, such as steel frameworks, some parts may be more highly stressed than others and may yield first. The *elastic load* is the highest load at which no plastic yielding occurs anywhere. In a complex structure this may be very small, e.g. only about $YA(\rho/c)^{1/2}$ for localized yielding at the tip of a notch of radius ρ and depth c. However, localized yielding—sometimes even of a whole member in a framework—does not collapse the structure so long as the yielded part is fully *contained* by other elastic parts round it, which restrict its plastic deformation to elastic orders of magnitude. For *plastic collapse* yielding must become sufficiently widespread to permit large displacements of the loading points.

Consider the redundant, statically indeterminate, pin-jointed frame of

Fig. 5.2(b). From Equations (5.6), the load W to deflect the loading point of the frame is

$$W = \frac{AE}{L}(1 + 2\cos^3 \theta)\delta, \tag{10.1}$$

where A is the cross-sectional area of a bar and L $(=l_2)$ and δ $(=\delta l_2)$ are the initial length and elongation of the centre bar. The *elastic load* W_e is reached when

$$\delta = \delta_e = \frac{Y}{E}L. \tag{10.2}$$

Neglecting work-hardening, any extra load beyond W_e must be carried entirely by the inclined bars, since the central bar can take no more. The frame is *statically determinate* with respect to this extra load $W - W_e$, i.e. is like the frame of Fig. 5.2(a), and the load–deflexion relation in this range $(W > W_e)$ is

$$W = AY + \frac{2AE \cos^3 \theta}{L}\delta. \tag{10.3}$$

Provided θ is small, the deflexion is of the same order as that below the elastic load, since the plastic deformation of the central bar is *contained* at this stage. The inclined bars start to yield when $(\delta/L)\cos^2 \theta = Y/E$, i.e. when

$$W = W_p = AY(1 + 2\cos \theta). \tag{10.4}$$

This is the *collapse load* W_p, since there is no longer any purely elastic connexion from the supports to the central loading point. Figure 10.1 shows these elastic (OA), contained plastic (AB) and fully plastic (BC) ranges.

A framework can sometimes collapse plastically *before* all its members become plastic. For example, if the load W in Fig. 5.2(b) were applied obliquely to the loading point, bearing more heavily on one inclined bar than the other, two bars could become plastic before the third, and the frame could then rotate (as a *mechanism*) about the upper hinge point of the remaining stiff bar, the other bars stretching plastically.

Figure 10.1 also shows some unloading effects for a material that has the *same* yield stress Y in tension and compression (i.e. has no Bauschinger effect) and does not buckle in compression. At first all bars contract elastically and an unloading line parallel to AO is followed. The centre bar then becomes plastic in compression when its stress state has changed from $+Y$ to $-Y$, i.e. when it has contracted elastically by $2\delta_e$. The *elastic range* of the frame under pulsating load is thus always $2\delta_e$. The frame can unload purely elastically from D and arrive at E in a state of self-stress. If re-loaded from E, it can now withstand a load W_1 *without further* plastic deformation. This effect is called *shakedown*. The self-stress remaining from the first loading, superposed on *any* of the elastic states between zero load and W_1, has produced a total stress everywhere below the yield stress. Had the frame been taken as far as G, however, the centre bar would have become plastic in compression after reaching

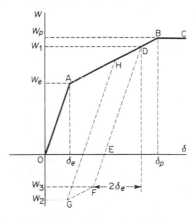

Fig. 10.1.   Load–deflexion diagram for the frame of Fig. 5.2(b).

F. On re-loading from G, the central bar would become plastic in tension again at H, and, in the absence of work-hardening, each cycle HDFGH would produce alternating plastic deformation. There is now no shakedown, because the range of load variation, from W_1 to W_2, exceeds $2W_e$. We notice that shakedown depends on the *range* of load and not on the absolute values of the extreme loads, provided the collapse load is not reached. Thus, after loading to D, a shakedown state would exist for load cycles between F and D.

After loading to D, the elastic load of the frame in compression is reached at F. Since $-W_3 < W_1$, the frame shows a *Bauschinger effect*, even though its material does not. Bauschinger effects in solids are generally believed to have a similar cause; e.g. one grain, perhaps because of its crystal orientation, slips at a lower load than its neighbours, and the self-stress so set up causes it to slip back under a reduced load in the reverse direction.

Shakedown is important in structures subjected to pulsating loads. When the shakedown range is exceeded, the small plastic strains produced repeatedly in each cycle may accumulate to produce *incremental plastic collapse* or, if continued for many cycles, may lead to *fatigue failure* (cf. Section 11.10). In engineering design there are thus three important loads, *elastic, collapse* and *shakedown*. The elastic load is important in precision apparatus where dimensional stability is vital, but is unduly conservative for redundant load-bearing structures. The collapse load is important where the working load is fairly steady (e.g. a steel-framed building) or where it is unlikely to be applied more than once or twice (e.g. a crash helmet). Where the working load is applied many times (e.g. an aircraft wing), however, the shakedown load should not be exceeded.

10.2 PLASTIC DESIGN

In *plastic design* the members are proportioned so that in every member the collapse load exceeds the working load by the same *plastic safety factor* (e.g. 1·75 for a steel-framed building). Plastic design has many advantages over *elastic design*, in which the aim is to keep everything elastic. Apart from economizing in material, it is more rational, accurate and simple. It is impossible to make an *exact* elastic analysis of a real structure, which would allow for all the detailed irregularities, misfits of construction, self-stresses, complexity of joints, settling of foundations, etc. In fact, the elementary *Strength of Materials* approach to engineering design, in which elastic design is simplified by ignoring complicated details, already relies extensively on plastic effects. The conventional elastic load used in elastic design is not the true elastic load (which is almost zero, because of notches, self-stresses, etc.), but the value at which the spreading plastic zones first produce measurable overall deflexions.

In some problems a form of plastic design has always been used. For example, a rigorous elastic analysis of a riveted or welded lap joint is hardly feasible. If we idealize the joint to a notch, as in Fig. 10.2, we see

Fig. 10.2. A joint regarded as a notch.

that the elastic load is very small, because of the stress concentration. The practical strength is calculated by assuming that failure occurs by plastic shearing of the whole rivet in the plane of the joint. If A is the area of this zone and k the shear yield stress of the material, the shear strength or collapse load of the joint is kA per rivet. The sheared zone must of course be ductile enough to yield completely before fracture.

It is useful to consider this problem in terms of dislocations, with k as the "propagation stress" for the passage of a dislocation across the sheared zone. Figure 10.3 shows a simplified example. At sufficiently small loads the dislocations created at the notch cannot glide far, because the nominal stress σ (i.e. load divided by notched sectional area) is small compared with k. At higher loads, σ approaches k and the dislocations spread further (diagram 2). The self-stress from these dislocations largely cancels the concentrated applied stress at the notch, so that the net stress there remains at about k. As σ reaches k, dislocations begin to leak out of the far end

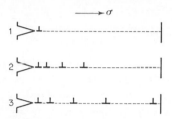

Fig. 10.3. Stages 1, 2 and 3 in the spread of yield from a sharp notch by the propagation of dislocations.

of the shear zone (diagram 3). There is then a steady-state flow of dislocations from the notch to the far end. Each dislocation dissipates a total energy kbA during its traverse. The applied load σA moves a distance b for each dislocation traverse. Hence, equating work done, *general yielding* (i.e. plastic collapse) occurs when $\sigma = k$ and the collapse load is kA. Since plastic collapse is determined from the work done by the dislocations, self-stresses cannot affect the collapse load, because they are self-balancing across the zone.

The dislocations here are of course not actual crystal dislocations (which move in a much more complicated way) but merely a convenient representation of the plastic strain gradient along the sheared zone. Provided k is measured as a macroscopic property of the bulk material, all the complications of the actual crystal dislocations, such as orientation, multiple slip and grain boundary effects, are automatically allowed for and a description in terms of "macroscopic" dislocations with a "glide stress" k is valid.

Bent beams collapse plastically by the formation of *plastic hinges*. Figure 10.4 shows the distribution of longitudinal stress across a simple rectangular beam of depth h and thickness b. Curve 1 represents the elastic state of Section 5.4, in which the stress varies linearly from the extended to the compressed side. The elastic limit is reached when the stress in the outer fibres equals $\pm Y$ (curve 2). The bending moment at this stage, from Section 5.4, is $Ybh^2/6$. This is not the limiting strength of the beam, however, since the inner fibres are still elastic. Further increase of load produces curves 3 and 4, in which more outer fibres yield and support the constant stress Y. Until curve 4 is reached, there are always some elastic fibres, near the neutral surface, to give stiffness. Hence, plastic collapse occurs at curve 4. The *fully plastic moment*, M_p, at this

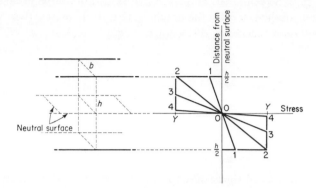

Fig. 10.4.   Longitudinal stress in a beam of lateral dimensions b and h.

stage is easily found, since we have two forces of magnitude $Ybh/2$ at distances $h/4$ above and below the neutral surface; hence,

$$M_p = 2\frac{Ybh}{2}\frac{h}{4} = \tfrac{1}{4}Ybh^2,\qquad(10.5)$$

which is 1·5 times the limiting elastic moment.

The plastic moment of an I-beam is only about 1·15 times the elastic moment, because the web is thin. However, when applied to redundant structures, plastic design leads to further economies. Consider the built-in beam in Fig. 10.5, of length $3L$, under a load W at a distance L from one

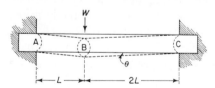

Fig. 10.5.   Regions A, B and C where plastic hinges form in a beam.

end. The largest bending moment initially is at A, and a *plastic hinge* (cf. curve 4 of Fig. 10.4) forms there when this bending moment reaches M_p. It can be shown that $W = 2 \cdot 25 M_p/L$ at this stage. Although plastic rotation at A is now possible, the beam cannot move because the rest of it is still stiff. Extra load beyond this value of W has to be supported elsewhere than at A. The next place to form a plastic hinge is B. The beam cannot collapse, however, until a third hinge is formed, at C. This collapse load is $3M_p/L$, i.e. about one-third higher than the load at which the first hinge formed.

There are two general methods for calculating the collapse loads of such structures: (1) the *stress-equilibrium* method, which in Fig. 10.5 would involve calculating the bending moments at A, B and C due to W, and equating them to M_p; and (2) the *plastic-work* method, in which the work done by a small displacement of W is equated to that done in all the rotating hinges. To apply this second method, suppose that C rotates by the small angle θ. Then, A and B rotate by 2θ and 3θ, respectively. The work done by W is $2L\theta W$. That done in each hinge is M_p times the angle turned. Equating these, $W = 3M_p/L$. More general methods for finding the collapse modes and loads of complex frameworks are given in books on plastic design. It is necessary of course for the joints to be strong enough not to collapse before the fully plastic moment is reached in adjoining sections of the beams. The collapse load is then independent of the properties of the joints, and can be calculated purely from the properties of the beams.

The stress-equilibrium and plastic-work methods are applicable to all problems of plastic collapse. The first insists that all forces shall be in equilibrium and that the stress shall nowhere exceed the yield stress; the second that the displacements of all points of the structure shall be *geometrically compatible* with one another. In an exact calculation based on the actual mode of collapse these should of course give the same answer. In complex structures, however, an exact treatment is not always possible. The two methods then become useful as the basis of *limit analysis*. Thus, if we do not know the actual mode of plastic collapse, we may *guess* a mode. This can be as simple as we please, provided the elements of the body deform compatibly with one another and provided the plastic zones spread far enough to ensure collapse. The collapse load for this guessed mode can then be calculated by the plastic-work method. This must be an *upper limit* to the true collapse load, because the body will certainly not endure it without collapsing. Alternatively, we may ignore geometric modes of collapse and, instead, guess a *distribution of loads* in the elements of the body at the point of general collapse. This distribution can be as simple as we please, provided the loads are everywhere in equilibrium and provided the yield stress is nowhere exceeded. The collapse load calculated from this guessed distribution must be a *lower limit* to the true collapse load, because *freeing* the system from the necessity to satisfy the compatibility condition cannot make the system *harder* to deform. Formal proofs of these theorems of limit analysis are given in books on plastic design.

As a simple example, we shall apply the theorems to a wide notched plate strained in tension, as in Fig. 10.6(a). To obtain a lower limit to the collapse load W, we guess a stress state in which all the material below ABC is on the point of yielding, as in diagram (b), and all that

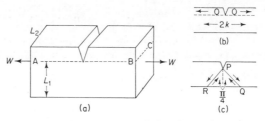

Fig. 10.6.   A notched plate in tension.

above is completely unstressed. We shall see in Section 10.7 that this *plane-strain* yield stress is $2k$, where k is the shear yield stress. The lower limiting collapse load is thus

$$W = 2kL_1L_2. \tag{10.6}$$

To find an upper limit, we guess, in diagram (c), a deformation mode in which shear occurs along two planes PQ and PR running from the tip of the notch, at 45° to W, everything outside these planes behaving rigidly. The shear force on the plane PQ, when plastic sliding occurs, is $kL_1L_2\sqrt{2}$, and a displacement δ along PQ moves W through a distance $\delta/\sqrt{2}$. Equating the work done, we have $(kL_1L_2\sqrt{2})\delta = W\delta/\sqrt{2}$, so that $W = 2kL_1L_2$. The upper and lower limits coincide in this calculation, so that Equation (10.6) gives the true collapse load.

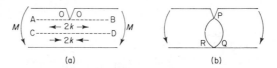

Fig. 10.7.   A notched plate in plane bending.

As a second example, we consider the **same** notched plate to be bent by couples M, as in Fig. 10.7. A guessed *stress state* that satisfies the necessary conditions is shown in diagram (a). Above AB there is no stress; CD divides the remaining material into two equal regions of thickness $\frac{1}{2}L_1$, fully plastic in tension and compression, respectively. The plastic moment, $\frac{1}{2}kL_1{}^2L_2$, is a lower limit to the true collapse moment of the notched beam. A guessed *collapse mode* is shown in diagram (b). Here two circular arcs PQ and PR, of length L and radius r, act as shear lines, and the two arms of the bar rotate rigidly by sliding round them as plastic hinges. Let each arm rotate by the small angle θ. A slip displacement $r\theta$ then occurs along a length L of each arc, against the shear stress k. The plastic work done in the two arcs is thus $2kLL_2r\theta$. Equating this to the external work $2M\theta$ gives $M = kLL_2r$. We are still free to choose values of either L or r. As we make one of them small, the other becomes large; the minimum is $Lr = 0\cdot69L_1{}^2$. Hence, an upper limit to the collapse moment is $0\cdot69kL_1{}^2L_2$. Closer estimates give the collapse moment as about $0\cdot63kL_1{}^2L_2$.

10.3 SHAKEDOWN

A *shakedown state* is a state of self-stresses which, when added to a variable applied stress, gives a total stress that satisfies the equilibrium conditions and does not at any time exceed the yield stress anywhere. In this state a body responds purely elastically to the variations of applied stress, even though it may earlier have deformed plastically to get into the state. The possibility of a shakedown state existing in the material depends upon the range of variation of the applied stress. This is because the shear yield stress k always opposes plastic shear on any slip line, whatever the direction of shear; thus, an element of a slip line can withstand a change of applied stress on it of $2k$, i.e. from $+k$ to $-k$, without slipping in the reverse direction. Provided the local change of applied stress is less than $2k$ everywhere, it is possible in principle to design a state of self-stress that can keep the total stress everywhere within the limits $\pm k$, i.e. a shakedown state is then possible. There is a *shakedown theorem* in applied mechanics, which proves that, if a shakedown state is possible, then shakedown will occur. This theorem is useful in engineering design, since it makes it unnecessary to calculate the precise shakedown state, which is often difficult to do, and replaces it by the simpler problem of merely investigating whether shakedown is possible. The physical basis of the theorem can be seen by thinking about inhomogeneous deformation. If one part of a body is initially overstressed, it adjusts its dimensions by plastic yielding in such a way as to set up self-stresses which partly cancel the applied stress on it. These self-stresses form the shakedown state.

It is difficult to keep the variation of stress less than $2k$ everywhere, because grooves and other stress concentrators produce large stresses at their tips. The inability to reach a shakedown state locally at such places is a major cause of *fatigue failure* (cf. Section 11.10) in bodies under variable loads. Although macroscopic shakedown, e.g. between one bar and another in a framework, can usually be arranged, it is almost impossible to ensure microscopic shakedown without making the range of working loads impracticably small. Fatigue failure is thus an almost unavoidable penalty of economic design in structures under variable loads.

Shakedown is used in the *autofrettage* process for strengthening thick-walled pressure tubes. Referring again to Fig. 5.24, it can be shown that the *elastic pressure p_e* at which the inner boundary first becomes plastic is given by

$$p_e = k\left(1 - \frac{r_0^2}{r_1^2}\right) \qquad (10.7)$$

and that the *collapse pressure p_c* at which the entire tube becomes plastic and bursts (neglecting work-hardening) is given by

$$p_c = 2k \ln\left(\frac{r_1}{r_0}\right). \qquad (10.8)$$

After loading to an intermediate pressure p ($<2p_e$), the tube unloads *elastically*, since the material at the inner boundary can accept a change of shear stress from $+k$ to $-k$ without yielding. The *shakedown pressure* is thus the smaller of the two values $2p_e$ and p_c. The working pressure p of thin-walled tubes is limited by plastic collapse ($p = p_c$) and that of thick-walled tubes by reversal of plastic deformation at the inner boundary on unloading ($p = 2p_e$).

10.4 GENERAL EQUATIONS OF PLASTICITY

We turn now to the general stress–strain analysis of plastic bodies. Plastic properties are far more complicated than elastic properties, and a manageable theory is possible only by making various simplifying assumptions, thus:

(1) The material deforms plastically at constant volume; hydrostatic stress does not alter the yield stress or the work-hardening properties. These assumptions are well supported by experiments on metals. They are not valid for soils, however, and a separate theory is needed for these (*soil mechanics*).

(2) The material is mechanically homogeneous and isotropic, is irreversible in its plastic deformation, is insensitive to strain rate, and has no Bauschinger or creep effects. The assumption of isotropy means that microstructure is ignored and that macroscopic shear occurs along lines of maximum shear stress, irrespective of the crystal structure. This is reasonable, provided the element of plastic volume is large compared with the grain size and a yield stress measured on macroscopic samples is used. The theory can be extended, where necessary, to anisotropic materials.

(3) The elastic strains can be neglected and the material regarded as a hypothetical rigid–plastic solid, when the plastic strains are large (e.g. many mechanical working processes). There are several important types of problems where this assumption cannot be made.

(4) Work-hardening can be neglected. This is reasonable for fully hardened materials and also for problems of hot-working, but not otherwise. Neglect of work-hardening means of course that *instabilities* such as tensile necking appear in the theory.

Consider first the *yield criterion*. We can apply various combinations of principal stresses σ_1, σ_2, σ_3, to the material. When does it yield under these various stress states? This is important even in elastic design, since it determines how much load, say, a pressure vessel can stand before elastic failure. We suppose that the more the stress state deviates from pure hydrostatic stress, the more likely it is to produce yield. In Fig. 10.8(a) any point on the line ON, equidistant from the three stress axes, represents a pure hydrostatic stress state. Yielding occurs at stress states a certain distance from this line. These states lie on a *yield surface*, with ON as

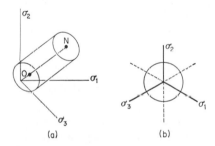

Fig. 10.0. The von Mises yield criterion

axis, such as the circular cylinder shown in diagram (a) and in projection in diagram (b). If yielding is independent of hydrostatic stress, as in metals, this yield surface is *cylindrical*. The conditions of isotropy (equal yield stresses along all three axes) and no Bauschinger effect (equal yield stresses along σ_1 and $-\sigma_1$, etc.) require the surface to have sixfold symmetry with respect to the stress axes.

The simplest figure with these properties is the circular cylinder. This is the basis of the *von Mises* (also *Huber*) yield criterion. The equation of this cylinder is

$$(\sigma_1 - \sigma_2)^2 + (\sigma_2 - \sigma_3)^2 + (\sigma_3 - \sigma_1)^2 = \text{constant}. \qquad (10.9)$$

To find the constant, we twist a thin-walled circular tube plastically (cf. Section 5.3). The principal stresses acting in the tube wall are

$\sigma_1 = -\sigma_2 = k$, where k is the _shear yield stress_ of the material; the principal stress σ_3 acting through the wall is zero. Substituting these values, we obtain

$$(\sigma_1 - \sigma_2)^2 + (\sigma_2 - \sigma_3)^2 + (\sigma_3 - \sigma_1)^2 = 6k^2, \qquad (10.10)$$

which is the _von Mises yield criterion_. To find the tensile yield stress Y in terms of this, we put $\sigma_1 = Y$, $\sigma_2 = \sigma_3 = 0$, to obtain

$$Y = \sqrt{3}k. \qquad (10.11)$$

An alternative to this is the _Tresca_ (also _St. Venant_, _Mohr_ and _Guest_) yield criterion, represented by the hexagonal cylinder in Fig. 10.9(a). The assumption here is simply that plastic shear occurs when the maximum shear stress reaches the critical value k, irrespective of other stresses

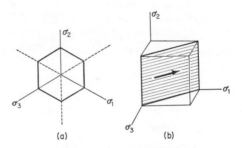

(a)   (b)

Fig. 10.9. The Tresca yield criterion.

present. Figure 10.9(b) reminds us that planes of maximum shear stress lie at 45° to the principal stresses. The shear stress acting on the shaded plane in the direction shown is $\frac{1}{2}(\sigma_1 - \sigma_3)$. Hence, assuming that $\sigma_1 > \sigma_2 > \sigma_3$, yielding occurs (by shear on these planes) when

$$\sigma_1 - \sigma_3 = 2k. \qquad (10.12)$$

This is the _Tresca yield criterion_. To find the tensile yield stress Y, we put $\sigma_1 = Y$, $\sigma_2 = \sigma_3 = 0$, to obtain

$$Y = 2k \qquad (10.13)$$

in place of Equation (10.11).

The Tresca criterion is simple and is the isotropic equivalent of the law of resolved shear stress in a crystal. The sharp corners of its yield surface sometimes raise mathematical difficulties, however, and it was to avoid these that von Mises proposed his criterion, which predicts a different ratio of tensile to torsion yield stress. Ductile metals, in fact, mostly follow von Mises' criterion rather better than Tresca's. This is due to the mode of crystal slip in a polycrystal, where the mutual constraints of the grains force slip to occur on several planes and directions at different orientations to the macroscopic axes of principal stress.

The shear stress for the plastic deformation of a *soil* (e.g. sand, gravel, clay) depends on *pressure*, because of the loose granular nature of the material (cf. Section 7.6). The more densely the grit particles are packed together, the harder it is for them to slide over one another. *Coulomb* treated this as a simple frictional resistance, proportional to pressure. Thus, if sliding occurs along a plane through the material, across which there is a normal pressure p, the *shear resistance* s of this plane contains a term of the type μp or $p \tan \phi$, where μ is the *coefficient*, and ϕ the *angle*, *of internal friction*. Thus, *Coulomb's yield criterion* is

$$s = k + p \tan \phi = k + \mu p, \tag{10.14}$$

where k is the shear strength due to cohesive forces between the grains. Since soils lack tensile strength, this formula does not apply when p is a tension. Two extreme cases are soft wet clay, where there is good cohesion but $\phi = 0$, in which case the formula reduces to Tresca's criterion $s = k$; and cohesionless powders such as dry sand, for which $k = 0$ and $s = p \tan \phi$. When $k = 0$, ϕ is equal to the *angle of repose*, the steepest angle a sand-hill or earth-bank can sustain without slipping. For dry sand $\phi \simeq 30°$. When $k > 0$, a sheer cliff can be sustained, at least temporarily, which is important for the modelling properties of clay. When $k = \phi = 0$, as in water-logged soil with grains supported by the buoyancy of the water, the material is then a fluid.

The plastic properties of clays are determined by compressing short circular cylinders (e.g. *length* $= 2 \times$ *diameter*) axially between flat ends. Failure commonly occurs by sliding on a plane, often called a *slip line*, at an angle θ, as in Fig. 10.10. From Equations (4.58) and (4.61), the axial pressure p_1 produces on this plane a normal pressure $p_1 \cos^2 \theta$ and

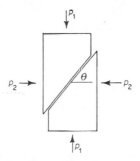

Fig. 10.10.   Shear failure in compression.

a maximum shear stress $p_1 \cos \theta \sin \theta$. In this uniaxial test $p_2 = 0$. Coulomb's criterion then becomes

$$p_1 \cos \theta \sin \theta = k + \mu p_1 \cos^2 \theta. \qquad (10.15)$$

Yield occurs first on the plane whose θ maximizes $\cos \theta \sin \theta - \mu \cos^2 \theta$, i.e. for which

$$\mu \tan 2\theta = -1; \quad \text{or} \quad \theta = \frac{\pi}{4} + \frac{\phi}{2}. \qquad (10.16)$$

Thus, slip occurs on the plane of maximum shear stress, $\theta = 45°$, when $\mu = 0$; on $\theta = 67 \cdot 5°$, when $\mu = 1$; and on $\theta \simeq 90°$, when $\mu \simeq \infty$. To determine k and μ the *triaxial compression test* is used, which is similar to that of Fig. 10.10 except that the specimen is enveloped in thin rubber and subjected to hydrostatic water pressure. Let p_1 be the total axial pressure and p_2 the lateral pressure at failure. We mark p_1 and p_2 along the axis of normal stress in a *Mohr circle diagram* (cf. Fig. 4.18), as in Fig. 10.11, and draw a semicircle of diameter $p_1 - p_2$ through these points.

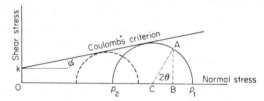

Fig. 10.11. Coulomb's criterion in a Mohr circle diagram.

The experiment is repeated at various pressures and semicircles are drawn through each pair of points so obtained. The tangential line to these semicircles represents Coulomb's criterion; it has a slope ϕ and an intercept k on the axis of shear stress. Any point such as A which subtends an angle 2θ at the centre C of its semicircle represents the state of stress on a plane of slope θ, since

$$OB = OC + CB = \tfrac{1}{2}(p_1 + p_2) + \tfrac{1}{2}(p_1 - p_2) \cos 2\theta$$

$$= p_1 \cos^2 \theta + p_2 \sin^2 \theta, \qquad (10.17)$$

which is the normal stress, produced by p_1 and p_2, on the plane; and

$$AB = \tfrac{1}{2}(p_1 - p_2) \sin 2\theta = (p_1 - p_2) \sin \theta \cos \theta, \qquad (10.18)$$

which is the shear stress on the plane.

We turn now to consider the *stress–strain relations* of plastic bodies.

Because plasticity is a form of *flow*, the property related to stress, particularly in non-hardening solids, is *strain rate*, not strain itself. This is like fluidity, but an important difference, at least in ideal plasticity, is that all rates of flow can occur at the same yield stress. We have therefore to imagine the applied force driving the material at some *externally determined* rate, e.g. set by the cross-head speed of a testing machine. The stresses in the material merely determine the *ratio* of strain rates ($\dot{\varepsilon}_1$, $\dot{\varepsilon}_2$, $\dot{\varepsilon}_3$) along their respective principal axes. In an isotropic body the principal axes of stress and strain rate coincide. For a *plastic–rigid* solid ($v = 0.5$) we thus expect, by analogy with Equation (5.9),

$$\frac{\dot{\varepsilon}_1}{\sigma_1 - \frac{1}{2}(\sigma_2 + \sigma_3)} = \frac{\dot{\varepsilon}_2}{\sigma_2 - \frac{1}{2}(\sigma_3 + \sigma_1)} = \frac{\dot{\varepsilon}_3}{\sigma_3 - \frac{1}{2}(\sigma_1 + \sigma_2)} = \text{constant}, \quad (10.19)$$

these being the *Lévy–Mises* equations of plastic flow. More general equations which take elastic strain into account have been developed by *Prandtl* and *Reuss*. Since the *deviatoric stresses* (cf. Section 4.9), σ'_1, σ'_2, σ'_3, are defined by relations of the type

$$\sigma'_1 = \sigma_1 - \frac{1}{3}(\sigma_1 + \sigma_2 + \sigma_3) = \frac{2}{3}[\sigma_1 - \frac{1}{2}(\sigma_2 + \sigma_3)], \quad (10.20)$$

Equations (10.19) can also be written as

$$\frac{\dot{\varepsilon}_1}{\sigma'_1} = \frac{\dot{\varepsilon}_2}{\sigma'_2} = \frac{\dot{\varepsilon}_3}{\sigma'_3} = \text{constant}, \quad (10.21)$$

the value of the constant depending on the external driving speed and *not* (unlike the coefficient of viscosity) on the properties of the material.

As a simple example, we consider again the pressure vessel of Fig. 5.3(a) and conclude, since $\sigma_1 = 2\sigma_2$ and $\sigma_3 \simeq 0$, that when such a cylinder is expanded plastically by internal pressure, its length remains constant.

10.5 PRESSURE OF SOIL AGAINST A WALL

An old but important application of plasticity theory concerns the horizontal pressure exerted by a soil against a vertical retaining wall which prevents it slipping down under its own weight. In Fig. 10.12 let AB be the level top surface, BC the wall, and p_2 the value at depth h of the horizontal pressure needed to prevent regions such as PBQ from sliding down slip lines such as PQ. All volume elements such as RS at the same depth h are subjected to the same horizontal pressure p_2 and the same vertical pressure $p_1 = \rho g h$, where ρ is the density of the medium and g the acceleration due to gravity. Since the direction of incipient slip of PBQ is down the slip line, the shear resistance k of the medium provides a force acting from Q to P on the face PQ of PBQ. The mechanical state

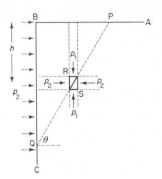

Fig. 10.12. Pressure on a vertical wall to prevent slip.

is thus similar to that of Fig. 10.10, with $\theta \geqslant 45°$ and p_2 smaller than p_1. Equations (10.17) and (10.18) give Coulomb's law as

$$(p_1 - p_2) \sin\theta \cos\theta = k + \mu(p_1 \cos^2\theta + p_2 \sin^2\theta), \quad (10.22)$$

and, since $\mu = -\cot 2\theta = (\tan^2\theta - 1)/2\tan\theta$, this can be reduced to

$$p_1 = 2k\tan\theta + p_2 \tan^2\theta \quad (10.23)$$

and, hence, to

$$p_2 = \rho gh \tan^2\left(\frac{\pi}{4} - \frac{\phi}{2}\right) - 2k\tan\left(\frac{\pi}{4} - \frac{\phi}{2}\right). \quad (10.24)$$

For cohesionless soils ($k = 0$), a finite retaining pressure is required to hold any vertical face, however shallow. For dry sand, $p_2 \simeq 30h$ lb/ft^2. For sand saturated with water, the pressure is increased, by the weight of the water, to about $90h$ lb/ft^2. For cohesive soils such as clay, Equation (10.24) predicts a *negative* retaining pressure to a certain depth below the top surface. This means that the cohesive strength k of the material is able itself to support a vertical face to this depth. We could not depend on this in practice of course, because the long-time creep strength of plastic clay is much smaller than its short-time yield strength.

Equation (10.24) describes what is called the *active Rankine state of plastic equilibrium*, in which the retaining pressure is just sufficient to prevent downward slip. There is, however, the complementary *passive Rankine state* in which this pressure is just large enough to produce *upward* slip against the force of gravity and the shear resistance of the slip lines. Here $p_1 < p_2$ and $\theta < 45°$. An analysis similar to the above then leads to Equation (10.24) again, but with all the negative signs made positive, for the horizontal pressure p_2 at which upward slip begins. This is much larger than the retaining pressure p_2 for downward slip. These considerations are important in the design of retaining walls with soils at different levels on the two sides, since the *active pressure* of soil banked up behind a wall has to be balanced against the *passive resistance* of soil being pushed forward in front of the wall.

10.6 TWO-DIMENSIONAL PLASTICITY

The problem of Fig. 10.12 is one of *plane strain*; i.e. if X_3 is the co-ordinate axis perpendicular to the diagram, then all particles move only in directions perpendicular to X_3 and their movements are independent of their X_3 co-ordinate. Many modes of plastic deformation occur under at least approximately plane-strain conditions, e.g. sheet rolling, lathe turning and deformation at the roots of long lateral grooves across tensile bars.

Since $\dot{\varepsilon}_3 = 0$ in plane strain, we have, from Equations (10.19),

$$\sigma_3 = \tfrac{1}{2}(\sigma_1 + \sigma_2). \tag{10.25}$$

Thus, σ_1 and σ_2 are the extreme principal stresses, the direction of maximum shear stress is perpendicular to X_3, and the plane of maximum shear stress is parallel to X_3. Tresca's criterion is

$$\sigma_1 - \sigma_2 = 2k, \tag{10.26}$$

and we see, by substituting Equation (10.25) into Equation (10.10), that for plane strain this is also von Mises' criterion.

Plane stresses exist in thin sheets and membranes stretched at their edges, as in Fig. 10.13. The principal stress σ_3 perpendicular to the sheet

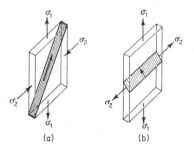

Fig. 10.13. Planes and directions of maximum shear stress in plane-stress distributions.

is zero, and the stresses σ_1 and σ_2 within the sheet are constant through the thickness. When σ_1 and σ_2 have opposite signs, the maximum shear stress is $\tfrac{1}{2}(\sigma_1 - \sigma_2)$ and acts along a line parallel to the sheet on a plane perpendicular to the sheet. When they have the same sign, the maximum shear stress is $\frac{1}{2}\sigma_1$ (assuming σ_1 is the numerically larger stress) and acts along a line and on a plane at $45°$ through the sheet.

Since $\sigma_3 = 0$, the yield surface becomes two-dimensional, as shown in Fig. 10.14. When σ_1 and σ_2 have opposite signs, the Tresca criterion becomes $\sigma_1 - \sigma_2 = 2k = Y$ and is represented by the straight lines CD and FA in the diagram. When they have the same sign, the criterion

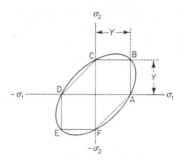

Fig. 10.14.   The von Mises yield ellipse and the Tresca yield hexagon for plane stress.

becomes $\sigma_1 = 2k = Y$ or $\sigma_2 = 2k = Y$ (since $\sigma_3 = 0$), as represented by the lines AB, BC, DE and EF. The von Mises criterion is

$$\sigma_1{}^2 - \sigma_1\sigma_2 + \sigma_2{}^2 = 3k^2 = Y^2, \tag{10.27}$$

which is the ellipse in Fig. 10.14. When the two yield criteria are expressed in terms of Y, the hexagon inscribes the ellipse.

10.7 SLIP-LINE FIELDS IN PLANE STRAIN

Plane-strain problems of rigid–plastic solids are solved by constructing *slip-line fields*. Using Equations (10.25) and (10.26), assuming $\sigma_1 > \sigma_2$, and defining $-p = \sigma_3 = \sigma_2 + k = \sigma_1 - k$, we can write the principal stresses as

$$\sigma_1 = -p + k,$$
$$\sigma_2 = -p - k, \tag{10.28}$$
$$\sigma_3 = -p,$$

and represent them diagrammatically as in Fig. 10.15, to show that they consist of a *hydrostatic pressure p* superposed on a *pure shear stress k* directed $90°$ from σ_3 and $45°$ from σ_1 and σ_2. The lines of maximum shear stress are the *slip lines* along which plastic sliding occurs. The two directions of maximum shear stress generate two families of slip lines

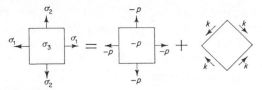

Fig. 10.15.   Plane-strain state of stress in a plastic volume element.

which intersect orthogonally, as shown by the lines α and β in Fig. 10.16. It is important to distinguish between the two families, and the rule is that, in an anticlockwise rotation about the point of intersection, starting from an α line, the line of algebraically highest principal stress (σ_1) is crossed before the β line is reached.

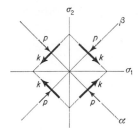

Fig. 10.16.   Notation for slip lines.

Uniform stresses give straight slip lines. Figure 10.17 shows a simple example, a flat slab compressed between flat, frictionless, plates. Because frictional forces are absent along the interfaces, the directions of the principal stresses and the slip lines are as shown. The algebraically highest principal stress is σ_1 ($=$ zero), and this identifies the α and β lines as shown.

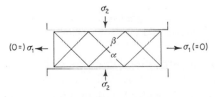

Fig. 10.17.   Slip-line field for plane-strain compression.

If we now make $\sigma_1 > 0$, $\sigma_2 = 0$, and $\sigma_3 = \frac{1}{2}\sigma_1$, the same slip-line field describes tensile elongation under plane-strain conditions along the σ_1 axis. This differs from simple tension in that deformation is now suppressed along the σ_3 axis. The hydrostatic stress is

$$p = -\tfrac{1}{2}(\sigma_1 + \sigma_2) = -\tfrac{1}{2}\sigma_1,$$

whereas for simple tension ($\sigma_2 = \sigma_3 = 0$) it is $-\frac{1}{3}\sigma_1$. The *plane-strain yield stress*, i.e. the value of σ_1 at yield when $\sigma_2 = 0$, is $\sigma_1 = \sigma_2 + 2k = 2k$. On Tresca's criterion this is identical with the tensile yield stress Y, but on von Mises' criterion it is larger, $1 \cdot 15\ Y$, which denotes a hardening due to the fact that the material is no longer free to deform in the X_3 direction.

10.8 PLASTIC RUPTURE

Tensile deformation without work-hardening is unstable. Thus, if Fig. 10.17 is interpreted as plane-strain tensile deformation under a positive stress σ_1, not all the infinite number of slip lines would be used. Slip could start on one of them, by chance, and then continue preferentially on it, owing to loss of cross-sectional area there, so producing complete "sliding-off" and a "chisel-edge" rupture, as in Fig. 10.18(a). Fractures

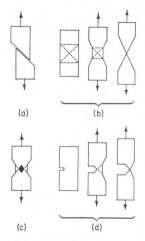

(a)   (b)

(c)   (d)

Fig. 10.18. Tensile ruptures by plane-strain plastic deformation: (a) sliding-off; (b) necking; (c) internal and external necking; (d) necking from a notch.

approximating to this are often observed in hard ductile materials (e.g. precipitation-hardened aluminium alloy plates), particularly where there is adiabatic softening (cf. Section 9.6), and they generally occur suddenly. Since they require an almost infinite shear strain in the active slip line, however, they are suppressed by a trace of work-hardening. If there is a little work-hardening, but not enough to prevent necking, the rupture often develops symmetrically on two diagonal slip lines, as in diagram (b). Here the plastic deformation is spread continuously through a thick zone of initial length equal to the thickness of the plate. At any instant only the thin layers in the slip lines themselves are plastic, but material is continually passed through each slip line by its displacement along the other slip line. Since each slip line is cut and equally displaced by the other, the overall shear strain in each is *unity*. In a fully ductile material this leads to a "knife-edge" rupture, the angle of the knife edge being $2 \tan^{-1} 0.5$ ($=53.2°$).

There are many variants of these forms of rupture. Diagram (c) shows how a cavity may grow by a process of "internal necking", and diagram (d) shows how a notch may spread, while maintaining a sharp tip, by slip along lines from its tip. The slip-line field in this case is also that of Fig. 10.6(c). Closely related is the process of cutting a plate with a frictionless knife, an elementary form of *machining*, as in Fig. 10.19. At first

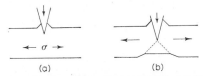

(a)   (b)

Fig. 10.19. Cutting with a frictionless knife.

the material piles up at the sides of the blade, as in diagram (a); but when the knife sinks deeper, the horizontal stress σ produced by it becomes large enough to push the plate apart in this direction, and the slip-line field of diagram (b), similar to that of Fig. 10.18(d), then becomes active. A striking aspect of this, which we can see by cutting a slab of well-lubricated clay, is that the underside *rises* to meet the advancing knife-edge.

Because slip occurs on lines at 45° to the tensile axis in Fig. 10.18, the tensile elongation due to necking rupture (assuming fully ductile material) is equal to the initial width of the region where the neck forms (i.e. from cavity to surface in diagram (c)). As mentioned in Section 9.6, in a long thin specimen this gives only a small overall strain. If the material contains many elongated cavities spaced an average distance w, as in Fig. 10.20,

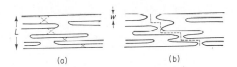

(a)   (b)

Fig. 10.20. Ductile fibrous ruptures.

necking can occur on slip lines between cavities, such as those shown in diagram (a), leading to the *fibrous* rupture of diagram (b). Such ruptures are commonly observed when ductile metals and alloys are broken, the cavities usually being nucleated on foreign inclusions and elongated during the period of uniform deformation before the ultimate tensile stress is reached. The elongation during necking is then reduced by the internal necks, from L to w, and the reduction of area during necking, measured from the *external* neck, is reduced to the order of w/L. When $w \ll L$, the material thus appears macroscopically "brittle" once the ultimate stress is reached, even though it may be microscopically ductile in the sense that individual internal necks are deforming to 100% reduction of area. The overall tensile ductility of metals containing foreign inclusions or of porous metals thus depends critically upon the capacity for work-hardening.

The growth of cavities by internal necking can also promote rupture during sliding-off, as shown in Fig. 10.21. The faces of the shear zone

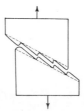

Fig. 10.21.   Sliding-off rupture promoted by internal necking.

move apart as they slide, so that this is no longer pure sliding, complete parting taking place during a slip displacement of order of the cavity spacing along the shear direction. Many shear fractures are of this type. Cavity-forming inclusions are deliberately introduced into some metals to reduce their ductility and so improve their ability to be machined (*free-cutting metals*).

The *cup-and-cone* fracture in a round tensile bar of a ductile metal is more complex than those we have so far described, because of the axial symmetry. Qualitatively, it follows the same principles. Fracture starts in the centre of the neck by the growth and coalescence of plastic cavities, so forming the fibrous "cup", and ends as an annular rim of sliding-off failure, similar to Fig. 10.21, so forming the "cone". Figure 10.22 shows these stages.

All these various forms of ductile fracture are brought about by *plastic strain concentration* in regions of reduced cross-sectional area. If these

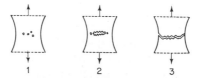

Fig. 10.22.   Stages 1, 2 and 3 in the formation of a tensile cup-and-cone fracture.

regions work-harden strongly, these strain concentrations can be dispersed and the material then becomes macroscopically ductile. There is a limit, however, at about the ultimate tensile stress U, to the extent to which work-hardening can compensate for differences of cross-sectional area. For example, in Fig. 10.18(d) we expect general yield to occur even in un-notched regions provided that

$$\frac{U}{Y} > \frac{A_u}{A_n} \tag{10.29}$$

approximately, where A_u and A_n are the un-notched and notched cross-sections. The importance of the *ultimate/yield* ratio is seen here; the material is *insensitive* to shallow notches but *sensitive* to deep ones, the critical depth increasing with this ratio. Similar criteria (sometimes numerically modified; e.g. if general yield involves simple extension and necking involves plane-strain extension, as in some forms of notched tensile bars) can be developed for other types of necking failure. A *large* volume fraction of spherical or irregular pores renders even a strongly work-hardening material brittle, in tension, by localized failure in the narrow waists between pores. Sintered powders thus have poor ductility until sintering is nearly complete.

Work-hardening is important to prevent fracture in mechanical working processes. Even in predominantly compressive processes such as rolling, there are usually some regions, e.g. along the sides, where tensile stresses may exist locally and may start strain concentration fractures once the work-hardening capacity of the material is exhausted. This capacity can be restored when cold-working, as well as making the material softer to work, by annealing at stages during working.

10.9 CURVED SLIP LINES

When the stress varies through the plastic region, in a material of constant k, the slip lines are *curved*; although the maximum shear stress, unlike p, cannot vary in magnitude, it can vary in direction. Consider part of a slip-line field with radial straight α lines and concentric circular β lines, as in Fig. 10.23. The stress distribution is determined from the mechanical equilibrium of the included element of material. The balance of moments about O gives

$$k(r + dr)^2\, d\phi - kr^2\, d\phi + p(r + \tfrac{1}{2}\, dr)\, dr - (p + dp)(r + \tfrac{1}{2}\, dr)\, dr = 0,$$

i.e.

$$dp - 2k\, d\phi = 0, \tag{10.30}$$

which gives the change dp along a β line in terms of the angle turned, $d\phi$. The equilibrium of forces gives $dp = 0$ in the radial direction. A similar derivation for curved α lines and radial β lines gives $dp + 2k\,d\phi = 0$, the

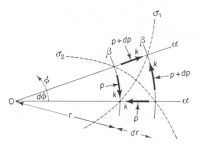

Fig. 10.23. Stresses on an element of material in a fan-shaped slip-line field.

change of sign coming from the different relation of α and β to the principal stress. Summarizing, we arrive at *Hencky's relations*,

$$p + 2k\phi = \text{constant along an } \alpha \text{ line},$$

$$p - 2k\phi = \text{constant along a } \beta \text{ line},$$

$$(10.31)$$

where ϕ is the *anticlockwise* rotation along the line. It is of course possible to combine the two derivations in a more general analysis, with both α and β lines curved, so that one gets simultaneously the two Equations (10.31).

Once the form of the slip-line field is known, these relations enable the stresses through the plastic region to be determined. This is the basis of the standard method of solving plane-strain problems in plastic–rigid materials. The slip-line field is first guessed, from the boundary conditions of the problem and from experience with related problems, the stress distribution is then determined and the validity of the solution examined. To verify the solution completely, it is essential to determine the stress distribution throughout the *whole* body, not merely in the plastic region, to ensure that the stress equilibrium and yield conditions are not violated anywhere; and also to determine the strain velocity field, to ensure that the compatibility conditions are satisfied.

10.10   INDENTATIONS AND NOTCHES

We press a flat frictionless punch into a plate under plane-strain conditions, as in Fig. 10.24(a). The elastic load is very small, and the sharp edges of the punch create small plastic zones beneath them almost as soon as the punch is loaded. The punch does not yet sink in, however, because most of the material beneath is still rigid. The plastic zones are *contained* by the surrounding rigid material at this stage. As the load is raised, the plastic zones grow and spread towards the centre of the punch. When they meet, the whole of the material under the punch has become plastic, and the punch can begin to sink in by sliding this material round the slip lines and heaping it up just outside its edges (cf. Fig. 10.19(a)).

The collapse load at which the punch first sinks in can be found without even constructing the slip-line field. Since no shear stresses act along the

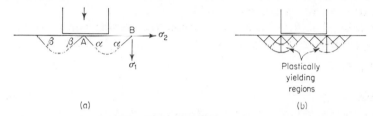

(a)   (b)

Fig. 10.24.   Plastic punching.

surface of the plate, this is a principal stress plane and the slip lines meet it at $\pi/4$. Hence, following the α line from A to B, the angle turned cannot be less than $\pi/2$. Thus, from the first of Equations (10.31) a hydrostatic pressure p under the punch gives a pressure $p - \pi k$ at the free surface. From Equations (10.28), the stresses at the free surface are $\sigma_1 = 0$, $\sigma_2 = -2k$, $\sigma_3 = \frac{1}{2}(\sigma_1 + \sigma_2) = -k$. Hence, the hydrostatic pressure in the free surface is k and that under the punch is found from $p - \pi k = k$, i.e. $p = (1 + \pi)k$. The normal stress under the punch is then

$$\sigma_2 = -p - k = -(2 + \pi)k. \tag{10.32}$$

In contrast to elasticity, this is *uniform* over the whole face of the punch. The compressive stress on the punch to press it in is thus $(2 + \pi)k$, i.e. $2 \cdot 57 Y$ on Tresca's criterion and $2 \cdot 82 Y$ on von Mises'. The resistance of the material to the flat punch is almost three times its resistance to uni-axial stress, so that the material of the punch needs to be three times stronger, if it is not to collapse first. This large *plastic constraint* effect is due to the fact that the slip lines have to *curve* to reach the nearest free surface; this lengthens them unduly compared with the width of the punch. The material offers a resistance to sliding proportional to the length of this slip path.

The slip-line field is also shown in Fig. 10.24. The lines run straight from the surface to meet straight lines of the other family radiating from the edges of the punch. They then curve as circular arcs through $\pi/2$, so forming a fan under each edge similar to that of Fig. 10.23.

The above theory in various modified forms is important for foundations of buildings. Plastic clays at ground level can, in fact, support pressures of about $3Y$. Usually a safety factor of 3 is imposed, which permits the design pressure to go up to Y (typically 15–30 psi). As developed above, the theory is not strictly applicable to indentations by, say, columns of equiaxed cross-sections, where the conditions are not plane-strain. The theory of axially symmetrical problems, although less developed than plane-strain, has shown that the indenting pressures are not much different; thus, a circular flat-ended punch will indent a Tresca rigid–plastic solid at a pressure of $2\cdot84Y$.

The slip-line field for a deep, symmetrical tensile notch can be found by regarding each half-piece, in Fig. 10.25, as a flat punch acting on the

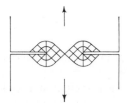

Fig. 10.25. Slip-line field for a deep, symmetrical tensile notch.

other, inwards or outwards. This is essentially the same problem as that of Fig. 10.24. The *nominal yield stress* (i.e. tensile yield load/notched sectional area) is then about $3Y$. This considerable strengthening of the notched section is again due to *plastic constraint*, i.e. to the support given by the rigid shoulders of material on either side. In fact, unless the notches are very deep, the collapse mode shown in Fig. 10.25 will not operate, because the specimen can yield more easily using *straight* slip lines from each notch, as in Fig. 10.18(d), even though these lines span a wider cross-section. We notice that, unlike elastic notches, the stresses in a plastic notch do not rise far beyond the general yield stress.

The slip-line field of Fig. 10.25 is equally applicable to long, flat *internal* notches, transverse to the tensile axis. Such cavities are often formed in metals at weak or weakly adherent foreign inclusions flattened by rolling or other processes of working. Worked materials thus often break in a weak, seemingly brittle, manner when stretched in directions transverse to those of working.

Since the shear stress in the plastic zone is constant $(=k)$, most of the stress $3Y$ in the notched region of Fig. 10.25 is *hydrostatic tension*. This high ratio of tensile/shear stress, which did not exist for the notch of Fig. 10.18(d), is important because it sometimes makes normally ductile materials brittle. Suppose that the tensile stress for *brittle cracking* (cf. Section 11.6) in uniaxial tension is F. We then have *Orowan's classification*: (1) if $F > 3Y$, the material is *simply ductile*; (2) if $F < Y$, the material is *simply brittle*; (3) if $Y < F < 3Y$, the material is *notch brittle*, i.e. ductile in plain tensile tests but brittle in notch tests with large plastic constraint factors.

Slip-line fields for various types of notch have now been determined. In bent bars they produce *plastic hinges*, as in Fig. 10.26. The bars rotate

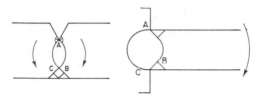

Fig. 10.26. Slip-line fields for a notched bar and for a cantilever in bending.

by sliding round the arcs AB and AC. Such slip-line fields have been confirmed experimentally on strain-ageing mild steel where the Lüders bands, which follow the active slip lines, can be afterwards revealed by chemical etching.

10.11 INDENTATION HARDNESS, ABRASION AND POLISHING

Hardness is resistance to indentation. It is determined by pressing a hard indenter such as a *diamond pyramid* (Vickers test), a *diamond cone* (Rockwell test) or a hard *steel ball* (Brinell test) into a flat surface of the material under a standard indenting force and measuring the size of indentation produced. Experiments have shown that the mean pressure under the indenter is about $3Y$, as expected. Some typical results are shown in Table 10.1. The *Vickers hardness number* H_v is defined as the

TABLE 10.1

Measured Yield Stresses Y and Vickers Indentation Pressures P, in kg mm^{-2} [a]

Material	Y	P	P/Y
Lead	2·1	6·7	3·2
Aluminium	12·3	39·5	3·2
Copper	27	88	3·3
Steel	70	227	3·2

[a] D. Tabor, *The Hardness of Metals*, Oxford University Press, London, 1951.

load divided by the *pyramidal* area of indentation, in kg mm^{-2}. From the geometry of the Vickers pyramid, $H_v = 0.927\,P$, so that

$$H_v \simeq 3Y, \qquad\qquad\qquad (10.33)$$

with Y in kg mm^{-2} ($= 1422$ psi). When the material is not work-hardened by the indentation, Y is the initial stress. The deformation in a Vickers indentation is equivalent to a tensile strain of about 8%, so that in a work-hardening material Y refers, not to the initial yield stress, but to a value after about 8% strain.

A disadvantage of the Brinell test (apart from deformation of the steel ball when used on hard materials) is that, unlike the Vickers pyramid or Rockwell cone, the geometry, and hence the strain, of the indentation varies with the depth of indentation. For consistent results the diameter of the indentation should be held at 0.3–0.6 of that of the ball. Under these conditions fairly soft materials ($H_v < 400$) have almost equal Brinell and Vickers numbers.

Mineralogists and geologists find *Mohs' hardness scale* useful for judging minerals. When two pieces of different minerals are scraped together, the harder will scratch the softer. In Mohs' scale 10 minerals are arranged in order, so that each can scratch all those of lower hardness number and is scratched by all those of higher. The chosen minerals and their Mohs hardnesses are: diamond (10), corundum (9), topaz (8), quartz (7), orthoclase (6), apatite (5), fluorite (4), calcite (3), gypsum (2) and talc (1). Many *gemstones* have hardnesses near the top of this scale (>7), and are thus not scratched by blown sand and silica dust in the air. Glass has a Mohs hardness of about 5.5, and so can be cut by tool steel (7) and tungsten carbide (8). *Abrasives* used for cutting, drilling, grinding and polishing, e.g. *emery* (impure sapphire or corundum), *carborundum* (silicon carbide) and *boron carbide*, have hardnesses greater than 9.

Even on normally brittle substances, abrasive cutting is mainly a process of ploughing plastic grooves by dragging hard angular particles across the surface. The large hydrostatic pressure under the indenting particle prevents brittle fracture and allows the material to yield plastically. Fracture occurs only when the indenter moves on and leaves behind internal stresses able to splinter the material along its track. Mohs hardness, in fact, correlates well with indentation hardness, each unit of increase on Mohs' scale corresponding to about 60% increase in indentation hardness, as shown in Fig. 10.27.

An indenting particle can detectably scratch a material only 20% softer than itself. To avoid *friction welding* (see Section 10.11) and severe mechanical *wear*, however, an indenter should be much harder than this. Diamond and sapphire have low coefficients of friction, against either

themselves or metals, and so are useful for cutting tools, wire-drawing
dies and bearing jewels. The importance of indentation hardness for wear
resistance is seen in the *hard-surfaced* metals that have been developed,
in which a very hard surface layer is built on to a softer ductile backing.
Various *carburizing* and *nitriding* treatments have been developed to produce hard carbide and nitride particles in the surface of steel. *Hadfield's
austenitic steel* hardens rapidly in its surface when plastically abraded.

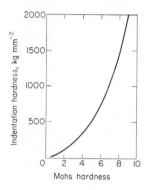

Fig. 10.27. Comparison of Mohs and indentation hardnesses.

Steel is also made wear-resistant by *hard chromium plating*, the hardness
in this case being apparently due to finely dispersed chromium oxide
particles in the metal. Hard surfaces are also made by *shot-peening*, in
which a surface is bombarded and work-hardened by a shower of hard
steel shot.

Polishing, produced by rubbing the surface with a fine polishing powder
on a soft cloth, is not abrasive cutting but *smearing*; friction produces local
hot spots which melt, rub along the surface, then freeze. It depends on
melting points rather than hardness. For example, despite their greater
hardnesses, glass and speculum metal (33% Sn in Cu; m.p. 1020°K) can
be polished by litharge (PbO; m.p. 1160°K), whereas soft nickel (m.p.
1728°K) cannot. Steel is usually polished with refractory oxide powders
such as chromium and aluminium oxides.

The brittle fracture strength, though subsidiary to yield strength in
determining the ability of a material to abrade, sometimes determines the
useful life of an abrasive. Thus, carborundum is harder than corundum
but is also more brittle, and so tends to crush to a fine powder against
hard substances. Large diamond crystals chip rather easily, and *carbonado*,
a cluster of diamond fragments, is preferred for rock drills. The resistance
of diamond to abrasion varies about five hundredfold, according to the
crystallographic plane and direction abraded. The octahedral plane, being

densely packed, is almost impossible to abrade. Their hard octahedral faces enable the grains of diamond powder to abrade soft faces on a diamond. Localized oxidation and transformation to amorphous carbon at frictional hot spots, and localized spalling fracture due to thermal stresses at frictional hot spots, also contribute to the abrasion of diamond.

10.12   PLASTIC ADHESION AND FRICTION

Solid surfaces, even when mirror smooth, usually consist of hills and valleys hundreds of atomic spacings high. When such surfaces are placed together, they touch only at the tips of their tallest asperities, so that the true area of contact is a small fraction of the apparent area and the asperities carry all the contact pressure. With elastic materials, such as rubber and diamond, this pressure brings many atoms into contact by *elastic* flattening of the tips; it is found that the true area of contact then increases as $W^{2/3}$, where W is the compressive load between the surfaces.

With ductile metals and many other materials, the true area (measured, for example, by the electrical resistivity of the contact) increases in direct proportion to W. This, and other observations, has led to the *Bowden–Tabor* theory, in which the tips crush down *plastically* until broad enough

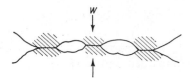

Fig. 10.28.   Plastic junctions (shaded) formed between asperities when ductile solids are pressed together.

to support the load, as in Fig. 10.28. These *plastic junctions*, with a yield pressure P, increase their area of contact A until

$$W = PA. \tag{10.34}$$

Thus, the *true* area A is proportional to W and *independent* of the total area of the surface. From Equation (10.32), we expect

$$P \simeq 5k, \tag{10.35}$$

where k is the (work-hardened) shear strength of the junctions.

The bonding of the atoms brought together by the plastic collapse of the junctions produces *pressure welding*, a process long utilized in the blacksmith's forge for joining white-hot wrought iron bars by hammering together. Unless the welding is done at high temperatures, the force to separate pressure-welded surfaces is usually less than the joining pressure, because some of the junctions are broken by self-stresses when the pressure is removed.

Pressure welding is important in electrical make-and-break contacts. Pressure is needed to give a large contact area to pass the current, but this also encourages "sticking". Silver is a good contact metal because it has a weak affinity for sulphur and oxygen; these elements on its surface prevent sticking without greatly increasing the contact resistance. Moreover, should the temperature rise, by resistance heating in the junctions, the sulphide and oxide decompose and the resistance falls. For very light contacts (e.g. 10 g load) a more noble metal (e.g. Pd, Pt, Ir) is usually necessary. Having such clean surfaces, these metals must also be fairly hard to avoid sticking by pressure welding; thus, pure gold is too soft. Graphite is also suitable for light contacts because it remains fairly clean and does not weld. For heavy-duty contacts, where electrical arc erosion is a problem, silver reinforced with a hard refractory material (e.g. W, WC or Mo) is often used.

Sliding friction can have many causes. We have already discussed elastic hysteresis, which is important in rubber (cf. Section 6.5), wood and some polymers. When a metal is plastically abraded by a lubricated slider, the friction is due mainly to the plastic work done in ploughing the grooves along its surface. If the surface is dry, however, or if the lubricating film breaks down under the applied pressure, *adhesive friction* occurs. Plastic junctions form and their resistance to shear failure gives a frictional force. We expect that the sliding force F should be simply

$$F = Ak \tag{10.36}$$

and hence, in view of Equation (10.34), that the coefficient of sliding friction,

$$\mu = \frac{F}{W}, \tag{10.37}$$

should be independent of the area of contact A, which, in fact, is experimentally observed. However, from Equation (10.35), we also expect

$$\mu \simeq \frac{k}{P} \simeq \frac{1}{5}, \tag{10.38}$$

whereas $\mu \simeq 1$ is commonly observed. This can be explained from the general plasticity equations. The junctions are already incipiently plastic under the pressure W itself, so that a negligible tangential force is needed to start them sliding plastically. Hence, $\mu \simeq 0$ initially. From the Lévy–Mises equations, however, the initial direction of movement is mainly *inwards*, not tangential, because the major force acts inwards. Initially, therefore, the process is rather like that of Fig. 10.21 acting in reverse, and the plastic junctions grow rapidly in size. The tangential force must then be increased to maintain the plastic state and, as F/W gradually increases, so also does the tangential component of sliding. However, so long as W continues to act, some further inward movement and junction growth also continues, so that this process eventually leads to *complete seizure*, i.e. to $\mu \gg 1$. This happens when very clean metals slide over each other under pressure in a vacuum. Under less stringent conditions, however, oxides and other contaminant films weaken the interface against shear. If $s\ (<k)$ is the interfacial shear strength, the junctions grow only until $F = As$, and the mode of deformation then changes from plastic crushing and shearing of the junctions to tangential sliding along the interface. The coefficient of friction becomes

$$\mu = \frac{F}{W} = \frac{sA_1}{5kA_0}, \tag{10.39}$$

where A_0 and A_1 are the areas of contact before and during sliding. If $s \ll k$, then $A_1 \simeq A_0$ and μ is very small. The value $\mu \simeq 1$ is obtained when $s \simeq k$.

When soft metals slide over each other, the junctions are weak but have large area, and so the friction is high; with hard metals the junctions have small area but are strong, and again the friction is high. In *sliding bearings* a film of soft metal is smeared between two hard ones, and in this way weak junctions of small area are obtained, thus reducing the friction. *Bearing bronzes* (e.g. 10% Sn and 10% Pb in Cu) consist of a network of soft lead—the lubricant metal—running through a hard bronze matrix and give a low friction. *Porous bearings*, usually prepared from partly sintered copper with lubricating oil, graphite or polytetra-fluorethylene in its pores, are based on the same principle. The *white metal* or *babbitt* bearings consist of lead or tin stiffened by alloy additions such as antimony and copper. They rely on lubricating oil for low friction; the tin or lead provides a thick pad of fusible metal which can readily melt and smear round to a cooler part of the bearing, so preventing seizure. Where pressures are very high and hydrodynamic and boundary lubrications are no longer effective, solid lubricants are used, e.g. phosphate coatings, graphite, molybdenum disulphide, boron nitride, talc, soapstone and pyrophillite. Lead inclusions act as a built-in lubricant during the machining of free-cutting brass; manganese sulphide likewise in steel.

When the surface of a plastic body is subjected to a frictional force, the slip lines no longer approach that surface at 45°. In Fig. 10.29 we consider again the problem of Fig. 10.17, but this time with friction between the slab and the compression plates. When $\mu = 0$, we have $P (= -\sigma_2) = 1\cdot15\,Y$ on von Mises' criterion. The material flows symmetrically outwards from the centre line AB, so that the frictional stress μP acts in the direction shown. Consider the equilibrium of forces on

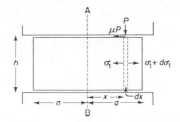

Fig. 10.29. Compression between rough plates.

the element of material between x and $x + dx$. The horizontal forces are balanced when $h\,d\sigma_1 = 2\mu P\,dx$. The plane-strain yield condition $\sigma_1 - \sigma_2\ (=\sigma_1 + P) = $ constant, gives $dP = -d\sigma_1$. Hence,

$$dP/P = -(2\mu/h)\,dx,$$

which integrates to

$$P = A \exp\left[(2\mu/h)(a - x)\right], \tag{10.40}$$

where $A = 1\cdot15\,Y$ since $\sigma_1 = 0$ at $x = a$. The normal pressure P thus rises exponentially towards the centre, giving a "friction hill", as in Fig. 10.30.

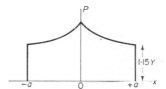

Fig. 10.30. Normal pressure between rough compression plates.

The extra pressure in the centre of the slab is mainly hydrostatic, the slip-line field being distorted by the frictional forces to the form shown in Fig. 10.31. The slip lines no longer approach the surface at 45°, and in the limit where sliding occurs in the surface layers of the slab itself rather than along the interface, they run parallel to the surface. No matter how large P becomes, the tangential friction cannot exceed k, since this is the shear strength of the material itself. Hence, μP (with constant μ) represents the friction only when $\mu P < k$; when P exceeds k/μ, the friction is constant at the value k.

Equation (10.40) shows that when $a \gg h$ the pressure in the centre of the slab becomes extremely high. This raises a problem in the rolling of thin sheets. Various measures are taken to overcome it: e.g. highly polished rolls flooded with lubricating oil to reduce μ to about 0·05;

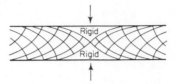

Fig. 10.31.   Slip-line field in a plastic slab between rough plates.

working rolls of small diameter mechanically supported by large "backing" rolls; tension applied to both ingoing and outgoing sheet to reduce P by increasing σ_1.

The analysis of Fig. 10.29 can also be applied to a slab cemented to plates pulled apart by a tension P. In the limit $a \gg h$ this describes a soldered joint in tension. Although the tensile stress P may become very large towards the centre of the joint, the deviatoric stress in the thin layer of metal in the joint remains small, most of the stress being hydrostatic tension. If the joint does not break, it can thus carry a tensile stress far above the ordinary tensile strength of the solder. This is clearly analogous to the behaviour of viscously glued joints (cf. Section 7.10).

10.13   DEFORMATION BY CREEP

The problem of calculating the deformation by creep of, say, a turbine blade remains largely unsolved, despite its importance, because the creep rate is such a complicated function of strain, stress and time. Geometrically simple problems have been solved by simplifying the creep curve; in particular by ignoring the accelerating creep (the specimen being assumed to have failed if this stage is reached) and by either ignoring transient creep or by approximating it to an initial, time-independent strain. The steady-state creep rate $\dot{\varepsilon}$ is then assumed to vary with stress σ as

$$\dot{\varepsilon} = A\sigma^n, \tag{10.41}$$

where A and n are determined experimentally.

A simple but important problem is the relaxation of stress in a bolt under constant strain ε_0. We have $\varepsilon_0 = \varepsilon_e + \varepsilon_c$, where ε_e and ε_c are the elastic $(=\sigma/E)$ and creep strains, respectively, with $\varepsilon_e = \varepsilon_0$ at time $t = 0$. Then

$$\dot{\varepsilon}_e + \dot{\varepsilon}_c = \frac{\dot{\sigma}}{E} + A\sigma^n = 0, \tag{10.42}$$

so that the time for the stress to relax from σ_0 to σ, for $n > 1$, is given by

$$t = \frac{1}{(n-1)AE}\left(\frac{1}{\sigma^{n-1}} - \frac{1}{\sigma_0^{n-1}}\right), \tag{10.43}$$

as shown in Fig. 10.32. Since transient creep is neglected, these curves overestimate the relaxation time initially but improve at longer times. The total transient strain ε_i can be allowed for approximately by relaxing

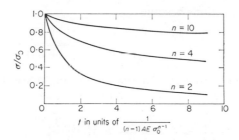

Fig. 10.32. Stress relaxation due to creep.

the stress from σ_0 to $\sigma_0 - E\varepsilon_i$ at $t = 0$. In practical problems it is usually necessary to allow for the elasticity of the joint as well as that of the bolt. The relaxation time is then increased, since there is more elastic strain to relax.

Bending of beams and torsion of shafts in creep have been analysed. An important problem, particularly in aircraft, is *creep buckling* in compressed struts and sheets, which can occur under loads far smaller than those estimated in Section 5.6. As a simple illustration, consider a Newtonian material, i.e. $n = 1$ in Equation (10.41). For small deflexions the stress in a bowed strut under constant end load is proportional to the deflexion. Since the creep rate is proportional to this stress, the rate of deflexion $\dot{y}$ is proportional to the deflexion y, i.e.

$$y = y_0\, e^{\alpha t}, \tag{10.44}$$

so that in each unit of time t_0, where $\exp(\alpha t_0) = 2$ and $t_0 = 0.693/\alpha$, the strut *doubles* its deflexion. The rate of collapse thus increases rapidly with time. In non-Newtonian materials, where the creep rate increases more rapidly with the stress, total collapse by buckling occurs very suddenly after a certain time under load. Methods of estimating these *buckling times* have been developed.

Problems of steady-state creep under multiaxial conditions can be analysed by the general methods of plasticity theory, provided the constants in Equations (10.9) and (10.21) conform to Equation (10.41). Thus, for uniaxial stress ($\sigma_1 = \sigma$; $\sigma_2 = \sigma_3 = 0$) Equations (10.21) become

$$\frac{\dot{\varepsilon}_1}{\sigma} = -\frac{\dot{\varepsilon}_2}{\frac{1}{2}\sigma} = -\frac{\dot{\varepsilon}_3}{\frac{1}{2}\sigma} = A\sigma^{n-1}. \tag{10.45}$$

For multiaxial stresses the stress σ on the right side of this is replaced by an "equivalent stress" Y defined by the von Mises function

$$2Y^2 = (\sigma_1 - \sigma_2)^2 + (\sigma_2 - \sigma_3)^2 + (\sigma_3 - \sigma_1)^2, \tag{10.46}$$

so that the flow equations become

$$\frac{\dot{\varepsilon}_1}{\sigma_1 - \frac{1}{2}(\sigma_2 + \sigma_3)} = \frac{\dot{\varepsilon}_2}{\sigma_2 - \frac{1}{2}(\sigma_3 + \sigma_1)} = \frac{\dot{\varepsilon}_3}{\sigma_3 - \frac{1}{2}(\sigma_1 + \sigma_2)} = AY^{n-1}. \tag{10.47}$$

As a simple example, we consider once more the thin-walled tube under internal pressure p. Since $\sigma_1 = 2\sigma_2$ and $\sigma_3 \simeq 0$, then

$$\frac{\dot{\varepsilon}_1}{\frac{3}{4}\sigma_1} = \frac{\dot{\varepsilon}_2}{0} = -\frac{\dot{\varepsilon}_3}{\frac{3}{4}\sigma_1} = A\left(\frac{\sigma_1\sqrt{3}}{2}\right)^{n-1}. \tag{10.48}$$

Since $\sigma_1 = pr/\delta r$, this becomes

$$\dot{\varepsilon}_1 = -\dot{\varepsilon}_3 = A\left(\frac{pr}{\delta r}\right)^n \left(\frac{3}{4}\right)^{(n+1)/2}, \tag{10.49}$$

with $\dot{\varepsilon}_3 = 0$. The strains due to steady-state creep in, for example, a boiler tube can thus be evaluated. Solutions of more complicated problems have been derived by similar methods.

Under multiaxial tension the rate of creep along the main tensile axis may be considerably reduced. This can affect ductility under creep conditions. Experiments have shown that the time to rupture by grain boundary cracking (cf. Section 8.9) is determined mainly by the highest tensile stress, irrespective of the other stresses. Designs based only on allowable creep strains may thus lead to failure by creep rupture, unless the effect of tensile stress on the rupture life is taken into account.

The effects of fluctuating temperatures and stresses are important in some problems. We saw in Section 4.7 that intergranular strains of order $\Delta\alpha\Delta T$, set up in polycrystalline materials of thermal expansion anisotropy $\Delta\alpha$ by temperature changes ΔT, can produce plastic deformation. Such thermal cycles can thus render the material *spontaneously plastic*, with an oscillating plastic strain rate

$$\dot{\varepsilon}_i \simeq [\Delta\alpha\Delta T - (Y/E)]/\tau, \tag{10.50}$$

where τ is the cycling time. Under a small externally applied stress σ *such a material behaves like a Newtonian fluid, whatever its intrinsic plastic properties*, because its yield strength Y has already been overcome by the intergranular stresses; the applied stress, being small compared with these (i.e. $\sigma \ll Y$), merely perturbs the existing motion in the material slightly. By the kind of argument used in Chapter 7, this perturbation is proportional to the stress which produces it. From the Lévy–Mises equations, we thus expect the external creep rate $\dot{\varepsilon}$ to be given by

$$\frac{\dot{\varepsilon}}{\sigma} \simeq \frac{\dot{\varepsilon}_i}{Y}, \qquad (10.51)$$

provided $\sigma \ll Y$. The coefficient of Newtonian viscosity for the external creep is then

$$\eta = \frac{\sigma}{\dot{\varepsilon}} \simeq \frac{Y}{\dot{\varepsilon}_i} \simeq \frac{Y\tau}{[\Delta\alpha\Delta T - (Y/E)]}. \qquad (10.52)$$

Suppose, for example, that $\tau = 10^5$ sec ($\simeq 1$ day), $\Delta\alpha = 25 \times 10^{-6}$ per °K, $\Delta T = 400°$K, $Y/E = 4 \times 10^{-3}$. Then $\dot{\varepsilon}_i \simeq 6 \times 10^{-8}$ sec^{-1}. An applied stress of only $10^{-2}Y$, for example, could then produce an external strain rate of 6×10^{-10} sec^{-1}. At this rate a bar, say 50 in. long and 1 in. diameter, could bend to a deflexion of 1 in. at its centre in about 1 month. Thermally anisotropic polycrystals may thus creep significantly even under their own weight, unless the temperature is held constant. Similar effects are possible in multiphase solids where grains of different substances have different expansion coefficients.

From its derivation, it is clear that this type of creep is not limited to the case where the self-stresses are produced by thermal strains. Any other source of spontaneous plasticity (e.g. neutron irradiation growth in uranium; mechanical vibrations; electromagnetically induced oscillations) could also produce it. Indeed, the Newtonian flow of a true liquid might be interpreted similarly, since small regions deform spontaneously by thermal motion and the applied stress merely perturbs this motion slightly (cf. Section 7.1.)

FURTHER READING

PHILLIPS, A., *Introduction to Plasticity*, Ronald Press, New York, 1956.

PRAGER, W., *An Introduction to Plasticity*, Addison-Wesley, New York, 1959.

HILL, R., *The Mathematical Theory of Plasticity*, Oxford University Press, London, 1950.

PRAGER, W. and HODGE, P., *Theory of Perfectly Plastic Solids*, Wiley, New York, 1951.

NADAI, A., *Theory of Flow and Fracture of Solids*, McGraw-Hill, New York, 1950.

HOFFMAN, O. and SACHS, G., *Introduction to the Theory of Plasticity for Engineers*, McGraw-Hill, New York, 1953.

JAEGER, J. C., *Elasticity, Fracture, and Flow*, Methuen, London, 1956.

BOWDEN, F. P. and TABOR, D., *The Friction and Lubrication of Solids*, Oxford University Press, London, 1950.

TABOR, D., *The Hardness of Metals*, Oxford University Press, London, 1951.

ELEY, D. D., *Adhesion*, Oxford University Press, London, 1961.

BRIDGMAN, P. W., *Studies in Large Plastic Flow and Fracture*, McGraw-Hill, New York, 1952.

OROWAN, E., *Rept. Progr. Phys.*, **12** (1948–9).

OSGOOD, W. R. (ed.), *Residual Stresses in Metals and Metal Construction*, Reinhold, New York, 1954.

JOHNSON, A. E., "Complex-stress Creep of Metals", *Met. Rev.*, **5**, 447 (1960).

LUBAHN, J. D. and FELGAR, R. P., *Plasticity and Creep of Metals*, Wiley, New York, 1961.

PROBLEMS

10.1   A plane frame is made from three steel bars, each 1 in^2 cross-section and 10 in. long, pin-jointed together at their lower ends and pin-jointed to a circular ceiling at their upper ends, so that the central bar is vertical and the others inclined at 60° to it. The steel has a Young's modulus of 3×10^7 psi, a uniaxial yield stress 6×10^4 psi, and yields at constant stress. A vertical load is suspended from the lower end. Sketch the load displacement diagram and find the values of load and displacement at the elastic load and the plastic collapse load.

[9×10^4 lb, 0·02 in.; 12×10^4 lb; 0·04 in.]

Does shakedown occur if the load is allowed to vary from zero up to any point below that for plastic collapse?

10.2   Before going into service, engineering components such as boilers are often deliberately overloaded. It is claimed that, in addition to testing the strength of the structure, this has the advantage of reducing unfavourable initial self-stresses and of inducing favourable self-stresses at stress concentrators. Explain this.

10.3   A steel bolt with Young's modulus $E = 3 \times 10^7$ psi holds two rigid plates together at a high temperature. Experiments have shown that the steel obeys Equation (10.41) with $n = 3$ and that $\dot{\varepsilon} = 2·8 \times 10^{-8}$ per hour under a stress of 4000 psi. The bolt is initially stressed to 10,000 psi. What is the stress in it after 1 year?   [2050 psi]

10.4   When a solid circular shaft of elastic–plastic material is twisted, radial lines in it remain straight. Show that the shaft becomes fully plastic at a torque 1·333 times that at the elastic limit.

10.5   A uniform beam with built-in ends carries a uniformly distributed load. Locate the positions of plastic hinges in it at the point of plastic collapse and show that the collapse load is at least one-third greater than that at which yielding begins.

10.6 An 8 in. by 8 in. T-section beam of uniform thickness 1 in. is bent about an axis perpendicular to the web. Show that the fully plastic moment is $1 \cdot 87$ times the elastic moment.

10.7 Show, by writing down the elastic strain energy of an isotropic elastic solid in terms of principal stresses and strains, and by separating off the dilatational part, that the von Mises yield criterion is equivalent to the assumption that yield takes place when the elastic shear strain energy reaches a critical value.

10.8 A thin-walled circular tube of radius r and wall thickness t which obeys von Mises' criterion yields plastically under an internal pressure p. This pressure is removed and the tube is then twisted by couples applied to its ends. Show that it yields when these couples are equal to $\pi r^3 p$. If the pressure is only half released, show that yielding occurs when couples of $0 \cdot 866$ of the above value are applied.

10.9 A thick-walled circular tube, of inner and outer radii r_0 and r_1, respectively, is brought to the point of general plastic yielding by an internal pressure. Noting that the directions of the principal stresses are determined by the symmetry of the problem, show that the slip lines, for a plastic–rigid solid, are spirals, the polar co-ordinates of a point r, θ on any one of them being determined by the relation $r = A \exp(\pm \theta)$, where A is a constant. Sketch the slip-line field and deduce Equation (10.8) for the collapse pressure.

10.10 A free single crystal of uranium, when irradiated in a nuclear reactor, elongates spontaneously along one crystal axis at a rate of 4×10^{-9} sec^{-1} and contracts equally along a perpendicular axis, so keeping its volume constant. Its yield strength is 5×10^4 psi. The same metal, as a poly-crystal with random crystal orientations, is similarly irradiated while subjected to an externally applied tensile stress of 100 psi. Estimate approximately its rate of creep. $[8 \times 10^{-12}$ sec$^{-1}]$

chapter **11**

Fracture of Solids

*In this chapter we shall consider how cracks form and grow
in solids. The ideal breaking strength, as determined from
the atomic bond strength, can be realized in specially prepared
solids, but everyday samples usually break at far lower stresses.
In brittle solids this is because sharp cracks act as stress
concentrators; in ductile solids because plastic deformation
occurs and enables the cross-section to diminish by localized
plastic straining. In many forms of fracture the processes of
yielding and breaking are closely linked. Cleavage cracks
in some crystalline solids are formed from strain
concentrations produced by deformation twins or slip bands,
so that the material breaks as it yields. Nevertheless, in a
notched sample the overall fracture stress may lie
below that for general yielding, because the localized plastic
deformation at the root of the notch is accommodated within
purely elastic surroundings. The fracture toughness of a
material, which is important for determining its bulk strength,
depends primarily upon the ductility at the roots of notches.
Fatigue fracture in metals is a process brought about by
alternating plastic deformations localized in the roots of
notches and in certain slip bands. Some progress has been
made in understanding it, but many problems remain.*

11.1 BRITTLE FRACTURE

Many solids, e.g. glass, appear brittle when pulled or bent at low temperatures. While still elastic they suddenly break, by the growth of a sharp crack across the line of tension, and the broken bits can be fitted together afterwards to re-form the original shape. Ordinary window glass usually breaks at about 10,000 psi; cast-iron at about 30,000 psi. Brittle solids commonly break at about $E/1000$, the ideal strength $E/10$ being approached only in specially prepared bodies such as fibres. Because it is difficult to make and test tensile specimens, brittle solids are often tested by bending. The highest tensile stress in the surface is determined, from the bending moment and the geometry of the cross-section, and the value of this at fracture is called the *modulus of rupture*. In highly brittle solids this is approximately equal to the tensile strength, but it is often somewhat larger, by a factor of about 2, in less brittle ones. Simple compression tests on short cylinders, similar to those on clays (cf. Section 10.4), are also made. Brittle solids often have useful strengths in compression, e.g. 10,000 psi (stone), 50,000 psi (glass) and 100,000 psi (cast-iron), and so are used in buildings, masonry dams, etc. to support compressive thrusts. Their relatively high strengths in compression, compared with those in tension, are due to the hydrostatic component of the compressive stress, which tends to hold them together. In compression they mainly break by sliding and crumbling along a path of shear failure such as that of Fig. 10.10.

Brittle solids are weak because sharp notches, corners of surface steps and various flaws, produce intense and localized concentrations of applied

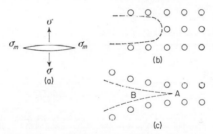

Fig. 11.1. Cracks and notches: (a) as concentrators of tensile stress; (b) a simple notch; (c) a simple crack.

stress. For a narrow notch or slit in a plate, as in Fig. 11.1(a), the applied tensile stress σ is increased to

$$\sigma_m \simeq 2\sigma\sqrt{\frac{c}{\rho}} \qquad (11.1)$$

within a distance of order ρ round the tip, of radius ρ, where $2c$ is the length of the crack, or c if a surface crack (cf. Equation (5.78)). When σ_m is equal to the ideal tensile strength σ_t (Equation (8.13)), we have

$$\sigma \sim \sqrt{\frac{\gamma E \rho}{4bc}} \qquad (11.2)$$

for the nominal breaking strength, with γ as the surface energy per unit area of crack surface.

Since $\sigma \to 0$ as $\rho \to 0$, can a sharply cracked body have any strength at all? For a *notch*, made by removing a slice of material as in Fig. 11.1(b), the atomic spacing clearly sets a lower limit to ρ; but what about a true crack, as in diagram (c), the faces of which converge tangentially? A. A. Griffith first recognized that, even with the sharpest crack, there is always some strength, because surface energy is created when the crack spreads and the energy for this must come from work done by the applied stress. For a crack at rest in an elastic solid there are three energies to consider: (1) external work, (2) surface energy of the crack, (3) elastic energy of the body. The first drives the crack forward. The second opposes this. The third can drive or oppose, according to the type of loading. Figure 11.2 shows two extreme types: *dead loading*, in which the body supports a constant force f and is free to elongate; and *fixed grips*, in which the displacement u is held constant. Both give the same criterion for the growth of the crack. Let f be applied when the crack has a certain length and let f elongate the body by u. Then, P represents the mechanical state and the

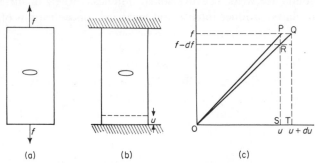

(a)   (b)   (c)

Fig. 11.2. Loading systems: (a) dead loading; (b) fixed grips; (c) load elongation diagram.

area OPS the elastic energy in the body. Now let the crack slightly increase its length at constant f, so that the body elongates to $u + du$ (P → Q). The work done is PQTS, i.e. $f\,du$. The elastic energy has increased by only OQT − OPS, i.e. $\tfrac{1}{2}f\,du$. Hence, the remaining external work $\tfrac{1}{2}f\,du$ is free to become energy of the extra crack surface. Starting from P again, now let the crack increase its length at constant elongation u (P → R). No external work is done, but the elastic energy falls from OPS to ORS; i.e. $\tfrac{1}{2}u\,df$ is given to the crack surface. In the limit of small PQ and PR, $\tfrac{1}{2}u\,df =$ OPR = OPQ = $\tfrac{1}{2}f\,du$, so that the energy given to the crack is the same in both cases, even though the elastic energy *increases* under dead loading and *decreases* under fixed grip conditions.

To estimate the elastic energy of the crack, we consider a small crack of length $2c$ transverse to a tension σ applied by fixed grips. Very near the crack faces this stress is relaxed to zero. Very far from them it is unchanged. Hence, roughly, a region of radius c round the crack is relieved of its elastic energy, $(\sigma^2/2E)\pi c^2$ per unit thickness of the body. The exact value obtained by integrating the actual strain field is precisely twice this. When this crack is on the point of growth under the stress σ, the decrease of elastic energy just balances the increase of surface energy, i.e.

$$\frac{d}{dc}\left(\frac{\pi c^2 \sigma^2}{E}\right) = \frac{d}{dc}(4\gamma c), \tag{11.3}$$

the 4 appearing because two surfaces grow at each end of the crack. Thus,

$$\sigma = \sqrt{\frac{2\gamma E}{\pi c}} \simeq \sqrt{\frac{\gamma E}{c}}. \tag{11.4}$$

This is the *Griffith stress* for the growth of a sharp crack. The same formula applies with minor numerical variations to cracks under plane-strain and plane-stress conditions, surface and internal cracks and penny-shaped cracks.

Equations (11.2) and (11.4) show that surface energy sets a lower limit $\rho \simeq 8b/\pi$ to the *effective* sharpness of a crack or notch. Sharper ones spread at the Griffith stress; blunter ones need a larger stress to set them off, but the stress for growth drops to the Griffith value as they start growing and become sharp. At the tip of a true crack, e.g. between A and B in Fig. 11.1(c), there are atomic bonds in every stage of elongation and rupture, so that the structure of the crack itself ensures the existence of bonds stressed to the limit σ_t, whatever the applied stress. Hence, Griffith's criterion is *sufficient*, as well as necessary, for the spreading of true cracks in tension. The atomic bond strength σ_t, in fact, does *not* determine the strength of a solid containing true cracks, except indirectly through its link with surface energy and Young's modulus (although it determines the width of the transition, A B in Fig. 11.1). The real resistance to crack propagation comes from the surface energy; as the crack advances by one atomic spacing, each atomic bond at the crack front takes up the strain that previously belonged to its predecessor, and the sum of all these increments of strain, from A to B, is equivalent to the complete rupture of one bond, the work required to do this being the surface energy.

Griffith's theory is based on the ordinary elastic theory of infinitesimal deformations, and so does not apply to highly deformable elastic bodies. Rubber is brittle, but it strains elastically so much at low stresses that slits in it distort into fairly harmless blunt shapes before the breaking stress is reached. Rubber owes its great wear resistance to this elastic resilience and notch insensitivity. For example, it indents without fracturing when rolled over a stone on a road; and then recovers its original shape as it moves on.

Griffith's formula has been confirmed by measuring the strengths of glass containing sharp cracks of known lengths; e.g. 300 psi for a 1 in. crack. The stress concentrators responsible for the usual strength of glass ($\simeq 10,000$ psi) are small surface scratches, about 10^{-4} cm deep. They can be detected by "decorating" them with droplets of sodium condensed on to the surface from vapour. Prolonged etching with hydrofluoric acid, to remove the cracked layer and so produce a "flawless" surface, can raise the strength of bulk glass almost to that of fibres. The brittleness of rock-salt crystals is also due mainly to surface flaws. When immersed in warm water to dissolve their surface, these crystals become ductile (the *Joffé effect*). Their brittleness usually returns on drying. This is because some salt crystallizes out from the evaporating water layer on the wet surface and creates stress-concentrating ledges. If this salt water is removed, the ductile condition can be preserved in a dry crystal.

The importance of surface stress concentrators is recognized in practical materials. Glass for windscreens is "toughened" by a heat treatment that induces *compressive* self-stresses in the surface which oppose the growth of surface cracks. Pottery is strengthened by *glazing* its surface with a glass of low thermal expansion coefficient, which goes into compression when cooled from the glazing temperature. The strong glass filaments in fibreglass are protected, by lubricant and resin, from touching and scratching one another.

The fracture strengths of brittle substances are sensitive to chemical effects, which range from simple lowering of the surface energy by adsorption (cf. Section 8.9) to the formation of notches by dissolution. Mica cleaves in air at about one-third of its cleavage stress in vacuum; *Orowan* showed that this could be explained in terms of adsorption on the crack faces, which lowers the surface energy from 4500 erg cm^{-2} (vacuum) to about 400 erg cm^{-2} (air) and which thus, from Griffith's formula, lowers the strength by about three. Glass can be broken by the prolonged action of a stress too small to break it immediately (*delayed fracture*; also called *static fatigue*). The sudden, spontaneous shattering of glassware appears to be due to the same effect, in this case due to the prolonged action of self-stresses left in it by faulty heat treatment. It seems that the surface cracks are initially too short to satisfy Griffith's criterion, unless their surface energy is lowered by adsorption. They thus grow slowly, at a rate controlled by the migration of the adsorbing substance, until large enough to satisfy Griffith's criterion with the full value of the surface energy, at

which point sudden complete fracture occurs. Surface-active agents are
sometimes used to ease the drilling of hard rocks. For example, if a little
aluminium chloride is added to the water used when drilling quartz rock,
the speed of drilling may be raised by about 60% and wear on the drill
reduced about 40%. Adsorption sometimes has other effects besides
altering surface energy. The fracture stress can be raised if the adsorbed
substance binds together the faces of the crack; in dry clay and coal, on
the other hand, the adsorbed molecules appear to reduce the strength by
prising the crack faces apart.

11.2 STRONG FIBROUS AGGREGATES

Striking evidence for the very large ideal strengths of solids is provided
by the properties of very fine filaments of dislocation-free (e.g. metal
whisker crystals) or intrinsically hard (e.g. oxide whiskers; glass fibres)
materials, as given in Table 8.2. The process of drawing a glass fibre
elongates cracks initially present along the fibre, where they become
harmless. Any cracks subsequently formed transverse to this axis are
short and separated by long, almost flawless lengths of the fibre.

Although these high strengths set an ideal to which designers of engineering materials aspire, it seems unlikely that *large* flawless bodies will
ever be developed for engineering use, although they have been made in
the laboratory and shown to have great strength. Apart from the difficulties and expense of manufacture, there would always be a risk of flaws
forming on them during service, e.g. by the impact of blown specks of
grit or by chemical attack. The more practical approach is to admit the
existence of cracks and, as with dislocations, try to block their progress
with obstacles. The best obstacle is a *weak* perpendicular interface, as
shown in Fig. 11.3. The load not supported by the members A and B,

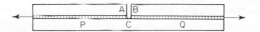

Fig. 11.3. A crack obstructed by a weak interface.

because of the crack between them, is then transmitted to the adjoining
member C in diffused form, over a long interface PQ, instead of being
concentrated at the tip. A bundle of parallel fibres joined by a soft adhesive has this property. If there is a transverse notch across some fibres,
the force released from the cut fibres is transmitted to the others as a
shearing force through the layers of adhesive. If this adhesive has a low
resistance to shear, so that one fibre can slide to some extent along another,
the force is transmitted across a long interface. In this way advantage can

be taken of the high strengths of brittle fibres without suffering the notch sensitivity normally associated with brittleness.

This is, in fact, the structure of some of the strongest and most reliable materials, both natural and man-made. As regards natural materials, we recall the fibrous proteins such as whalebone and horn; wood and cane; and inorganic materials such as jade (fibrous silicate). Among the man-made materials we have fibreglass, reinforced plastics, drawn metal wire, wrought iron (soft, but resistant to cracks), spun fibres and rope.

In fibreglass, and other aggregates based on fibres such as asbestos and cellulose, the fibres are glued together compactly by a substance of lower elastic modulus, usually a resin. A relatively weak interface between glue and fibre is necessary to interrupt cracks, but the fibres must be long in relation to their diameter, so that the total shear force transmitted through this interface is sufficient to exercise the tensile strength of the fibre. The product of interfacial area and shear strength of the interface should thus be comparable with the tensile breaking load of the fibre. The strength of a parallel bundle prepared in this way is then of order of the average strength of the individual fibre multiplied by the fraction of the cross-section occupied by fibres. This is a one-directional strength, but two- and three-directional strengths can be achieved by assembling the fibres randomly in plane and solid "mats". Isotropic two- and three-dimensional strengths, respectively one-third and one-sixth the one-directional strength, are possible. Fibreglass rods and tubes containing about 75% glass have bend strengths of about 200,000 psi. Made up into isotropic plane mats, tensile strengths of about 60,000 psi are reached. An advantage of many of these fibrous aggregates is their lightness, being composed of elements such as oxygen, silicon, carbon and aluminium. They thus have high strength–weight ratios.

The principle of increasing the fracture strength and crack resistance (i.e. *toughness*) of a brittle material by interrupting the continuity with a softer crack-resistant substance, is also used in tungsten carbide cutting tools. Small, hard, brittle tungsten carbide crystals are cemented together by a thin film of cobalt, which wets them and forms a strong, continuous bond. Being ductile, it prevents cracks from spreading, but is too thin to collapse plastically.

11.3 PROPAGATION OF BRITTLE CRACKS

Once a brittle crack grows beyond its Griffith length, Equation (11.3) is no longer satisfied. The crack becomes unstable and accelerates, and the excess work done by the applied load becomes kinetic energy as the crack faces fly apart. The kinetic energy cannot increase faster than energy is supplied from external work, and this sets an upper limit to the speed of an elastic crack, which has been shown to be about $0.38\ c_0$, where c_0 is the longitudinal elastic wave velocity (Equation (6.8)). At this crack speed an atomic bond in the crack-front is strained to its limit in a time of order one atomic vibration, which is the highest rate the material can be strained in tension. Table 11.1 shows that measured crack speeds agree approximately with the theoretical value.

TABLE 11.1
Theoretical and Observed Maximum Crack Velocities (10^5 cm sec^{-1})

	Steel	Glass	Cellulose acetate	Irradiated polyethylene	Silicone rubber
Theoretical ($0.38\ c_0$):	2·0	2·0	0·4	0·05	0·008
Observed:	1·0–1·8	1·5–2·0	0·3	0·05–0·09	0·002–0·008

One method of measuring crack speeds is to inject ultrasonic waves into the specimen while the crack is running. These produce undulations in the path of fracture, which can be seen as "ripple" markings on the fracture surface, indicating successive positions of the crack-front. From their spacing and from the ultrasonic wave frequency the crack speed can be found. This technique was suggested by the observation of *Wallner lines* on fracture surfaces, i.e. ripple marks caused by the interaction of the crack with stress waves radiated from it and reflected back from nearby surfaces. Such stress waves can completely shatter brittle materials by starting secondary fractures.

Many other features are seen on fracture surfaces. New cracks sometimes nucleate ahead of the main crack, not in its own plane, and steps are produced where they join the main front, as shown in Fig. 11.4. They form parabolic markings which point back to a smooth region, called the *mirror area*, where the original fracture started (diagram (b)), and they give the fracture surface a seashell-like appearance from which the name *conchoidal*, applied to fractures in glass and brittle polymers, derives. Sometimes a new crack-front may be completely overtaken by the main crack, giving the ripple structure of diagram (c). Related to these effects is the branching or *bifurcation* of the crack, which is familiar in glass. None of these effects appear until the primary crack has run smoothly a certain distance from its source and gathered enough speed and kinetic energy to set off new fractures in the stress wave at its tip. At high crack velocities

nearly equal tensile stresses exist over a wide arc about the tip, which
encourages branching.

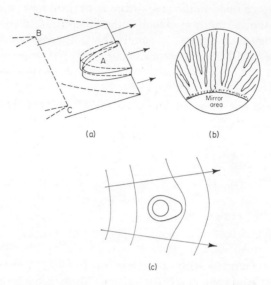

(a)   (b)

(c)

Fig. 11.4. Steps on fracture surface: (a) due to new crack nucleated at A when crack-
front was at BC; (b) radiating back to original source of fracture; (c) new crack-front
completely overtaken by main crack-front.

Most brittle crystals fracture by *cleavage* along certain crystal planes
(there are exceptions, e.g. quartz). Almost perfect crystal faces can be
exposed, on some crystals, if a sharp knife is placed along the edge of the
intended cleavage plane of a firmly supported specimen and then struck,
sharply and lightly, with a hammer. There is very little crushing and
cutting by the knife, because the brittle crack runs on ahead of it. Layer
crystals, such as mica, talc and graphite, weakly bonded between layer
planes, can be split easily along these planes. Many cubic crystals cleave
well, usually on close-packed planes across which there are fewest bonds
(e.g. octahedral planes in diamond). Most cubic ionic crystals, such as
MgO, NaCl and LiF, cleave on cube planes and it is probable that low
surface energy is the determining factor. This picks out widely spaced,
close-packed planes across which Young's modulus is small as the most
likely cleavage planes. B.C.C. transition metals cleave on cube planes,
and most hexagonal metals cleave on basal planes. F.C.C. metals and
alloys, unless weakened on certain planes by chemical effects (e.g. stress-
corrosion), do not cleave.

11.4   GENERAL CONDITIONS FOR FRACTURE

We turn now from specifically brittle fractures to more general aspects. To spread a crack or notch, two distinct conditions have to be satisfied. First, the *thermodynamics* must be favourable, i.e. the free energy released by the fracture must be large enough to make the fracture. This includes Griffith's criterion as a special case. Second, it must be possible to operate an *atomic mechanism* of fracture. The difference between these conditions is shown by considering particular examples. In a cleavage such as that of Fig. 11.1(c) the atomic mechanism of fracture is ensured by the structure of the crack itself, but if the crack is too short, the thermodynamics requires it to *close*, not open. Conversely, a blunt notch in an elastic solid cannot grow at the Griffith stress, despite the favourable thermodynamics, because the stress at the tip is too small to break the atomic bonds there. Indeed, the Griffith criterion is equally well satisfied if the applied stress across the notch is *compressive*, even though the atoms are then being pushed together at the tip!

The thermodynamic condition is therefore *not always sufficient* to ensure fracture, when the applied stress has the entire task of parting the atoms at the tip. *It becomes sufficient, however, when other means of separating those atoms exist*; for example, by evaporation (possibly with condensation elsewhere on the notch or rest of the body), diffusion (along the surface of the notch or through the body) and chemical dissolution. Under these conditions blunt notches can grow, even under compressive stresses, provided they are long enough to satisfy Equation (11.3) (with a γ sometimes reduced by adsorption and by the release of chemical free energy). This is important both for high-temperature fractures (cf. Section 8.9) and for various types of *stress–corrosion* fractures in chemically active environments. When stressed metals are exposed to certain chemical agents (e.g. ammonia vapour on brass), they sometimes develop deep fissures along active slip planes (particularly those containing wide dislocations in metals and alloys of low stacking-fault energy) or along grain boundaries. When the chemical agent is sufficiently selective and aggressive, practically all the surface energy can be supplied by the free energy of the chemical reaction and the fracture may then spread under extremely low stresses. Organic polymers also suffer from stress–corrosion cracking. Their crack resistance is due to molecular strands which join the sides of the crack, like minute ropes, and so prevent all the load concentrating at the tip. Organic solvents attack these molecules and allow them to slip along one another, so releasing their load.

Even when chemical and other external sources of free energy are absent, the thermodynamic condition does not always reduce to Griffith's condition. There are other ways in which the processes of fracture can cause the applied load to do work, besides the elastic deformation which is the basis of Griffith's condition. For example, in fracture by *vacancy creep* the external work is done, not by elastic displacement, but by crystal growth on the grain boundaries (cf. Fig. 8.21). The applied stress is too small to satisfy Griffith's condition, and the elastic energy is thus unable to prevent the surface energy from spheroidizing the holes (by atomic migration). This stress is nevertheless large enough to satisfy the thermodynamic condition for growth by vacancy creep (if the inequality (8.45) is satisfied) and these spheroidal holes can grow. In this way fracture is possible at extremely low stresses.

Plastic ruptures such as those of Fig. 10.18 provide another example. Here the external work is done through plastic glide and the thermodynamic condition for this also differs from the Griffith condition. Surface energy is usually unimportant, because very much more energy is expended as plastic work along slip lines. The thermodynamic condition is then simply that the applied stress (allowing for work-hardening and plastic constraint effects) should equal the general yield stress. It is this large plastic work, for plastic rupture, that makes ductile materials *tough*. Even though usually softer than brittle materials, they absorb large amounts of mechanical energy before breaking, especially when work-hardening prevents plastic strain concentrations. They thus have good *shock resistance*, being able to absorb energy of impact by plastic distortion.

11.5   ELASTIC–PLASTIC FRACTURES

In *elastic–plastic* solids it is possible under certain conditions, as in the *tearing* of thin sheets such as aluminium foil, for plastic rupture to occur *below* the general yield stress. This is important since it appears to threaten the basis of plastic design. The essential condition is that the *plastic* displacement at the tip of the crack must be accommodatable by *elastic* displacement of more distant parts of the material. There is then no need for plastic zones to span the entire load-bearing cross-section. To illustrate this, we consider in Fig. 11.5 two simple forms of plastic rupture, under plane-strain and plane-stress conditions, in a fully ductile material. The argument can be generalized to other modes of fracture and less ductile materials. In the plane-strain example each incremental growth of the notch requires more edge dislocations to be injected into the slip lines, as shown. These push their predecessors forward, and the plastic front spreads, across the section, much faster than the crack front. Although the fracture may start from the initial notch at a low applied stress, it cannot spread far, unless the stress is raised, since the yield zones have to extend over a rapidly increasing proportion of the cross-section. They must, in fact, spread completely across before the crack has advanced far. The crack is therefore *stable* (i.e. requires an increasing stress to grow) below the general yield stress. We may call this a *non-cumulative* process,

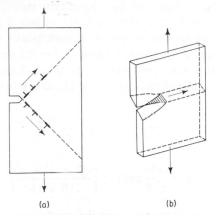

(a)   (b)

Fig. 11.5. Simple types of (a) plane-strain and (b) plane-stress ruptures in a fully ductile elastic–plastic solid.

implying that the dislocations formed in the earlier stages of crack growth do not contribute to the later stages of growth and more dislocations have to be created at every stage.

By contrast, the plane-stress process in diagram (b) is *cumulative*. Here the direction of sliding is transverse to the direction of crack growth, on a plane similar to that of Fig. 10.13(b), and the elastic–plastic distortion of the fracture is represented as a group of screw dislocations running along the shear plane. The important point is that each such dislocation produces an increment of sliding-off on *every* element of plane swept by it. The original set of dislocations, sufficient to produce complete sliding-off at the starting end of the plane, remains sufficient to slide all the rest of the plane off. It has merely to glide forward to accomplish this. If the sheet is wide and thin, only a small part of the shear plane is plastic at any instant and the crack can propagate unstably below the general yield stress.

Plastic zones in practice are usually less simple than those of Fig. 11.5, but the same principles apply. Unstable low-stress fractures are also possible in thick plates, where conditions approximate more to plane strain than to plane stress, when fracture nuclei are formed ahead of the main front, as in Fig. 11.6. Here the fracture occurs by necking of the bridges between cavities, and this deformation, because it lies along the path of the crack, is *cumulative*. The elongation of the bridges can, in fact, be represented formally by the advance of a group of edge dislocations along the path of fracture, as in Fig. 5.30(a).

How wide does a plate have to be to hold such a group of dislocations? This problem has been solved for a few simple cases, both by formal elastic–plastic theory and also by dislocation theory from solutions of Equation (5.89). Consider an internal crack of length $2c$ and suppose that dislocations run forward in its plane, symmetrically up to a distance R from each end, against a shear resistance k. Let the nominal applied shear stress be σ, where $\sigma < k$. (We ignore any effect of loss of cross-sectional area in the region immediately ahead of the crack-front.) It can then be shown that

$$\frac{c}{R + c} = \sin\left\{\frac{\pi(k - \sigma)}{2\sigma}\right\}$$ (11.5)

and that

$$u = \frac{2ck}{\pi^2 A} \cosh^{-1}\left[\frac{(R + c)^2 + c^2}{2(R + c)c}\right],$$ (11.6)

where u is the plastic displacement at the crack-front and $A = \mu/2\pi(1 - v)$ for edge dislocations and $\mu/2\pi$ for screws (μ = shear modulus, v = Poisson's ratio). Table 11.2 gives some numerical values for $A = \mu/2\pi$ and $k = \mu/1000$. In a cumulative fracture of the type in Fig. 11.6, u is of order

TABLE 11.2

Plastic Displacement and Length of Plastic Zone at the Tip of a Crack

$\dfrac{\sigma}{k}$:	0·1	0·5	0·75	0·9	0·95	0·99	0·999
$\dfrac{R}{c}$:	0·01	0·41	1·61	5·4	11·75	62·7	635
$\dfrac{10^4 u}{c}$:	0·185	4·41	12·1	23·6	32·4	52·9	82·3

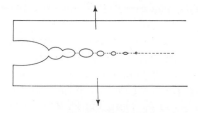

Fig. 11.6. Crack propagation by coalescence of fracture nuclei ahead of the crack-front.

of the spacing of the fracture nuclei. In a polycrystalline metal this might typically be about the grain size; e.g. $u \simeq 0 \cdot 01$ cm. Then a notch of length $2c \simeq 46$ cm could grow at one-half of the general yield stress in a wide plate. For much smaller notches and plates the fracture would not spread appreciably until the general yield stress was reached across the notched section. A convenient criterion for *low-stress* notch sensitivity due to cumulative fractures is

$$\frac{u}{c} \leqslant \frac{k}{\mu},\qquad(11.7)$$

which gives a fracture stress less than 70% of the general yield stress. As a second example, we now let Fig. 11.6 represent a vacancy creep fracture (cf. Fig. 8.21) and write $k \simeq \gamma/r_0 \simeq \mu b/8r_0$, using Equation (8.15), where r_0 is the initial radius of the holes. There is then little sensitivity to notches smaller than $c \simeq 8r_0 u/b$. If $r_0 \simeq 10^{-4}$ cm, $u \simeq 3 \times 10^{-4}$ cm, $b \simeq 3 \times 10^{-8}$ cm, then $c \simeq 8$ cm, so that once again long cracks and wide plates are needed. The condition for low-stress notch sensitivity in ductile materials thus lies beyond the range commonly encountered in plastic design, except in very thin sheets where u is limited by sheet thickness.

For small stresses, $\sigma \ll k$, Equations (11.5) and (11.6) reduce to

$$R = \frac{\pi^2}{8}\left(\frac{\sigma}{k}\right)^2 c \simeq \frac{\sigma^2 c}{k^2}\qquad(11.8)$$

and

$$u = \frac{\pi c \sigma^2}{2\mu k}.\qquad(11.9)$$

In the form

$$\sigma = \sqrt{\frac{2\mu u k}{\pi c}} = \sqrt{\frac{2\mu \gamma_p}{\pi c}},\qquad(11.10)$$

the second of these is very similar to Griffith's equation, with $\gamma_p = uk$ in place of the surface energy. Since uk is the plastic work required to produce unit area of fracture, this equation, in fact, is strictly analogous to Griffith's equation but with plastic work along the fracture path acting as *effective* surface energy. This is a natural way to interpret cumulative plastic fractures *when the applied stress is very small* and the plastic deformation is localized in a narrow zone ahead of the crack. The corresponding formula expressed in tensile stresses,

$$\sigma \simeq \sqrt{\frac{E\gamma_p}{c}},\qquad(11.11)$$

usually known as the *Orowan–Irwin formula*, was first derived from this analogy with Griffith's equation. For a propagating crack these formulae correctly give the stress only when the plastic deformation is cumulative. When the plastic zones spread above and below the plane of the crack, tending towards non-cumulative conditions, the propagation stress of even a long crack approaches the general yield stress.

11.6 THE DUCTILE–BRITTLE TRANSITION

Many solids can break in either a ductile or brittle manner. We have already noted the effect of hydrostatic pressure in suppressing fracture and making normally brittle substances ductile (cf. Section 9.1). Two other obviously important variables are strain rate and temperature. A block of pitch flows slowly like a fluid at room temperature but can be smashed into sharp pieces with a hammer. Glass and polymethyl methacrylate are worked to various shapes at elevated temperatures but are brittle at room temperature. The size and shape of the body are also important. Large thick pieces, e.g. of steel, may be brittle where small ones are ductile. Sharp notches are particularly effective in changing the fracture from ductile to brittle (notch-brittleness).

These effects are of great practical importance. The danger of brittle or semi-brittle fracture at points of stress or strain concentration, before general yielding sets in, undermines the basis of structural engineering design (cf. Section 10.2). Materials liable to sudden brittle fracture are avoided as far as possible in engineering members which have to support tensile stresses. In complex structures and irregular service conditions this is not always easy to arrange. Many steel ships have broken in two by brittle fracture round the hull, particularly all-welded ships where the continuity of metal across the welded joints allows a crack to pass without interruption from one plate to the next.

A difficulty about testing for brittleness in service is that service conditions may be more severe than those in the laboratory. A ship's plate may break in a highly brittle manner, but a tensile bar cut from the fractured region may elongate 30% and give a fibrous cup-and-cone fracture. Materials likely to be brittle in service are, however, usually brittle in simple tensile tests at low temperatures (e.g. about 100°K for mild steel;

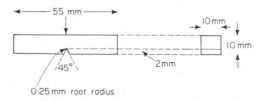

Fig. 11.7. The Charpy V-notch specimen.

room temperature for hard tool steels). To simulate more severe conditions various notch tests are made, e.g. the *Charpy V-notch impact* test, in which a small notched bar, shaped as shown in Fig. 11.7 and supported at its ends, is struck behind the notch by a heavy pendulum. The energy of fracture is measured from the amplitude of swing. The notch-brittleness of the specimen is also assessed from the appearance of the fracture, in particular from the "crystallinity", i.e. the proportion of crystal grains showing bright cleavage facets. A notch-tough steel typically absorbs 50–100 ft lb of energy in the Charpy test. When the impact energy of ship's steel is below about 15 ft lb at the temperature of service, there is a fair risk of brittle failure in service. Experience has also shown that service failure is likely when the test shows 70% or more of crystallinity at the service temperature.

Various other tests are also used; e.g. slow notch-bend tests for very large specimens and notch-tensile tests for hard, thin sheet. To determine the *crack-arrest temperature*, at which a steel plate can stop a running brittle crack by deforming plastically at its tip, the *Robertson test* has been developed. Here a large steel plate is loaded to a certain tensile stress and a fast brittle crack is started by a gun shot aimed at a sawcut at one end, locally cooled by liquid nitrogen. Mild steel can usually stop such a crack at temperatures above the 30 ft lb Charpy temperature for all stresses below the general yield stress. By contrast, at temperatures 20°C or more below this, brittle cracks are often able to run at stresses of 15,000 psi ($\simeq Y/3$) or less, i.e. in the range of engineering design stresses. Another severe test is the *Pellini test*, in which a heavy weight is dropped on to the centre of a wide steel plate, supported at its edges, which is allowed to deflect 5° before meeting a supporting stop. The fracture is started by a notched bead of brittle weld metal on the centre of the plate, and the plastic deflexion of the plate and appearance of the fracture are assessed. This test imitates the rough service conditions which a ship's plate may experience.

Figure 11.8 shows ductile–brittle transitions produced in several materials by testing at different temperatures. In the plain tensile tests they are indicated by reduction of area, or elongation, and in the Charpy test by impact energy. The large effect of a notch is seen in the mild steel curves. Most solids other than f.c.c. metals and alloys become brittle at low temperatures. Very roughly, b.c.c. transition metals become brittle in simple tension below about $0 \cdot 1 \, T_m$ (T_m = absolute melting point) and non-metallic solids and intermetallic compounds below about $0 \cdot 5 \, T_m$. Comparing this with Fig. 9.10, we see a correlation with yield strength. Intrinsically hard materials become brittle at low temperatures, in the range where their yield strength rises sharply.

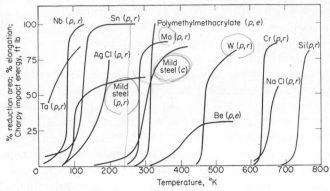

Fig. 11.8. Typical ductile–brittle transitions (p, plain tension; r, reduction of area; e, elongation; c, Charpy V-notch impact energy).

Many features of the ductile–brittle transition, including notch effects, can, in fact, be explained on the assumption that brittle fracture becomes possible when the yield stress exceeds a critical value. This is the basis of the *Ludwik–Davidenkow–Orowan* hypothesis, illustrated in Fig. 11.9.

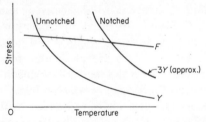

Fig. 11.9. Interpretation of plain bar and notched bar ductile–brittle transition temperatures.

Plastic flow and brittle fracture are assumed to be independent processes, each with its characteristic tensile stress, Y and F, respectively. As outlined in Section 10.9, the material is *simply ductile* at high temperatures where $Y < F$. For materials which behave in this way Y is known to vary rapidly with temperature, in the manner shown; and it is assumed that F is fairly insensitive, which is reasonable, because thermal activation plays little part in brittle crack propagation once the conditions outlined in Section 11.4 are satisfied.

The notch can raise the yield stress curve by *plastic constraint*, by increasing the *strain rate*, and by *work-hardening*. The transition temperature is then also increased, as shown in Fig. 11.9. Plastic constraint, discussed in Section 10.9, can raise the effective tensile yield stress to about $3Y$ near the tips of notches. Its effect in practice is shown by the existence of a *size effect.* When a thick steel plate is broken at a temperature near the top of the notch-brittle range, its fracture is brittle inside the plate but changes to a ductile "shear lip" near the external surface, where a large hydrostatic tensile stress cannot exist. If a thin plate is tested, so that plane-stress conditions prevail and the yield stress is hardly altered by plastic constraint, the ductile zone extends right through the plate and there is no brittleness.

The effect of strain rate on the yield stress was discussed in Section 9.3. At the tip of the V-notch in a Charpy impact test a strain rate of order 10^3 sec^{-1} is reached, which is about 10^6 times larger than that in a typical simple tensile test. Even higher strain rates are possible in the moving stress pulse of a running crack. Thus, at a small distance r from the path of an elastic crack, with speed v, the strain rate in the pulse is of order

$$\dot{\varepsilon} \simeq \frac{\text{maximum strain at } r}{\text{duration of pulse}} \simeq \frac{\dfrac{\sigma_t}{E}\sqrt{\dfrac{b}{r}}}{\left(\dfrac{r}{v}\right)}, \qquad (11.12)$$

where σ_t/E ($\simeq 0.1$) is the elastic breaking strain at the crack tip and b is the atomic spacing. If $v = 10^5$ cm sec^{-1} and $r = 10^4 b \simeq 2.5 \times 10^{-4}$ cm, then $\dot{\varepsilon} \simeq 4 \times 10^5$ sec^{-1}. The effect of strain rate on the ductile–brittle transition is demonstrated, not only directly by simple tensile tests at different rates, but also by the fact that, when a brittle crack slows down in a normally ductile material, it becomes ductile and showers of glide dislocations are released from its tip. This has been proved directly in LiF and MgO crystals and indirectly in Robertson tests on steel.

The effect of work-hardening at the tip of the notch comes from the fact that in fully ductile material very large plastic strains, of order unity, can develop there (cf. Section 10.8). These strains may start a ductile fracture or raise the yield stress to the level at which a brittle fracture can start. If a ductile fracture starts, the sudden increase in strain rate near the tip of the ductile crack may then raise the yield stress further and so promote a transition to brittle fracture. In fact, when a steel is broken at temperatures near the top of the notch-brittle range, there is usually heavy plastic deformation, work-hardening and ductile fracture around the tip of the notch.

11.7 NUCLEATION OF CRACKS BY GLIDE

There is much evidence to support the view, due originally to *Zener*, that in some crystalline solids cracks can be nucleated by plastic glide.

For example, in partly yielded tensile bars of steel, cleavage microcracks
have been observed within the Lüders bands, where yielding has occurred,
but not in the unyielded regions outside them. Some of the various

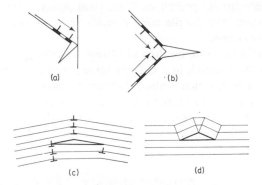

Fig. 11.10.   Cracks produced by glide processes: (a) single pile-up;   (b) double pile-up;
(c) split bend planes;   (d) twins.

mechanisms envisaged are shown in Fig. 11.10. In diagram (a) edge dislocations in a glide band pile up against an obstacle, such as a grain
boundary or twin band, and there produce a crack. We can regard these
glide dislocations as transforming into "cavity" dislocations, as in Fig.
5.30(a), and then advancing down the crack. Diagram (b) shows a variant
of this in which dislocations on two intersecting slip or twin bands coalesce
and produce a crack. Diagrams (c) and (d) show how lattice rotations
associated with bend planes and deformation twins can lead to cracks.
Cracks nucleated from slip band pile-ups have been observed in ionic
crystals (LiF and MgO) and at points where slip bands meet grain boundaries in polycrystalline ionic solids. Intersecting twins produce fracture
nuclei in steel at low temperatures (e.g. 100°K). Near room temperature,
fractures in steel have been seen to start at intersections of plastic shear
lines from notches. Diagram (c) fractures have been observed on basal
planes in zinc and beryllium; and diagram (d) fractures in graphite (also
on basal planes) and in calcite.

   The condition for brittleness by plastically induced fracture is that the
material breaks as it yields. We then expect the fracture stress to coincide
with the yield stress. This has been observed. In plain bar tensile tests on
soft steel at low temperatures the brittle fracture stress below the transition temperature often follows, not the level curve F in Fig. 11.9, but a
continuation of the steeply rising curve Y. It has also been shown that
the ductile yield stress of a steel, measured in compression, coincides with

the brittle fracture stress, measured in tension at the same temperature
(Fig. 11.11). At temperatures well below the ductile–brittle transition

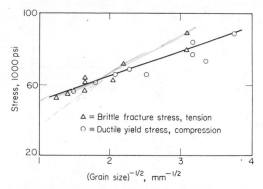

Fig. 11.11. Ductile yielding in compression and brittle fracture in tension of mild
steel of various grain sizes at 77°K (after J. R. Low, *Madrid Colloquium on Deformation
and Flow of Solids*, Springer-Verlag, Berlin, 1956).

point the nominal fracture stress often falls below the extrapolated Y curve.
This does not necessarily mean that the fracture has become *completely*
brittle, however, since localized plastic deformation can occur, for
example in regions of stress concentration, before the general yield stress
is reached.

An important feature of the glide mechanism of cracking is that it
enables running cracks to be created suddenly from avalanches of glide
dislocations, and so enables brittle fracture to occur at temperatures where
resting cracks might be immobilized by plastic blunting. A material may
thus be resistant to Griffith's mechanism of fracture, if it relaxes stresses
at notches by plastic yielding, yet it may still be brittle if this same yielding
nucleates running cracks.

The glide mechanism appears to be quite contrary to Fig. 11.9, since it
relates fracture directly to the yield process. It also raises other apparent
difficulties. Why is the ductile–brittle temperature so sensitive to hydrostatic tension, if the brittle cracks are created by glide processes insensitive
to hydrostatic stress? There is also the fact, of great practical importance
as a means of reducing brittleness by refining grain size, that variation of
grain size produces a ductile-brittle transition at constant temperature
quite as spectacular as the temperature transition. Figure 11.12 shows an
example. We expect, in steel, that both fracture and yield are triggered by
stress concentrations at dislocation pile-ups and therefore expect, in
agreement with Fig. 11.11, both the fracture and yield stresses to depend
on grain size in similar manner. This leaves no opportunity for a grain
size transition, however, since, if we then write Equation (9.12) in the

forms $F = \sigma_i + k_f d^{-1/2}$ and $Y = \sigma_i + k_y d^{-1/2}$, we have either fracture or yield at *all* grain sizes, according as k_f is smaller or larger than k_y.

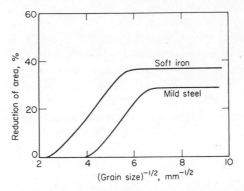

Fig. 11.12. Effect of grain size on the ductility of iron and steel at 77°K (after N. J. Petch, *J. Iron Steel Inst. (London)*, **173**, 25 (1953)).

To resolve these difficulties, we recall from Section 11.4 that *two* distinct general conditions have to be satisfied for fracture. The plastic glide provides the atomic mechanism of fracture, when the yield stress Y is reached, either locally or generally, but curve F in Fig. 11.9 represents the stress at which the thermodynamic condition is satisfied. This is a *permissive* condition for the growth of cracks, given that these are somehow formed, and it bears no obvious relation to the yield stress at which they form. In fact, if steel is strained to its yield point just above the simple brittle temperature, microcracks can then be seen in many of the yielded grains; the plastic deformation has formed cracks at a stress below that at which thermodynamics permits them to grow.

The embrittling effect of hydrostatic tension can then be explained in terms of the work done by this stress on the growing microcrack, which contributes to the thermodynamic requirement. As regards grain size, small grains produce short shear bands and, hence, short microcracks; large grains likewise produce long microcracks. We may thus expect a Griffith-like relation,

$$F = k_f d^{-1/2}, \qquad (11.13)$$

with the grain size d taking the place of c, as the thermodynamic condition for the growth of such microcracks. Since $Y = \sigma_i + k_y d^{-1/2}$, we have a grain size transition if $k_f > k_y$; then $F > Y$ when the grains are small, and $F < Y$ when coarse. We note that F is only a *permissive* fracture stress. The actual fracture stress is usually higher than F when the formation of microcracks depends on plastic yielding. These two stresses coincide at the critical grain size where $F = Y$.

11.8 FRACTURE TOUGHNESS

Tool steel and window glass are both brittle. They are comparably hard. Yet the first is much stronger than the second. It has greater *fracture toughness*. We are concerned here with resistance to cumulative fractures, brittle or semi-brittle, and we know from Section 11.5 that for these a crack spreads when a critical *fracture displacement u* is produced in each element of material at the crack-front. For a fully brittle cleavage *u* is the fracture elongation of an atomic bond across the crack faces; for semi-ductile fractures the value of *u* depends on the micromechanism of fracture, as discussed in Section 11.5. The larger is *u*, the more resistant is the material to the propagation of a crack. However, *u* by itself is not sufficient for fracture toughness; otherwise soft clay would be tough. We have also to consider the forces involved. Guided by Equations (11.10), we use γ_p, the *work of fracture*, as a measure of fracture toughness. This is of course closely related to *u*, since

$$\gamma_p \simeq \sigma_p u, \tag{11.14}$$

where σ_p is the stress, across the crack-front, at which the fracture displacement *u* takes place; for example, in Equations (11.10) we have $\sigma_p = k$. When the crack is long enough to give low-stress fracture (cf. Equation (11.7)), the overall strength of the body can then be expressed by Equation (11.11), with γ_p as the *effective surface energy*.

Soft steel at about 100°K typically has $\gamma_p \simeq 10^5$ erg cm^{-2}, about 50 times larger than the true surface energy. This difference is mainly due to the difficulty a cleavage crack has in cutting through grain boundaries. The orientation of the cleavage plane changes from one grain to the next, and this change occurs across a belt of plastically deformed and torn material near the boundary, the plastic work of which contributes to γ_p.

The value of γ_p allows us to distinguish two broad classes of brittle solids: *weak* ($\gamma_p \simeq$ true surface energy) and *strong* ($\gamma_p \gg$ true surface energy). Strong solids, such as hard steel, retain considerable strength even when their surfaces are heavily scratched; weak ones, such as glass and stone, do not.* The advance of primitive man from the stone age to the iron age was an advance in fracture toughness.

Many engineering components are made from materials capable of brittle fracture in service. What matters in practice is not brittleness itself but reliability against sudden fracture. This depends on fracture toughness and on the size and sharpness of notches and other stress concentrators. For a hard clock-spring these may be small scratches or foreign inclusions

---

* It has been shown recently that some plastic deformation occurs at the tips of cracks in glass [cf. D. M. Marsh, *Proc. Roy. Soc.* (*London*) (1954)].

364	The Mechanical Properties of Matter

in the metal; for a large turbine rotor they may be machined keyway grooves; for a steel ship they may be faulty welds or badly designed hatch openings in the deck. Toughness is a relative term, since a sufficiently large value of c, in Equation (11.11), can give a low-stress fracture even in a material with a large γ_p. We saw in Section 11.5 that even a fully ductile metal, in the form of a large thin sheet containing a long notch, can tear at a stress well below that for general yield.

Fracture toughness is determined from measured breaking strengths of specimens containing sharp notches (e.g. $\rho = 10^{-3}$ in. for steel) of various lengths. Equation (11.11) is then fitted to the results to give a value of γ_p. Figure 11.13 shows such results obtained from notched bars of alloy steel, up to 9 in. deep, slowly bent at room temperature. The scaled Charpy specimens (notch radius = 0·025 beam depth) approached general yield before breaking, but the large sharply notched bars broke well below their general yield stress. From Equations (11.11) and (11.7) (assuming $\mu/k \simeq E/Y \simeq 200$), we obtain $\gamma_p \simeq 100$ lb in^{-1} ($= 1·75 \times 10^7$ erg cm^{-2}) and $u \simeq 5 \times 10^{-4}$ in. Similar values of γ_p are obtained from sharply notched tensile bars of mild steel tested at temperatures near the top of the notch-brittle range.

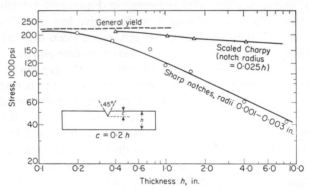

Fig. 11.13. Effect of size and notch sharpness on the bend strength of a Ni–Mo–V turbine steel at room temperature (J. D. Lubahn and S. Yukawa, *Am. Soc. Test. Mater. Proc.*, **58**, 661 (1958)).

We see that long notches (e.g. 1 in.) are necessary to weaken this metal appreciably, in strong contrast with glass. This introduces a *macroscopic size effect*, important in large engineering structures. The fracture displacement u at a given notch tip is accommodated only by *general* yield zones in small specimens. The full weakening effect of a long notch in a large body cannot then be reproduced in a small sample. Given only a small sample, however, we can determine u from it and then, as in Table 11.2, estimate the weakening effect of a long notch in a large body.

To what extent is γ_p sensitive to the sharpness of the notch? Obviously, if the notch is very blunt, γ_p is increased. At the other extreme, however, two observations show that there is a *microscopic size effect* which sets a lower limit to the *effective* sharpness of a notch, irrespective of the geometrical sharpness: (1) X-ray analysis of lattice distortion, and measurements of heat released, have shown that γ_p can approach 10^7 erg cm^{-2} along the track of a running brittle crack in steel; (2) experiments with extremely sharp notches and natural cracks have shown that the macroscopic strength does not decrease beyond a certain level as the crack is sharpened. We can represent this microscopic size effect roughly as follows. Let ε_f be the fracture *strain* in the root of a notch. We think of the root, of radius ρ, as a small tensile specimen with gauge length of order ρ, embedded in a larger specimen, and write the fracture displacement as

$$u \simeq (\rho + \rho_0)\varepsilon_f, \tag{11.15}$$

where ρ_0 is a measure of the effective limiting sharpness. Suppose that ε_f is approximately the simple tensile ductility and that ρ_0 is determined by the grain size, or spacing of inclusions or slip or twin bands. Then, typically, $\varepsilon_f \simeq 0{\cdot}3$, $\rho_0 \simeq 0{\cdot}003$ cm and $\sigma_p \simeq 4 \times 10^9$ dyn cm^{-2}, giving $u \simeq 0{\cdot}001$ cm and $\gamma_p \simeq 4 \times 10^6$ erg cm^{-2} for sharp notches ($\rho < \rho_0$). For steel, from the top to the bottom of the notch-brittle range of temperature, γ_p falls from about 10^7 to about 10^5 erg cm^{-2} and ε_f falls from about $0{\cdot}3$ to about $0{\cdot}01$.

The theory of fracture toughness is useful in engineering design in several ways. It re-emphasizes the familiar fact that unavoidable large stress concentrators should be made as blunt as possible. It shows that some tensile ductility, e.g. $\varepsilon_f \simeq 0{\cdot}05$, is necessary if highly stressed small parts for machinery, turbine blades, etc. are not to be weakened unduly by small scratches. It shows that reducing the yield strength of the material can actually *increase* the breaking strength, provided that the changes that decrease σ_p in Equation (11.14) also bring about a larger increase in u; in fact, the surface of hard steel sheet is sometimes deliberately softened, by decarburization, to strengthen the sheet against fracture. It also shows that thin sheets can be tougher than thick ones, because they enable plane-stress rather than plane-strain conditions to develop in the plastic zone ahead of the crack-front, so reducing the hydrostatic tension, changing the fracture to a more ductile mode such as that of Fig. 11.5(b), and so raising γ_p.

It should also be possible to produce safe engineering designs with notch-brittle materials, by determining the fracture toughness and then controlling the maximum size of stress concentrators in the material, so that fracture cannot occur at the working stress. Recent engineering design has tended towards this procedure. There are difficulties, however. In a large and complicated structure it is difficult to detect and eliminate all large notches and other stress concentrators. Self-stresses, e.g. thermal stresses from welding, may bring parts of the structure almost up to the general yield stress. Embrittlement may occur at the roots of notches, owing, for example, to chemical action from the surroundings, or to localized plastic strains caused, for example, by thermal stresses from welding. The service loads may be irregular, as when a ship is buffeted by high seas.

Careful radiographic inspection, to seek out notches, and stress relief annealing, to reduce self-stresses, are helpful for improving the safety of such structures. Where utmost safety is essential, however, the service temperature should never fall below the crack arrest temperature, as measured, for example, in the Robertson test. Then, even if a crack should start, it could not propagate in a brittle manner. This will result in a great increase in safety, since the minimum lengths for the unstable propagation of ductile cracks are very large. The physical conditions which determine the crack-arrest temperature are complicated. At the top of the notch-brittle range, the "brittle" crack is already partly ductile and its effective surface energy γ_p is large ($\simeq 10^7$ erg cm^{-2}), not only when it starts but also when it is running at speed. The displacement u associated with this plastic work could hardly be accommodated in the manner of Fig. 11.5 except at the general yield stress, and it is probable that this semi-brittle crack propagates by the nucleation of cleavages in some grains ahead of the main front and by the plastic tearing of the bridges between them. Then, as in Fig. 11.6, a cumulative fracture is obtained even under plane-strain conditions. Small cracks ahead of the main front have, in fact, been observed in broken steel plate. Where they join the main front they produce "chevron" markings, similar to the markings of Fig. 11.4, as shown in Fig. 11.14. Near the sides of the plate a large hydrostatic tension cannot

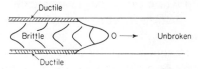

Fig. 11.14. Appearance of a crack in steel plate, showing chevron marks in brittle zone, ductile shear lips, tunnelling crack front, and cracks nucleated ahead of front.

exist. The material is much tougher under these approximately plane-stress conditions and breaks there by a ductile "shear lip". As a result, the centre of the crack, where conditions approximate to plane strain, *tunnels* on ahead of the sides and a curved crack-front is obtained. Conditions at the crack-front, where new cracks are nucleated in the region ahead, are then fairly similar to those in a V-notch impact specimen, and the crack arrest temperature is roughly indicated by the V-notch impact transition temperature.

11.9 MICROSTRUCTURE AND BRITTLENESS

An important property of a material determining its brittleness is its intrinsic hardness. All crystalline solids in which dislocations move fast only at high stresses appear capable of brittle cleavage at low temperatures or under impact loading. Their dislocations cannot run fast enough to relax the stress of a high-speed crack. The importance of notches for revealing the brittleness latent in such materials has been emphasized, but it should be remembered that sharp, elongated inclusions in the microstructure can also act as notches. Graphite flakes make grey cast-iron fairly brittle and weak in tension. *Nodular* cast-iron is tougher and stronger, because the graphite is rearranged into spheroidal shapes. Similarly, sintered ceramics such as alumina can be made much less fragile by ensuring that no unsintered holes remain.

Grain size is important for the reasons discussed in Section 11.7. Coarse-grained zinc is brittle at room temperature, but fine-grained zinc can be deformed by up to 90%. Tungsten is too brittle to be cold-drawn to wire until its grain size has first been refined by hot working. Low-carbon steels sometimes grow coarse grains when recrystallized and are then brittle (*Stead brittleness*). Brittleness due to coarse grains is also a problem in large steel forgings and plates. Conversely, strength and toughness can be gained in most materials by grain refinement. Mild steel is *normalized*, i.e. air-cooled after heating to about 900°C, to refine its grain size and lower the transition temperature. Grain-refining additions, mainly aluminium, are also often added for the same purpose. In hard alloy steels strength is combined with toughness by developing a *tempered martensite* structure in which carbide particles are finely dispersed through the grains. These particles interrupt the crack and increase the effective surface energy of fracture.

Hardening without refinement of microstructure generally raises the transition temperature. Strain-ageing in low-carbon steels has a particularly adverse effect (*strain-age embrittlement*), and for this reason nitrogen-bearing steels are generally avoided in applications where brittle fracture would be a serious hazard. Mechanical working at 150–300°C is also avoided, because the simultaneous work-hardening and strain-ageing hardens the metal so much that it becomes brittle (*blue brittleness*). Manganese usefully lowers the transition temperature of steel, partly by suppressing strain-ageing and partly by refining the grain size.

Chemical embrittlement was mentioned in Section 8.9. Many substances can segregate to grain boundaries or crack faces and produce weak interfaces for fracture. There are innumerable examples of this, many of great practical importance. Even f.c.c. metals and alloys can be made to show ductile–brittle transitions (intercrystalline cracking) in this way; e.g. by a trace of antimony in copper. The brittleness of steels after *overheating* is due to sulphides segregated in their grain boundaries. Nickel–chromium and other alloy steels often show *temper brittleness* after heating to 500–550°C, owing to segregation to grain boundaries. Hydrogen embrittles copper by reacting with oxygen to form pockets of steam along the grain boundaries. It embrittles steel by reducing the surface energy of cracks and also by nucleating cracks by gas pressure (*hair-line cracking*). To avoid this, large steel castings are now often made under vacuum. Embrittlement of metals by attack from molten metals and by stress–corrosion cracking is a frequent cause of failure. Molten lead, for example, reduces the effective surface energy for intercrystalline cracking in copper to such a low value (about 200 erg cm^{-2}) that the Griffith condition can be satisfied very easily for the propagation of such cracks. The Griffith stress is then usually too small to produce plastic deformation in the grains bordering the crack, and an extremely brittle fracture is obtained.

11.10 FATIGUE FAILURE

If we bend a ductile metal wire to and fro many times, it eventually breaks. This is a simple example of *fatigue failure*, in which a given load or deformation, applied and removed repeatedly, gradually breaks the material. Many service failures of engine components, e.g. shafts, gears, connecting rods and springs, are due to metal fatigue. A small crack forms, usually at a point of stress concentration, and then slowly spreads,

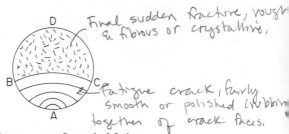

Fig. 11.15. Appearance of a typical fatigue crack.

often over a period of millions of stress cycles, in a direction across the main tensile stress. The final break occurs suddenly, when the unbroken part of the cross-section has been sufficiently reduced in area to fail immediately by ductile or brittle fracture. The fatigue crack itself seems to be brittle but this is merely because the alternating stress pushes the crack faces together as often as it pulls them apart. Fatigue failure is easily recognized from the appearance of the fracture. Two distinct zones can usually be seen. The first (e.g. ABC in Fig. 11.15) is the fatigue crack itself, rather smooth, sometimes polished by the rubbing together of the crack faces, sometimes discoloured by corrosion. The second (e.g. BCD) is the final sudden fracture, rough and fibrous or crystalline. The fatigue crack often shows roughly concentric rings. These mark successive positions of the crack-front and they focus back to the source (e.g. A) of the fracture.

The *fatigue strength* is determined by *endurance tests* in which specimens are repeatedly stressed under controlled conditions and their *fatigue life* or *endurance*, i.e. the number N of cycles to fracture, measured. The stress is usually oscillated sinusoidally at a few kilocycles per minute. Except at high frequencies ($> 10^4$ cycles min^{-1}) or at temperatures where creep or ageing occur during each cycle, the fatigue strength is usually insensitive to frequency. Figure 11.16 defines the main features of the stress cycle:

Fig. 11.16. Stress cycle in fatigue test.

the *maximum* and *minimum stresses*, $\sigma_{max.}$ and $\sigma_{min.}$, which may each be either positive or negative; the *mean stress* σ_m; the *stress amplitude* S; and the *stress range* R. In *reversed stress* tests $\sigma_m = 0$ and $\sigma_{max.} = -\sigma_{min.} = S$. In *pulsating stress* tests $\sigma_{min.} = 0$ and $\sigma_{max.} = 2\sigma_m = 2S$. The most popular tests are *push–pull* tests on tensile specimens and *rotating beam* tests in which a bending moment is applied through loaded bearings to a rotating shaft.

Provided σ_m is small compared with the general yield stress, the important variable is the stress amplitude S. Figure 11.17 shows typical S,N curves. Plain carbon and low-alloy steels show a definite *fatigue limit*, at about 10^6 cycles, and at stresses below this limit their fatigue life becomes virtually infinite (in non-corrosive environments), whereas most other metals and alloys appear never to reach such a limit, at least up to lives of order 10^9 cycles. The *fatigue strength* is defined as the stress amplitude S that produces failure in a given number of cycles, e.g. $N = 10^7$. The fatigue life decreases tenfold for about 15–30% increase in stress (above the fatigue limit, where this exists). At high amplitudes, where appreciable

plastic deformation occurs during each cycle, stress is no longer a satis-
factory measure of the intensity of the fatigue cycle. Under these conditions (which are important for studying fatigue produced by thermal

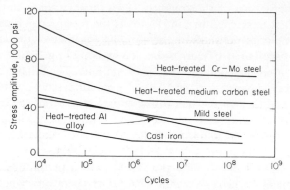

Fig. 11.17. Typical endurance curves under reversed bending (based on data from
Am. Soc. Metals Handbook (1948)).

strains resulting from large temperature oscillations, e.g. in steam power
plant) fatigue tests at _constant plastic strain amplitude_ ε_p are useful;
these have shown that many metals obey the _Coffin–Tavernelli_ relation

$$N \simeq \frac{C}{\varepsilon_p^{2}}, \tag{11.16}$$

where C is a constant, in the range $N < 10^5$.

The effect of mean stress on fatigue can be shown by plotting the stress
amplitude S for a given life against the mean stress σ_m, as in Fig. 11.18.
The line OC, of unit slope, marks the edge of the region BOC in which the
stress changes sign during the cycle. The line joining the ultimate tensile

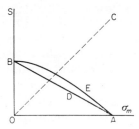

Fig. 11.18. Effect of mean stress on stress amplitude for a given endurance.

stress, A, to the fatigue strength for reversed stress, B, determines the
possible combinations of $\sigma_{max.}$ and $\sigma_{min.}$ for a given life. Results on most
materials usually lie between the straight line ADB (the _Goodman line_)

and the parabola AEB (the *Gerber parabola*), the latter being followed closely near B, where σ_m has almost no effect on fatigue strength.

Any study of fatigue must take account of notch effects, because fatigue is notch-sensitive and because the fatigue crack itself is a notch. The *effective stress concentration factor* k_f is the ratio of fatigue strengths measured on un-notched and notched specimens. It usually differs from the *elastic stress concentration factor* k_t, expected from a purely elastic analysis of the notch, and a *notch sensitivity* factor

$$q = \frac{k_f - 1}{k_t - 1} \qquad (11.17)$$

is defined to indicate the extent to which the material is sensitive ($q = 1$ and $k_f = k_t$) or insensitive ($q = 0$ and $k_f = 1$). We expect $q \simeq 1$ for large blunt notches under small stresses, since the material at the roots of these is large enough to behave as a macroscopic fatigue specimen. This is observed. The notch sensitivity decreases as the dimensions of the notch are uniformly scaled down. It decreases more quickly in soft steel than hard, so that there is little gain in using hard steel rather than soft under notch-fatigue conditions. For notches such as keyways and fillets ($k_t < 5$) q drops from 0·9 to 0·5, approximately, as the root radius decreases from 0·1 to 0·01 in. in an average steel.

To remove surface notches, fatigue specimens are usually highly polished. The fatigue strengths of engineering parts are often improved by introducing compressive stresses into the surface layers, e.g. by surface rolling, shot-peening and nitriding (on steels). Corrosive agents which attack the surface, producing etch pits and grooves at slip bands and grain boundaries, greatly reduce the fatigue strength, particularly of notch-sensitive materials (*corrosion fatigue*).

11.11   FORMATION AND GROWTH OF FATIGUE CRACKS

As the name "metal fatigue" implies, fatigue failure is mainly a problem of ductile materials. Brittle solids tend to break either all in one cycle or not at all. There are many indications that the fatigue crack forms and grows by plastic deformation. Under combined stresses the fatigue strength follows von Mises' relation, Equation (10.9). The fatigue strength is roughly proportional to the ultimate tensile stress, such that usually $B \simeq \frac{1}{2}A$ (for $N \simeq 10^7$) in Fig. 11.18. Slip bands can be seen on the surfaces of specimens fatigued after polishing, in greater or lesser abundance according to the amplitude of oscillation. Even at stresses below the fatigue limit, a few slip bands are still seen in grains favourably oriented for slip under the applied stress, or near stress concentrators. Under stresses that produce fatigue failure, some of these slip bands gradually become strongly marked and ineradicable by light polishing (*"persistent"* slip bands). Fatigue cracks have begun to form along them. If this cracked surface layer is removed by deep polishing, the fatigue life of the specimen is renewed. Strain-ageing solutes, such as nitrogen and carbon in steel and magnesium in aluminium, able to pin dislocations in motion and bring them to rest, thereby halting the slip, usefully improve the fatigue strength and give the metal a definite fatigue limit. It is also possible in such alloys to raise the fatigue strength by *coaxing*; that is, by gradually increasing the amplitude of applied stress at a temperature where strain-ageing can occur during straining.

It is important to notice that slip can continue repeatedly to and fro on slip bands under alternating stress, whereas under monotonic stress it would cease. This happens when shakedown cannot be achieved on some slip bands (cf. Section 10.3); if the local value of applied shear stress acting on the band oscillates over a range exceeding $2\sigma_i$ (where the "friction" stress σ_i of the band may be raised by work-hardening), the band continues to slip to and fro. Near sharp stress concentrators this happens even when the overall stress varies weakly, since the local stress on the bands is then intensified by the notch effect. Thus, except in strain-ageing materials, some oscillating slip and fatigue is almost always possible.

Oscillating slip bands in internal grains do not seem to produce fatigue cracks. A free surface, usually the external surface, appears to be necessary, probably as a place to deposit material pushed out of active slip bands to make space for cracks. This may be why sharp, deep notches have less effect on fatigue strength than elasticity theory would suggest. Active slip bands which meet the free surface behave quite remarkably. Thin tongues of material, called *extrusions*, grow rapidly out of them, and at the same time narrow crevices, called *intrusions*, spread down them into

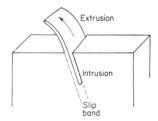

Fig. 11.19.   One form of extrusion and intrusion on a slip band.

the body, as shown in Fig. 11.19. Extrusions and intrusions have been seen on many materials, under various conditions. In a long fatigue test they usually form during the first 5–10% of the fatigue life. The exact manner of their formation is still not established. They may form by a process in which the oscillating dislocations do not precisely retrace their path up and down the slip plane but jump from one face of the plane to the other, e.g. by cross-slip, at the ends of their path. Alternatively, they may form simply statistically, by random slips on a packet of planes, so that some planes by chance slide out of the packet. A third possibility is that soft material in the slip bands is squeezed out plastically during the compressive phase of the cycle and not sucked back again during the tensile phase.

An obvious practical consequence of this is that hardening the surface of the material, to suppress these plastic processes, improves the fatigue strength. Many surface treatments not only induce compressive surface stresses in the material but also harden the surface layers. The adverse effect of a soft surface layer is also known. Thus, a soft coat of pure aluminium reduces the fatigue strength of a hard aluminium alloy, as also does a soft decarburized layer on steel.

There seems little doubt that the fatigue crack starts from an intrusion (or in some cases at a similar notch formed along a grain boundary). As it spreads down the slip band into the material, the incipient fatigue crack slows down and at a depth of 10^{-3}–10^{-2} cm becomes almost "dormant". In tests lasting 10^6–10^7 cycles, over three-quarters of the total fatigue life may be spent in this dormant phase. If the applied stress is sufficiently small (the crack having been started from a sharp notch, for example), the crack may even cease growing altogether (*non-propagating fatigue cracks*). What happens during the almost dormant stage of growth is not yet understood. The growth may be slow at this stage because the crack is too deep to allow discarded material to be transported easily from the tip to the surface; and yet too short to grow by the final rapid phase of fatigue-crack growth.

This final phase, which produces most of the region ABC in Fig. 11.15 and which takes up the final 10% or so of the total life in a long fatigue test, starts when the crack turns out of the slip plane into a path perpendicular to the tensile axis. The fracture surface in this phase becomes covered with *ripple markings*. Those shown in Fig. 11.15 are merely the few which happen to be outstandingly large and clearly visible, e.g. because of a temporary increase of applied stress. On a microscopic scale thousands of fine, parallel ripple lines, which mark successive positions of the crack-front, can be seen. It has been proved that one ripple forms during each cycle of stress, which enables the progress of the crack to be retraced unambiguously by counting ripples. It seems that each ripple is produced by alternating tensile and compressive plastic strain at the tip of the crack and that the process of crack growth is quite simple. During the tensile phase the crack-front advances slightly by ductile rupture (cf. Fig. 10.18(d)), presumably then being halted by work-hardening and by self-stresses from the dislocations pushed out in front of it. On the compressive stroke the tip closes up again, thus eliminating these dislocations, but the crack faces do not weld together. At the next tensile stroke a further increment of growth can occur, since the dislocations which brought the previous increment to a halt have now disappeared. The crack continues to spread in this manner, at an increasing rate, until the remaining cross-section becomes small enough to break completely in a single tensile stroke.

We have now completed our study of solids. It will have become all too obvious that the processes of fracture are extremely complicated and that, particularly for fatigue fracture, we are still far from understanding them. There are many opportunities for further research here.

FURTHER READING

OROWAN, E., *Rept. Progr. Phys.*, **12** (1948–49).

JAEGER, J. C., *Elasticity, Fracture, and Flow*, Methuen, London, 1956.

PARKER, E. R., *Brittle Behaviour of Engineering Structures*, Wiley, New York, 1957.

IRWIN, G. R., "Fracture", *Handbuch der Physik*, Vol. VI, Springer-Verlag, Berlin, 1958.

BIGGS, W. D., *The Brittle Fracture of Steel*, MacDonald and Evans, London, 1960.

TIPPER, C. F., *The Brittle Fracture Story*, Cambridge University Press, London, 1962.

AVERBACH, B. L., *et al.* (eds.), *Fracture*, Wiley, New York, 1959.

McLEAN, D., *Mechanical Properties of Metals*, Wiley, New York, 1962.

KENNEDY, A. J., *Processes of Creep and Fatigue in Metals*, Oliver and Boyd, Edinburgh, 1962.

ELEY, D. D., *Adhesion*, Oxford University Press, London, 1961.

CRANFIELD COLLEGE OF AERONAUTICS, *Proceedings of the Crack Propagation Symposium*, Cranfield, Bedfordshire, England, 1962.

PROBLEMS

11.1   Explain under what circumstances the fracture toughness (a) increases, (b) decreases with increasing thickness of a notched sheet.

11.2   Explain under what circumstances the fracture toughness (a) increases, (b) decreases with refinement of the microstructure of a solid.

11.3   A sheet of glass ($E = 9 \times 10^6$ psi; $\gamma = 0.003$ lb in^{-1}), $40 \times 20 \times \frac{1}{4}$ in., is scribed into two 20×20 in. squares with a sharp diamond scratch to a depth of 0·001 in. and is then broken along the scribed line by bending. What approximately is the smallest bending moment for this? [350 lb in.]

11.4 10^{15} particles of a perfect gas have collected at room temperature in a penny-shaped internal crack in an elastic solid, of Young's modulus 10^{12} dyn cm^{-2} and surface energy 1000 erg cm^{-2}. Roughly estimate the equilibrium values of gas pressure, crack radius and elastic displacement at the centre of the crack. [100 atm; 0·1 cm; 10^{-5} cm]

11.5 At 100°K the critical grain size for the ductile–brittle transition in simple tension of a steel is 0·005 cm and the tensile yield stress at this grain size is 0·0045 E ($E =$ Young's modulus). At one-half of this grain size the yield stress is 0·0055 E. At this smaller grain size the metal is hardened to increasing degrees by a process which increases the frictional resistance of the slip planes. At what tensile yield stress do you expect it then to become brittle again? [0·0084 E]

11.6 A steel with a tensile yield stress of $1·6 \times 10^5$ psi and a plastic fracture strain of 0·005 gave the following results in notch tests:

Notch length, in.:	2·5	2·5	1·0	0·4	0·4
Root radius, in.:	0·0125	0·002	0·005	0·010	0·002
Breaking stress, 10^4 psi:	6·0	4·0	6·4	12·8	9·4

Estimate the lower limit to the effective radius of the root. [0·005 in.]

11.7 In a standard Charpy V-notch test on a steel the energy absorbed is 10 ft lb. Assuming that this is practically all absorbed before fracture as plastic work done along hinges, as in Fig. 10.7, each of length 0·4 in., and that the shear yield strength of the metal is 6×10^4 psi, estimate approximately the fracture displacement at the root of the notch. A 20 in. deep slit with the same root radius as the Charpy notch is cut centrally in a 20 ft wide plate of the metal. Assuming that the tensile yield strength is twice the shear yield strength, estimate approximately the fracture strength of the notched plate and length of accommodation-yield zone. [0·012 in.; 9×10^4 psi; 16 in.]

11.8 Experiments on a sharply notched sheet of an aluminium alloy gave the following results:

Notch length, in.:	1·98	1·89	1·74
Breaking stress, 10^4 psi	3·50	3·58	3·74

It is proposed to make use of a large sheet of this material, in which there is a window opening, 20 in. square. There may be small sharp cracks round the opening. At what tensile stress do you expect the sheet to fail? [11,000 psi]

11.9 A thin rectangular crystal flake of mica of length L is cleaved along its central plane by a wedge, driven into it along one edge, the two half-crystals being separated by a fixed distance s at this edge. The crack runs to a distance l ($\ll L$) along the crystal and then stops. Treating the two cleaved half-crystals as elastic cantilever beams, freely loaded at their cleaved ends but built-in where they join at the crack-front, show that the crack is in *stable* equilibrium at the length

$$l = \left(\frac{3}{32} \frac{Es^2t^3}{\gamma} \right)^{1/4},$$

where E is Young's modulus, γ is surface energy per unit area, and t is the thickness of a beam.

11.10 When a ductile metal, e.g. in a railway carriage coupling, is repeatedly struck with blows each heavy enough to produce large plastic distortion (compared with elastic distortion), the total energy absorbed in fracturing it is almost independent of that absorbed in a single blow. However, when the blows are all light, so that small plastic deformation is produced by each, the total energy absorbed to fracture is increased. Explain these observations.

chapter 12

Fluid Mechanics

Fluids shear easily. The characteristic problems of fluid mechanics are therefore those where the viscous forces are small or absent. After briefly considering the equilibrium of static fluids, we shall examine the motion of fluids and see that kinetic energy plays a decisive part. The general equations of fluid flow are formidable, but fortunately many simplifying concepts can be introduced, e.g. streamline flow, irrotational flow, representation of velocity in terms of a potential, which, when combined with various approximations, e.g. neglect of viscosity and compressibility, enable a variety of problems to be solved. The concepts of circulation, vorticity and separation bring other problems, such as the behaviour of airfoils, within reach, but at this stage it becomes essential to take account of viscosity and then to discuss drag, turbulent wakes and related effects, in terms of the properties of boundary layers in which the viscous and inertial forces are comparably strong. The study of wave motion starts with the analysis of simple surface waves on water and then leads on, through tidal waves, to shock pressure waves in fluids and problems of supersonic flow and high-speed flight in rarefied atmospheres.

12.1 STATIC FLUIDS

There are no shear stresses in a fluid at rest. The only stress is the hydrostatic pressure p, i.e.,

$$\sigma_{11} = \sigma_{22} = \sigma_{33} = -p,$$

$$\sigma_{12} = \sigma_{23} = \sigma_{31} = 0. \tag{12.1}$$

Equations (4.69) for the equilibrium of forces along an axis X_i then simplify to

$$\frac{\partial p}{\partial x_i} - F_i \rho = 0, \qquad (12.2)$$

so that, in the absence of gravity or other body forces ($F_i = 0$), the pressure is constant throughout the fluid.

Consider the force of gravity. Let it act in the negative z ($=x_3$) direction. Thus, $F_z = F_3 = -g$ and $F_1 = F_2 = 0$. Equations (12.2) become

$$\frac{dp}{dz} = -\rho g, \qquad (12.3)$$

which gives the familiar *equation of hydrostatics*,

$$p = p_0 - \rho g z, \qquad (12.4)$$

when ρ is constant. *Archimedes' principle* then follows. The upthrust due to the fluid on a cubical volume element $dx\,dy\,dz$ of an immersed body is equal to the upward force $(p - \rho g z)\,dx\,dy$ on its bottom face, minus the downward force $(p - \rho g z - \rho g\,dz)\,dx\,dy$ on its top face. Writing $dV = dx\,dy\,dz$ and integrating over the whole immersed body, we see that the total upward *buoyant force*,

$$F = \rho g V, \qquad (12.5)$$

is equal to the weight of displaced fluid.

Consider now the earth's atmosphere. Let the mass of a molecule be m. The density is no longer constant and we write $\rho = mp/kT$, from the gas law. Substituting for ρ in Equation (12.3) and integrating, we obtain

$$p = p_0\, e^{-mgz/kT}, \qquad (12.6)$$

as in Equation (1.21). The *stability* of such distributions is interesting. If a test mass of air is moved upwards through the rest of the atmosphere, does it then become denser than its surroundings and so tend to sink back again? And conversely, if moved downwards? Although the atmosphere is stratified into layers of low density above layers of high density, our test mass of air changes its pressure to the local value as it rises or falls, and so changes its density also. In fact, omitting temperature changes, Equation (12.6) gives a *neutral* equilibrium because the change of pressure brings the test mass always to the same density as its surroundings. However, this change of pressure also adiabatically changes the temperature of the mass and this further changes the density (cf. Equations (1.33)–(1.35)). When this temperature change brings the mass to the same temperature as its surroundings as well as to the same pressure, the equilibrium remains neutral. This happens when the temperature of the atmosphere decreases by about 1°C per 100 m height, this critical temperature gradient being known as the *adiabatic lapse rate*. If the atmosphere is well stirred, as in windy weather, its temperature distribution approaches this equilibrium value. This contrasts sharply with incompressible fluids, which become *uniform* in temperature when stirred.

These considerations are important in meteorology. If the air at ground level is too warm, a *convective instability* exists and "bubbles" of turbulent air break away from ground level and float upwards, so bringing the actual temperature gradient back to the adiabatic lapse rate by turbulent mixing. These instabilities develop especially at points where the wind blows over projecting obstacles, and they make the wind more gusty. Thus, on a warm sunny day the wind freshens towards noon and dies away later to give a calm evening. In dull cloudy weather this cycle is suppressed and the meteorological state remains fairly constant. In cold, clear, still weather a *temperature inversion* may occur in which radiation cooling of the ground reduces the temperature of the nearby low-lying air below that of the upper layers. This dense cold air then runs down into valleys and does not mix appreciably with the air above, which is a hazard in populated industrial regions, because there is then almost no way for fumes and smoke to escape into the upper air.

12.2 STEADY FLOWS

We turn now from a resting fluid to one in *steady flow*. We choose co-ordinate axes which are independent of the fluid and we write u_i for the velocity components of an element of fluid in this co-ordinate system. We are thinking macroscopically of course, so that an "element" contains a large number of molecules and is a representative miniature sample of the bulk fluid. It is often convenient to choose co-ordinate axes that are themselves in motion. For example, for an aircraft in steady flight we would attach the axes to the aircraft itself. We always choose axes, however, which are *not* in accelerated motion; thus, a body moving with uniform velocity or at rest in this co-ordinate system has no force of inertia. If, subject to this restriction, we can find a set of axes for which $\partial u_i/\partial t = 0$ everywhere, the fluid is in steady flow.

Any flow can be represented instantaneously by a field of arrows, each of which indicates the velocity of an element of fluid at that instant. The lines formed by sequences of infinitesimal arrows, head to tail, are called *streamlines*. In steady flow, but not otherwise, the streamlines coincide with the *lines of motion* of the elements of the fluid. Streamlines show the simultaneous motion of many elements at a particular time, whereas lines of motion trace the paths followed by particular elements through the course of time.

In steady flow, as when water glides smoothly over a weir or air flows smoothly over the leading edges of an aircraft, the fluid never crosses from one streamline to another. A *tube* of streamlines, i.e. a *stream-tube* drawn through the boundary of an area such as A in Fig. 12.1, behaves like a frictionless pipe along which the fluid is flowing. Then, if the stream velocity is v_1 where $A = A_1$, and v_2 where $A = A_2$,

$$A_1 v_1 \rho_1 = A_2 v_2 \rho_2, \tag{12.7}$$

or, for an incompressible fluid,

$$A_1 v_1 = A_2 v_2. \tag{12.8}$$

The streamlines crowd together where the fluid flows quickly and spread apart where it flows slowly; i.e. still waters run deep. A diagram of streamlines can thus indicate both speed and direction of flow.

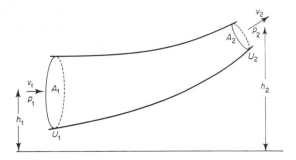

Fig. 12.1. A stream-tube.

Suppose that the fluid is *incompressible*. This is a less restrictive assumption than first appears. Even in gases, compressibility is usually negligible in flow problems at velocities less than about one-quarter that of sound waves in the gas (cf. Section 12.11). Consider the element of such a fluid in the stream-tube between A_1 and A_2 in Fig. 12.1. During an element of time dt the pressure p_1, at A_1, moves a distance $v_1 \, dt$ and so does the work $p_1 \, dV$, where $dV = A_1 v_1 \, dt = A_2 v_2 \, dt$. A similar argument applies at A_2. Hence, the net work done on this element is $(p_1 - p_2) \, dV$. Suppose that the loss of energy due to viscous forces in the fluid is negligible. This also is a reasonable assumption for a wide range of conditions (cf. Section 12.9). We thus assume the fluid to be *ideal*, i.e. have zero viscosity. All the work $(p_1 - p_2) \, dV$ must then become kinetic and potential energy of the fluid. A volume dV of the element with kinetic energy $\frac{1}{2}\rho v_1^2 \, dV$ at A_1

has been replaced by an equal volume with energy $\frac{1}{2}\rho v_2^2\, dV$ at A_2. The kinetic energy of the element has thus increased by $\frac{1}{2}\rho(v_2^2 - v_1^2)\, dV$. Similarly, the potential energy has increased by $\rho(U_2 - U_1)\, dV$, where U is the potential energy of unit mass of fluid due to an external force. If this force is gravity, then $U_2 - U_1 = (h_2 - h_1)g$, where h_1 and h_2 are the respective heights. Since energy is conserved we have

$$(p_1 - p_2)\, dV = \tfrac{1}{2}\rho(v_2^2 - v_1^2)\, dV + \rho(U_2 - U_1)\, dV,$$

i.e.

$$p_1 + \tfrac{1}{2}\rho v_1^2 + \rho U_1 = p_2 + \tfrac{1}{2}\rho v_2^2 + \rho U_2, \tag{12.9}$$

or

$$p + \tfrac{1}{2}\rho v^2 + \rho U = constant\ along\ a\ streamline. \tag{12.10}$$

This result (which holds for steady flows of incompressible ideal fluids) is *Bernouilli's theorem*. It expresses the most striking single feature of fluid dynamics, that the pressure in a fluid *increases* when the fluid slows down (at constant U), and vice versa. The force of the wind is thus caused by the increase in the pressure of the impinging air as it is brought to rest against an obstacle. Figure 12.2 shows how the streamlines divide to allow a fluid to flow round an obstacle. At the *stagnation point*, where the central streamline meets the nose of the obstacle, the fluid is brought to rest and its pressure there, p_1, exceeds that in the undisturbed stream (p,v) by the amount

$$p_1 - p = \tfrac{1}{2}\rho v^2 \tag{12.11}$$

called the *dynamic pressure*. Where the streamlines crowd together at the sides of the obstacle the velocity is high ($v_2 > v$) and the pressure low ($p_2 < p$).

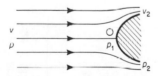

Fig. 12.2. Flow against an obstacle.

If a small hole is made in the obstacle at the stagnation point, to allow the pressure there to penetrate to a pressure gauge connected to a chamber inside the obstacle, the stagnation pressure can be measured. This is the basis of the *Pitot tube*, used, for example, for measuring aircraft speeds. In the example shown in Fig. 12.3, air enters a narrow open-ended tube B at the nose of a large closed-end tube A and its pressure p_1 is measured at the other end of B. Along the sides of A, away from the nose, are holes which allow the pressure inside A to reach the value p. This is also measured and, from the difference $p_1 - p$, the air speed v is found. Another

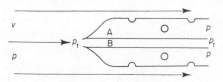

Fig. 12.3.   A Pitot tube.

device is the *Venturi meter*, shown in Fig. 12.4. Here the speed is found from the difference in manometer readings at the narrow and wide parts of the tube.

Bernouilli's equation shows that the pressure should become negative (i.e. hydrostatic tension) in regions of sufficiently high speed. In gases,

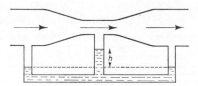

Fig. 12.4.   A Venturi meter.

negative pressures cannot occur and a modified theory, which takes account of compressibility, is necessary for high-speed flows (cf. Section 12.11). In liquids, negative pressures are limited by the onset of *cavitation*, usually at $p \simeq 0$ (cf. Section 2.5). There is thus an upper limiting speed, $v \simeq \sqrt{2p_0/\rho}$, where p_0 is the pressure at zero velocity, for an incompressible fluid in steady flow.

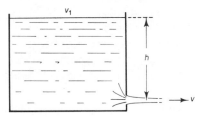

Fig. 12.5.   Flow from a reservoir.

We can now find the speed of a jet issuing from a small orifice in a reservoir of still fluid, as in Fig. 12.5. Suppose that the external pressure is virtually the same at the top of the reservoir and at the orifice, i.e. $p_1 \simeq p_2$; also that the reservoir is so broad, compared with the orifice, that $v_1{}^2 \simeq 0$. Equation (12.9), with $U_1 - U_2 = gh$ and $v_2 = v$, then gives

$$v = \sqrt{2gh}, \tag{12.12}$$

as if the fluid had fallen freely under gravity through the height h. Torricelli (1643), in fact, first derived this formula by analogy with Galileo's law of falling bodies.

The jet contracts in cross-sectional area as it leaves the orifice (*vena contracta*). This is because the fluid in the reservoir approaches the orifice radially and cannot instantly change direction as it enters the jet. All streamlines, which start from the surface of the reservoir and lead down to the orifice, have the same values of p_1, v_1 and U_1. The Bernouilli constant (cf. Equation (12.10)) thus has the same value on every streamline. Streamline flows of this type are said to be *irrotational*, and have special properties that we shall discuss later.

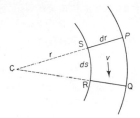

Fig. 12.6. Element of fluid between curved streamlines.

Consider now the change of pressure in a direction *across* the streamlines. In Fig. 12.6 we have an element of fluid PQRS with velocity v, bounded by two streamlines and two radial lines which meet at C (SC $= r$, SP $= dr$, SR $= ds$). The acceleration of the element towards C is v^2/r and, since the mass is $\rho\, dr\, ds$ per unit thickness, the centrifugal force on the element is $\rho(v^2/r)\, dr\, ds$. The pressures on the faces of the element give a radial force $-(\partial p/\partial r)\, dr\, ds$. The radial force due to the potential energy U of the element is $-\rho(\partial U/\partial r)\, dr\, ds$. Hence, balancing radial forces,

$$\frac{\partial(p + \rho U)}{\partial r} = \frac{\rho v^2}{r}. \tag{12.13}$$

We see that the pressure is constant across straight streamlines, when $U = $ constant. When the Bernouilli constant is the same on all streamlines, we can differentiate Equation (12.10) with respect to r to obtain $\partial(p + \rho U)/\partial r = -\rho v(\partial v/\partial r)$, so that Equation (12.13) then becomes

$$\frac{\partial v}{\partial r} + \frac{v}{r} = 0. \tag{12.14}$$

We shall see in Section 12.5 that this result means that the individual particles of the fluid do not rotate.

12.3   EQUATIONS OF FLUID FLOW

We shall now set up the equations governing the motion of fluids. Expressed in their most general form, these are somewhat formidable and insoluble, but we shall nevertheless proceed this way in order afterwards to see the relation between their various simplified and soluble forms.

First, we need some stress–strain relations. For the dilatational components these reduce to an equation of state,

$$\rho = f(p, T), \tag{12.15}$$

which gives the density as a function of pressure and temperature. In a wide range of problems it is sufficient to simplify this to $\rho = $ constant. For the shears we use Newton's law of viscosity. For incompressible viscous fluids this is expressed by Equations (7.38) and (7.39).

Next, we need the *equation of continuity*; what flows into a given region either flows out again or stays in. In Fig. 12.7 we show a volume element with sides δx_i ($i = 1, 2, 3$), fixed with respect to the co-ordinate axes, through which the fluid is flowing. The mass of fluid in the element

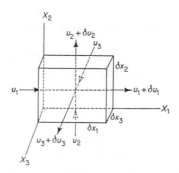

Fig. 12.7.   Flow through a fixed element of space.

changes at the rate $(\partial \rho / \partial t)\, \delta V$, where $\delta V = \delta x_1\, \delta x_2\, \delta x_3$. The influx through the face at $X_1 = 0$ is $\rho u_1\, \delta x_2\, \delta x_3$ and the efflux through the opposite face at $X_1 = \delta x$, is $\{\rho u_1 + [\partial(\rho u_1)/\partial x_1]\, \delta x_1\}\, \delta x_2\, \delta x_3$. The net influx is then $-[\partial(\rho u_1)/\partial x_1]\, \delta V$. Similarly for the other faces. Equating the total influx to the rate of increase of mass in the volume element, we obtain the equation of continuity,

$$\frac{\partial \rho}{\partial t} + \frac{\partial(\rho u_j)}{\partial x_j} = 0, \tag{12.16}$$

where $j (=1, 2, 3)$ is a *repeated suffix* and is to be summed as explained in Section 4.6. When $\rho =$ constant, this reduces to Equation (7.37).

Finally, we need the equations of equilibrium of forces. We have to introduce the forces of inertia and viscosity into Equations (4.69). The force of inertia is a body force and so joins F_i. Since it is *mass* × *acceleration*, we might think it simply equal to $\rho(\partial u_i/\partial t)$ per unit volume. This is wrong. We are now dealing with the acceleration of a *moving element of fluid*, whereas $\partial u_i/\partial t$ gives the variation of velocity with time at a *fixed point of space*. Strictly, we should have allowed for this effect in our previous analyses of deformation and flow, but it is only now, where the displacements of the medium can be very large, that we must take it into account. Let H be any quantity belonging to a particular moving element of fluid, (e.g. $H = u_i$ or $H = p$) and let it change by dH during a small interval of time dt. Now H is a function of both time and position; furthermore, during the time dt, our element of fluid has moved distances $dx_j (=u_j\, dt)$ along the three axes X_j. The total change of H is therefore contributed partly by the direct lapse of time, i.e. $(\partial H/\partial t)\, dt$, and partly by the movement of the element to a different place. We thus have

$$dH = \frac{\partial H}{\partial t}\, dt + \frac{\partial H}{\partial x_j}\, dx_j = \left(\frac{\partial H}{\partial t} + \frac{\partial H}{\partial x_j}\frac{\partial x_j}{\partial t}\right) dt,$$

i.e.

$$\frac{dH}{dt} = \frac{\partial H}{\partial t} + u_j \frac{\partial H}{\partial x_j}, \tag{12.17}$$

where j is a repeated suffix. Whereas $\partial H/\partial t$ tells us how H changes with time at a fixed point of space, dH/dt tells us how the value of H changes in a given moving element of fluid; dH/dt is called the *convective derivative*.

We take $H = u_i$ and add the three inertial forces $-\rho(du_i/dt)$ per unit volume to $F_i\rho$ in Equation (12.2). This gives the three *Euler equations* $(i = 1, 2, 3)$,

$$\frac{\partial u_i}{\partial t} + u_j \frac{\partial u_i}{\partial x_j} + \frac{1}{\rho}\frac{\partial p}{\partial x_i} - F_i = 0, \tag{12.18}$$

for the motion of a *non-viscous* fluid along the axes X_i. Slow viscous flow, in which the inertial forces are neglected, has already been described by Equations (7.42); if we now add inertial forces to these equations, we obtain the *Stokes–Navier equations* for incompressible, viscous, fluid flow,

$$\frac{\partial u_i}{\partial t} + u_j \frac{\partial u_i}{\partial x_j} + \frac{1}{\rho}\frac{\partial p}{\partial x_i} - F_i - \nu\nabla^2 u_i = 0, \tag{12.19}$$

where ν is the *kinematic viscosity* $(=\eta/\rho)$ and $\nabla^2 u_i$ is defined by Equations (7.41). To generalize to *compressible* viscous flow, we note that some of the tensile strain rate $\dot{\varepsilon}_{11}$ in Equation (7.38) is not necessarily due to shear. We subtract the contribution $\frac{1}{3}\dot{\Theta}$ due to the dilatation, and Equation (7.38) then becomes

$$\sigma_{11} = 2\eta(\dot{\varepsilon}_{11} - \tfrac{1}{3}\dot{\Theta}) - p. \qquad (12.20)$$

When this is substituted into Equation (7.40), the new term $-\frac{2}{3}\eta(\partial\dot{\Theta}/\partial x_1)$, which is $-\frac{2}{3}\eta\,\partial(\partial u_j/\partial x_j)/\partial x_1$, cancels two-thirds of the last term in Equation (7.40). The remaining one-third appears in the Stokes–Navier equations which then become

$$\frac{\partial u_i}{\partial t} + u_j\frac{\partial u_i}{\partial x_j} + \frac{1}{\rho}\frac{\partial p}{\partial x_i} - F_i - \nu\nabla^2 u_i - \tfrac{1}{3}\nu\frac{\partial^2 u_j}{\partial x_i\,\partial x_j} = 0. \qquad (12.21)$$

If the fluid is ionized, its particles may experience electromagnetic forces. Further terms have then to be added to these equations. The new effects which then appear (which are important in astrophysics and in flames and plasmas) are dealt with by the theory of *magnetohydrodynamics*. It is found, for example, that lines of magnetic force must move with the fluid, because charged fluid would induce infinite electrical currents if it crossed lines of force. Because the fluid is, in effect, "glued" to magnetic lines of force and because there is a tension along such lines, it behaves as if made of taut strings. These can be vibrated and will transmit transverse waves, called *magnetohydrodynamic waves*, by a motion more like the transverse vibrations of a solid than the flow of a liquid.

12.4 SOUND WAVES

The general equations can be greatly simplified when applied to the propagation of infinitesimal disturbances through an otherwise resting fluid. The displacements are everywhere small and so, in Equations (12.21), we can substitute the average density ρ_0 for its actual value at any point, ρ, and can drop the terms $u_j(\partial u_i/\partial x_j)$, which involve products of the small u components and their derivatives. We can also express the pressure p as a power series expansion in the density and drop the higher terms, i.e.

$$p = p_0 + c^2(\rho - \rho_0) + \text{higher powers of } (\rho - \rho_0), \qquad (12.22)$$

where p_0 is the average pressure and c^2 will be explained below. Let $F_i = 0$, and consider a plane wave moving along the X_1 axis, with $u_2 = u_3 = 0$ and with constant u_1 and p on the wave surfaces $x_1 = \text{constant}$. We drop the subscripts on u_1 and x_1. The equation of continuity then becomes

$$\frac{\partial\rho}{\partial t} + \frac{\partial(\rho u)}{\partial x} = \frac{\partial\rho}{\partial t} + \rho_0\frac{\partial u}{\partial x} = 0, \qquad (12.23)$$

where we have dropped the term $u_0(\partial\rho/\partial x)$, since $u_0 = 0$. Equation (7.41) reduces to $\nabla^2 u = \partial^2 u/\partial x^2$, and so Equations (12.21) become

$$\frac{\partial u}{\partial t} + \frac{1}{\rho_0}\frac{\partial p}{\partial x} - \tfrac{4}{3}v\frac{\partial^2 u}{\partial x^2} = 0. \tag{12.24}$$

We differentiate with respect to time and then eliminate p as follows:

$$\frac{1}{\rho_0}\frac{\partial}{\partial t}\left(\frac{\partial p}{\partial x}\right) = \frac{1}{\rho_0}\frac{\partial}{\partial x}\left(\frac{\partial p}{\partial t}\right) = \frac{c^2}{\rho_0}\frac{\partial}{\partial x}\left(\frac{\partial \rho}{\partial t}\right) = -c^2\frac{\partial^2 u}{\partial x^2}, \tag{12.25}$$

to obtain

$$\frac{\partial^2 u}{\partial t^2} - c^2\frac{\partial^2 u}{\partial x^2} - \tfrac{4}{3}v\frac{\partial^3 u}{\partial t\,\partial x^2} = 0. \tag{12.26}$$

When $v = 0$, this reduces to an ordinary wave equation (cf. Equation (6.5)) with wave velocity c, where

$$c^2 = dp/d\rho. \tag{12.27}$$

Newton first derived an expression for the speed of sound waves in air, but did not know that volume changes in the pressure wave occur *adiabatically*, there being not enough time for heat to flow. We have to use the adiabatic equation $p \propto \rho^\gamma$, where γ is the ratio of specific heats. Thus,

$$c^2 = \gamma p_0/\rho_0. \tag{12.28}$$

For air at NTP, $p_0 = 10^6$ dyn cm^{-2}, $\rho_0 = 0\cdot0013$ g cm^{-3}, so that $c = 33{,}000$ cm sec^{-1} ($=742$ mile/hr), which almost equals the mean speed of its molecules. Although the molecules themselves are deflected by collisions and diffuse slowly, their momentum goes forward, from one molecule to another, since it is conserved. This momentum, because it acts in the forward, not the transverse, direction, does not depend on random diffusion for its forward propagation, unlike that of Section 1.10. Since p/ρ is proportional to T, the speed of sound in a gas varies as $T^{1/2}$. For example, the speed of sound in air varies from about 760 mile/hr at sea level to about 680 mile/hr at 30,000 ft.

To examine the effect of viscosity, we try a solution $u = y\cos kx$ in Equation (12.26), where y is a function of time. This gives

$$\frac{d^2 y}{dt^2} + \tfrac{4}{3}vk^2\frac{dy}{dt} + c^2k^2 y = 0. \tag{12.29}$$

From the analysis of the similar Equation (6.41), we see that, when the damping is small, the wave travels with speed c and its amplitude decays as $\exp(-t/\tau)$, where

$$\tau = 3/2vk^2, \tag{12.30}$$

k being the wave number, $2\pi/\lambda$. The wave decays over a distance of order $c\tau$, approximately $\lambda^2 c/25v$. Ordinary sound waves are hardly affected, but shrill notes (e.g. 16,000 c/s; $\lambda \simeq 2$ cm, $\tau \simeq 1$ sec) do not carry much beyond about $\frac{1}{4}$ mile. Thus distant echoes sound mellow.

12.5 IRROTATIONAL FLOW

To derive Bernouilli's equation from the general equations, we put $v = 0$ and $\rho = $ constant (which together define an *ideal fluid*) and assume that F_i can be expressed in the form

$$F_i = -\partial U/\partial x_i, \tag{12.31}$$

where U is potential energy. We also have steady flow, i.e. $\partial u_i/\partial t = 0$. Equations (12.19) then become

$$\rho u_j \frac{\partial u_i}{\partial x_j} + \frac{\partial (p + \rho U)}{\partial x_i} = 0, \tag{12.32}$$

which can be integrated if we can change x_j into x_i. This is certainly possible when

$$\frac{\partial u_i}{\partial x_j} = \frac{\partial u_j}{\partial x_i}. \tag{12.33}$$

Assuming that this is so, and noting that $u_j(\partial u_j/\partial x_i) = \frac{1}{2}\,\partial(u_j u_j)/\partial x_i$ and that $u_j u_j = u_1{}^2 + u_2{}^2 + u_3{}^2 = v^2$, where v is the velocity, Equation (12.32) becomes

$$\frac{\partial}{\partial x_i}\,[\tfrac{1}{2}\rho v^2 + p + \rho U] = 0,$$

i.e.

$$p + \tfrac{1}{2}\rho v^2 + U = \text{constant.} \tag{12.34}$$

Since the velocity components u_i and u_j are displacements per unit time, we see from Equations (4.24) that Equations (12.33) imply that elements of fluid do not rotate during their motion. The flow is *irrotational*. A fluid can of course *circulate* irrotationally round closed paths. Figure 12.8 shows two circulatory flows. In the first, the particles move irrotationally by a sequence of horizontal and vertical translations, like passenger cars on a fairground wheel. In the second, they rotate, like the moon round the earth. Irrotational circulation requires the particles to slip past one another without setting up rotations, which is possible because, by definition, there is no friction between the particles of an ideal fluid.

To describe these different kinds of flow, two properties are defined, *circulation* and *vorticity*. Consider a closed curve S in a fluid. Roughly, the circulation is the length of this curve multiplied by the average speed of motion, if any, round it; precisely, it is the line integral, once round the circuit, of the tangential component of the velocity. In Fig. 12.9 the

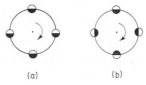

(a)   (b)

Fig. 12.8. Circulation: (a) irrotational; (b) rotational.

velocity v has the tangential component v_s at the point s. The element of line between s and $s + ds$ contributes $v_s\, ds$ to the integral. The circulation C round S is then

$$C = \int_S v_s\, ds = \int_S v \cos \alpha\, ds = \int_S u_i\, dx_i, \qquad (12.35)$$

where α is the angle between v and v_s, u_i and dx_i are the components of v and ds along co-ordinate axes X_i, i is a repeated suffix, and the subscript S denotes an integration once round the curve S.

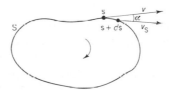

Fig. 12.9. Component of velocity round a circuit.

The circulation round macroscopic circuits does not tell us whether the fluid is moving as in Fig. 12.8(a) or (b). To investigate *rotation*, we must ask how individual particles are moving, i.e. take microscopic circuits. The flow is irrotational when the circulation is zero round every *infinitesimal* circuit. Consider the circulation C round PQRSP in Fig. 12.6. The velocity is zero along QR and SP; and $-v$ and $v + (\partial v/\partial r)\, dr$, respectively, along RS and PQ. The lengths of RS and PQ are ds and $(1 + dr/r)\, ds$, respectively, so that

$$C = \left(v + \frac{\partial v}{\partial r}\, dr\right)\left(1 + \frac{dr}{r}\right) ds - v\, ds = \left(\frac{\partial v}{\partial r} + \frac{v}{r}\right) dr\, ds, \qquad (12.36)$$

neglecting the third-order term. From this and Equation (12.14), we see that, when the Bernouilli constant is the same on all streamlines, the circulation vanishes round every infinitesimal circuit and the flow is irrotational.

For circular streamline flow, as in Fig. 12.10, we integrate Equation (12.14) to obtain

$$v = \text{constant}/r. \tag{12.37}$$

Thus, in irrotational circular flow the fluid is not spinning like a solid body, since its inner circles move faster than the outer ones. The circulation is simply

$$C = 2\pi r v = \text{constant}. \tag{12.38}$$

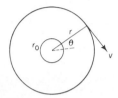

Fig. 12.10.   Circular streamline flow.

12.6   VORTICES, WAKES AND EDDIES

Equations (12.37) and (12.38) show that the circulation is the same round every circle centred on the origin in Fig. 12.10. Thus, even though the flow as a whole is irrotational, we must, as we reduce r to zero, in the limit reach an infinitesimal circle with a finite circulation, which is a *rotating element*. The centre of a region of circulation therefore lies *outside* the irrotational region. It is called a *vortex*. The important property here is the ratio C/A of the circulation C to the area A of the circuit, which is called the *vorticity*. In an irrotational region, irrespective of the circulation, the vorticity vanishes in every infinitesimal circuit. This is true in Fig. 12.10 except at the origin.

It follows that, when a fluid has circulation, its irrotational region is *doubly connected* (or *multiply connected*), in the sense that a closed circuit round which there is circulation cannot be shrunk to a point without passing out of the irrotational region, at the vortex. Such a fluid thus typically consists largely of irrotational regions through which there run vortex lines where there is rotation. Familiar examples are smoke rings, eddies, whirlpools, whirlwinds and sun spots.

There may even be no fluid at all at the centre of circulation. A familiar example is a *whirlpool* or *free vortex*, as shown in Fig. 12.11. The funnel remains empty because the centrifugal force due to the high velocities near the centre brings the pressure there down to atmospheric. Consider an

element of fluid which started at a distant point on the surface and is now
at a depth h and radius r on the funnel. The (atmospheric) pressure p is
nearly the same at both points. The velocity at the starting point ($h = 0$,
$r = \infty$) is zero, by Equation (12.37). Hence, Equation (12.12) applies.
Combining this with Equation (12.38), we obtain

$$r = \frac{C}{2\pi\sqrt{2gh}} \qquad (12.39)$$

for the radius of the funnel at depth h. Ideally, the direction of rotation
of whirlpools in bathtubs is determined, oppositely in the northern and
southern hemispheres, by the Coriolis force from the earth's motion. In
practice it is usually determined by adventitious effects such as circulation
introduced during the filling of the bath.

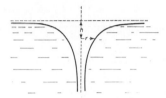

Fig. 12.11.   A free vortex.

Alternatively, there may exist a stationary or rotating cylinder of solid
or fluid along a vortex line. Since the fluid outside is frictionless, it can
slip freely along the surface of this cylinder. The vorticity has not been
eliminated, since the Bernouilli constant now differs on the two streamlines
which circle the cylinder along the two faces of the slip interface. This
interface, in fact, forms a *tube* of vortices, since the slip along it is a form
of *simple shear*, which is partly a rotation (cf. Fig. 4.13). A slip line in a
plastic solid is geometrically equivalent to a *sheet of vortices*, which we may
picture as a set of minute, parallel, "roller bearings" between the slipping
faces. Such *vortex sheets* exist along all solid surfaces over which an ideal
fluid slips.

The *strength* of a vortex is the circulation round any circuit which en-
closes that vortex alone. Such a circuit can be expanded or shrunk, or
moved along the vortex line, without any change of circulation, provided
it does not cross a vortex. The strength of a vortex line is thus analogous,
in its conservation properties, to the Burgers vector of a dislocation line.
There are further analogies. The total strength of a bundle of vortices is
the algebraic sum of the individual strengths. Vortex lines must either
form closed rings (e.g. smoke rings) or end at surfaces (e.g. whirlpools;
vortices trailing from wing tips of aircraft).

We shall now derive an important theorem about circulation in a moving fluid. Consider a closed circuit made up of a *fluid line*, i.e. a line always composed of the same particles of fluid. At a given time two neighbouring particles on this line have co-ordinates and velocities x_i, u_i, and $x_i + \delta x_i$, $u_i + \delta u_i$, respectively. The element δx_i of the line contributes $u_i \, \delta x_i$ to the circulation. We take the convective derivative of this,

$$\frac{d(u_i \, \delta x_i)}{dt} = \delta x_i \frac{du_i}{dt} + u_i \frac{d(\delta x_i)}{dt}. \tag{12.40}$$

Since $d(\delta x_i)/dt$ is the rate at which the second particle is moving relative to the first, it is equal to δu_i. The second term on the right-hand side is thus $u_i \, \delta u_i = \frac{1}{2} \delta(u_1{}^2 + u_2{}^2 + u_3{}^2) = \frac{1}{2} \delta v^2$. Equations (12.17), (12.18) and (12.31) enable us to transform the first term. Equation (12.40) then becomes

$$\frac{d(u_i \, \delta x_i)}{dt} = -\frac{1}{\rho}\left[\frac{\partial(p + \rho U)}{\partial x_i}\right] \delta x_i + \frac{1}{2} \delta v^2 = \delta\left(\frac{v^2}{2} - \frac{p}{\rho} - U\right). \tag{12.41}$$

The circulation along the fluid line thus changes at a rate equal to the line integral of $\delta(v^2/2 - p/\rho - U)$ round the fluid line. But each of the quantities v^2, p/ρ and U has the same value at the beginning and end points of the line integral, since these points coincide. Hence, the integral is zero. This is *Thomson's theorem*: *the circulation round a closed fluid line is constant with time*. It means, for example, that a vortex is indestructible, i.e. maintains itself at full strength as it drifts with the fluid. The theorem is of course restricted to ideal fluids and to body forces that obey Equation (12.31).

Every closed fluid line in a still fluid has no circulation. By Thomson's theorem it continues so even if this fluid is subsequently set in motion. This seems to contradict the familiar fact that, when a paddle or spoon is drawn sharply through still water, vortices are formed in its wake. We

(a)   (b)

Fig. 12.12. Flow over an edge: (a) without, (b) with, separation.

must remember, however, that we have introduced a sliding solid interface into the fluid. Consider the flow of an ideal fluid over a *sharp* edge, as in Fig. 12.12. The fluid cannot turn the sharp corner to slide down the trailing side without reaching infinite velocity, acceleration and negative pressure at the edge. A liquid, for example, would cavitate at this edge. To avoid such impossible conditions, the fluid does not follow the interface but *separates* away from the leeward side. It glides off the edge into the space beyond, and a *line of discontinuity* is formed along the streamline which separates, as in Fig. 12.12(b).

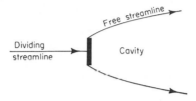

Fig. 12.13. Free streamlines formed by flow past a plate.

The separated region may be a *cavity*, as in the wake of a fast projectile fired into water. A *jet* from an orifice is essentially similar, but in reverse. The separated streamlines in such cases mark the free surface of the fluid and are called *free streamlines*. Their theory was developed by Helmholtz. Figure 12.13 shows an example. Neglecting gravity, the pressure in the cavity is constant and so is also constant along a free streamline. If $U =$ constant, Bernouilli's theorem then shows that the velocity is also constant along a free streamline.

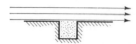

Fig. 12.14. Flow past a stagnant region.

The separated region may instead be a *wake*, filled with fluid. Kirchhoff and Rayleigh showed that this could be a static wake, filled with "dead" fluid, similar to a backwater on a river (Fig. 12.14). In place of the free streamline we now have a *streamline of discontinuity*, across which there is a *finite* change of velocity. These, analogous to slip lines in plastic flow, are familiar as the boundaries of rising smoke plumes from chimneys in still air. They are also the meteorological *fronts*, formed where different winds meet.

Since a discontinuous streamline is the continuation of a sliding interface, it is also a *vortex sheet*. The infinite rate of shear across this infinitesimally thin sheet is possible in an ideal fluid, because there is no friction, but in a real fluid the frictional forces cause the sheet to thicken into a *boundary layer* across which the velocity changes rapidly, but not discontinuously. Thomson's theorem is preserved, despite the injection of a vortex sheet into the fluid, thrown off the point of separation, because a fluid line in the stream always remains on the same side of this vortex sheet. In a sense, the fluid in the wake becomes part of the obstacle and the main stream slips along this enlarged but now smoothly contoured obstacle.

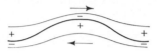

Fig. 12.15.   Unstable discontinuous streamline.

Discontinuous streamlines are unstable. In Fig. 12.15 we suppose such a streamline (heavily outlined) to be stationary, with the fluid flowing along it in opposite directions, above and below. Suppose that a small undulation develops in it. Because of the altered spacings of the neighbouring streamlines the pressure decreases on its convex side and increases

Fig. 12.16.   Formation of an eddy from an unstable discontinuous streamline.

on its concave side, so that the wave will grow in amplitude. As the wave protrudes further across the stream, it is sheared by the motion of the fluid and rolled up into an *eddy*, as in Fig. 12.16. This leads to the familiar *turbulent* flow behind sharp obstacles, as in Fig. 12.17, in which eddies form in otherwise stagnant fluid. These eddies grow larger as more of the

Fig. 12.17.   Eddies formed behind a sharp edge.

vortex sheet is peeled off the edge and fed into them. They may then break away and drift downstream, new ones taking their place.

In a real fluid the viscous forces enable eddies to form from solid surfaces even without sharp edges. As we shall see later, the viscosity lifts the vortex sheet off the interface and so helps it to separate.

12.7 THE ANALYSIS OF IRROTATIONAL FLOW

Even at their simplest (Euler's equations), the general equations of fluid flow are formidable. They contain four unknowns (u_1, u_2, u_3, p) and are non-linear. Restricting ourselves to the conditions under which Thomson's theorem is valid, we can, however, use this theorem to conclude that velocity fields with zero circulation are solutions of these equations. Zero circulation is equivalent to the condition that the velocity, like the body force, is derivable from a "potential", ϕ, i.e.

$$u_i = -\partial\phi/\partial x_i. \tag{12.42}$$

To see this, consider two points A and B in a fluid with no circulation. We go from A to B along a fluid line and evaluate $\int_A^B u_i \, dx_i$ along this line. We then return from B to A along *any* other fluid line and know that the integral of $u_i \, dx_i$ along it is, irrespective of its path, equal to $-\int_A^B u_i \, dx_i$ since we have completed a closed circuit in a fluid without circulation. Hence, since the integral is independent of the path between A and B, its value depends only on the positions of A and B. Thus, if we take A as origin, we can then attach to every other point such as B its own particular value of $-\int_A^B u_i \, dx_i$ which is a property of that point and not of the path to reach it. This integral is thus formally analogous to a potential. We denote it as

$$\phi = -\int_A^B u_i \, dx_i, \tag{12.43}$$

called the *velocity potential* of B relative to A. The change of ϕ with the position x_i of B then gives Equation (12.42).

Any function ϕ of the space co-ordinates x_i thus generates a velocity field that satisfies the equilibrium equations. It has, however, also to satisfy the equation of continuity (with $\rho = $ constant),

$$-\frac{\partial u_i}{\partial x_i} = \frac{\partial}{\partial x_i}\left(\frac{\partial\phi}{\partial x_i}\right) = \frac{\partial^2\phi}{\partial x_1{}^2} + \frac{\partial^2\phi}{\partial x_2{}^2} + \frac{\partial^2\phi}{\partial x_3{}^2} = \nabla^2\phi = 0, \tag{12.44}$$

i.e. ϕ has to satisfy *Laplace's equation*. This is a familiar equation in many branches of physics. It has only one unknown, ϕ. It is a linear equation, and so we can add solutions to make new ones (*principle of superposition*). Our problem is thus greatly simplified.

The simple solution $\phi = ax_1$ describes uniform flow with constant velocity $-a$ along the X_1 axis. The function $\phi = ax_1{}^2 + bx_2{}^2 + cx_3{}^2$ satisfies Laplace's equation when $a + b + c = 0$. For example,

$$\phi = a[x_1{}^2 - \tfrac{1}{2}(x_2{}^2 + x_3{}^2)], \tag{12.45}$$

which gives $u_1 = -2ax_1$, $u_2 = ax_2$, $u_3 = ax_3$, describes flow against a plate, as in Fig. 12.18. To find the streamlines, we note that the components dx_i of an element of a streamline must be proportional to the velocity components u_i, i.e.

$$\frac{dx_1}{u_1} = \frac{dx_2}{u_2} = \frac{dx_3}{u_3}, \tag{12.46}$$

because $dx_i = u_i\,dt$ along a streamline. Hence, in the plane $X_2 = 0$ we have $dx_1/dx_3 = u_1/u_3 = -2x_1/x_3$, so that

$$x_1 x_2{}^2 = \text{constant} \tag{12.47}$$

is the equation of the streamlines in this plane. The *equipotential lines*, $\phi = \text{constant}$, intersect these streamlines perpendicularly. From the velocities and Bernouilli's theorem, we can find the pressure. If $U = 0$ then $p = \text{constant} - \frac{1}{2}\rho a(4x_1{}^2 + x_2{}^2 + x_3{}^2)$.

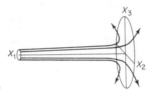

Fig. 12.18.   Flow against a plane.

Spherically symmetrical solutions have the form

$$\phi = \pm a/r, \tag{12.48}$$

where r is the radial distance from a point *source* $(+a)$ or *sink* $(-a)$ of fluid, such as might be simulated approximately by the orifice of a long thin pipe, immersed in a reservoir of fluid, through which fluid is slowly introduced or withdrawn. By superposition, we can then obtain the important solution

$$\phi = a\left(\frac{1}{r_1} - \frac{1}{r_2}\right) - v_0 x_1, \tag{12.49}$$

which represents an axially symmetric diversion in an otherwise uniform flow with velocity v_0, as shown in Fig. 12.19. The fluid which appears at the source A and disappears at the sink B, does not mix with the main stream, which diverts round the *dipole* AB as shown. The boundary C could equally well represent the surface of a frictionless obstacle in the stream. If the distance AB is reduced, the flow from A to B being increased in inverse proportion, this becomes in the limit a pattern of flow round a sphere.

Such patterns are symmetrical upstream and downstream of the obstacle. The fluid pressures on the leading and trailing sides of the obstacle are therefore equal and opposite. There is then no resultant force on the obstacle! This is *D'Alembert's paradox*, a clear contradiction with everyday experience. The point of course is that most real flows are not like that of Fig. 12.19 but are unsymmetrical, like those of Figs. 12.13 and

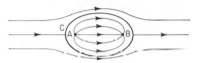

Fig. 12.19.   Source (A) and sink (B), equivalent to an obstacle (C) in a stream.

12.17, with wakes behind the obstacles. Despite its unreality, the theory developed above does describe the upstream flow rather well and is a stepping stone to better theories.

A powerful method is available for solving Equation (12.44) for *plane flows*, i.e. where $u_3 = \partial u_1/\partial x_3 = \partial u_2/\partial x_3 = 0$ (cf. Section 5.7). Laplace's equation

$$\frac{\partial^2 \phi}{\partial x^2} + \frac{\partial^2 \phi}{\partial y^2} = 0, \tag{12.50}$$

where $x_1 = x$ and $x_2 = y$, then resembles the wave equation except for the different sign. We saw in Section 6.1 that any function of the variable $x + ay$, where $a = \pm 1$, satisfies the wave equation. The same is true of Equation (12.50), provided now that $a^2 = -1$, i.e. $a = \pm i$, where $i = \sqrt{-1}$. General solutions are therefore any functions of the form $f(z)$ or $f(z^*)$, where $z = x + iy$ and $z^* = x - iy$. Despite the complex numbers, this is useful, because these functions can easily be transformed, i.e.

$$f(x + iy) = \phi + i\psi, \tag{12.51}$$

where ϕ and ψ are *real* functions (i.e. not containing i) of x and y, each of which by itself satisfies Equation (12.50). Thus, by choosing functions such as $(x + iy)^2$, $(x + iy)^{-1}$, $\ln(x + iy)$, etc. and changing these into the form $\phi + i\psi$, we can create many functions ϕ and ψ, each of which is a solution. Suppose, for example, that $f(z) = z^{-1}$. We transform it by multiplying it, numerator and denominator, by $f(z^*)$, i.e.

$$f(z) = \frac{1}{x + iy} = \frac{1}{x + iy}\frac{x - iy}{x - iy} = \frac{x - iy}{x^2 + y^2} = \frac{x}{x^2 + y^2} + i\frac{-y}{x^2 + y^2}, \tag{12.52}$$

so that

$$\phi = \frac{x}{x^2 + y^2} \quad \text{and} \quad \psi = -\frac{y}{x^2 + y^2}. \tag{12.53}$$

Substitution in Equation (12.50) shows that these are both solutions.

What do ϕ and ψ mean? We note first that $\partial f/\partial y = i(\partial f/\partial x)$, since x and iy have equal status in any function $f(x + iy)$.
Hence,

$$\frac{\partial \phi}{\partial y} + i\frac{\partial \psi}{\partial y} = \frac{\partial f}{\partial y} = i\frac{\partial f}{\partial x} = i\left(\frac{\partial \phi}{\partial x} + i\frac{\partial \psi}{\partial x}\right) = -\frac{\partial \psi}{\partial x} + i\frac{\partial \phi}{\partial x}. \tag{12.54}$$

Since $a = ib$ is impossible when a and b are both real numbers, this means that

$$\frac{\partial \phi}{\partial y} = -\frac{\partial \psi}{\partial x} \quad \text{and} \quad \frac{\partial \phi}{\partial x} = \frac{\partial \psi}{\partial y}. \tag{12.55}$$

These are the *Cauchy–Riemann* equations. Suppose $\phi + i\psi = -v_0(x + iy)$, where v_0 is a constant. Then $\phi = -v_0 x$, $\psi = -v_0 y$, $u_x = -\partial \phi/\partial x = v_0$ and $u_y = -\partial \phi/\partial y = 0$. We thus have uniform flow with speed v along the X axis. The lines $\phi =$ constant are then *equipotentials*, and the lines perpendicular to them, for which $\psi =$ constant, are the *streamlines*, ψ being called the *stream function*. We can prove this quite generally. Along any line where $\psi =$ constant we have

$$d\psi = \frac{\partial \psi}{\partial x} dx + \frac{\partial \psi}{\partial y} dy = 0 \tag{12.56}$$

and hence, from Equations (12.42) and (12.55),

$$\frac{dy}{dx} = -\frac{\partial \psi/\partial x}{\partial \psi/\partial y} = \frac{u_y}{u_x}, \tag{12.57}$$

which is the equation of a streamline (Equation (12.46)). We could, of course, equally well regard the lines $\phi =$ constant as streamlines and $\psi =$ constant as equipotentials, since both ϕ and ψ are solutions of Equation (12.50).

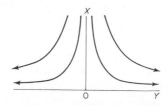

Fig. 12.20. Two-dimensional flow against a plane.

Functions of the type z^n provide interesting solutions. Uniform flow is given by $n = 1$, as shown above. For $n = 2$ we write $\phi + i\psi = \frac{1}{2}a(x + iy)^2 = \frac{1}{2}a(x^2 - y^2 + 2ixy)$, so that $\phi = \frac{1}{2}a(x^2 - y^2)$ and $\psi = axy$. The *streamlines* are then hyperbolas. This is two-dimensional flow against a plane. We see from Fig. 12.20 that the dividing streamline ($y = 0$) could be replaced by a frictionless wall, so that the pattern on one side also represents flow in a square corner. This can be generalized; the function

$$\phi + i\psi = \frac{a}{n} z^n, \tag{12.58}$$

where $n = \pi/\alpha$, represents flow in the angle α between two plane walls which meet at the origin, one of which lies along the X axis. Using polar co-ordinates r and θ, we have $z = x + iy = r(\cos\theta + i\sin\theta) = r\,e^{i\theta}$ and hence, $z^n = r^n\,e^{in\theta} = r^n(\cos n\theta + i\sin n\theta)$. Also, $\psi = (a/n)r^n \sin n\theta$, so that the radial lines along $\theta = 0$ and $\theta = \alpha$ are streamlines on which $\psi = 0$. Figures 12.12(a), 12.20 and 12.21 show examples of such flows. The velocity goes as r^{n-1} and so becomes infinite or zero at the origin, $r = 0$, according as $n < 1$ or $n > 1$. In reality, infinite velocities are avoided by *separation* of the flow, as discussed earlier. Equation (12.58) therefore does not apply downstream of obstacles for which $n < 1$. The true solutions in such cases involve *discontinuous* functions (cf. Fig. 12.13).

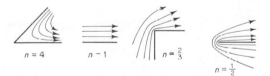

$n = 4$ $n = 1$ $n = \frac{2}{3}$ $n = \frac{1}{2}$

Fig. 12.21. Flow at corners and edges.

The function

$$\phi + i\psi = -v_0\left(z + \frac{r_0^2}{z}\right), \tag{12.59}$$

which gives (cf. Equation (12.52))

$$\psi = -v_0 y\left(1 - \frac{r_0^2}{x^2 + y^2}\right) = v_0\left(\frac{r_0^2}{r} - r\right)\sin\theta, \tag{12.60}$$

represents symmetrical flow round a circular cylinder of radius r_0 in a stream of undisturbed velocity v_0, as in Fig. 12.22. Because of the symmetry, there is no resistance to flow (D'Alembert's paradox).

Many similar problems have been solved. Figure 12.23 shows irrotational flow (without circulation) round an inclined plate. The stagnation points, where the pressure is high, are placed so that the fluid tends to turn the plate *broadside* to the flow, as may be seen in a falling leaf. Although there is a couple, there is no resultant force on the plate. It can, in fact,

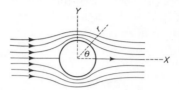

Fig. 12.22.   Flow round a cylinder.

be proved generally that in any irrotational flow of an ideal fluid without circulation there is no resultant force on an immersed obstacle, whatever its shape. The characteristic *lift* (force across the stream) and *drag* (force down the stream) on airfoils come from other effects.

Fig. 12.23.   Flow without circulation round a plate.

12.8   IRROTATIONAL FLOW WITH CIRCULATION

The argument for a velocity potential holds also for irrotational flows with circulation. We can thus apply Laplace's equation to such flows. Using the co-ordinates of Fig. 12.10, consider

$$\phi + i\psi = ia \ln z = ia \ln (r\, e^{i\theta}) = -a\theta + ia \ln r, \qquad (12.61)$$

i.e. $\phi = -a\theta$, $\psi = a \ln r$. The streamlines, $\psi = $ constant, are concentric circles and the equipotentials are radial lines from the origin. The velocity on a streamline is $v = -d\phi/ds = a/r$, where $ds\, (=r\, d\theta)$ is an element of a circle of radius r. The circulation is then $C = 2\pi a$. The superposition of this with Equation (12.59), i.e.

$$\phi + i\psi = -v_0\left(z + \frac{r_0^2}{z}\right) + \frac{iC}{2\pi} \ln z, \qquad (12.62)$$

represents a stream which is both flowing past and rotating round a cylinder, as in Fig. 12.24. Below the cylinder, where the circulatory and non-circulatory flows run against each other, the velocity is low and the pressure high. Above, where they run with each other, the velocity is high and the pressure low. The fluid thus exerts a force on the cylinder, lifting it upwards. This is the *Magnus effect*, the basis of airfoil action.

Figure 12.25 shows flows round an airfoil. The irrotational flow of diagram (a) gives neither lift nor drag. The sharp trailing edge of the airfoil produces a line of discontinuity which curls up into a *starting vortex*, diagram (b), as soon as the flow begins. This vortex eventually breaks away and drifts downstream. By Thomson's theorem, the total circulation

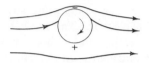

Fig. 12.24. Lift on a cylinder due to flow with circulation.

round a large circuit enclosing both airfoil and vortex remains at zero (assuming no circulation initially). Hence, to compensate for the circulation round the vortex, another circulatory flow is formed round the airfoil, diagram (c). Combined with the irrotational flow of diagram (a), this gives a flow with circulation and hence a lift, by the Magnus effect, on the airfoil, diagram (d).

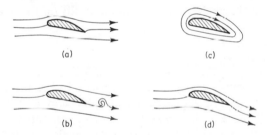

(a)   (c)

(b)   (d)

Fig. 12.25. Flows round an airfoil: (a) irrotational without circulation; (b) the starting vortex; (c) circulation; (d) combination of (a) and (c).

By the conservation rules of Section 12.6, the circulation cannot suddenly end at the wing-tips of an airfoil. Hence, *vortex trails* run backwards from these tips to the ends of the starting vortex; the airfoil, the vortex trails and the starting vortex thus form one large closed vortex ring. The trails are formed by the flow of air round the wing tips, from the underside of the airfoil where the pressure is high, to the topside where the pressure is low.

The lifting force L can be found from the *Kutta–Joukowsky theorem*: *in an irrotational two-dimensional flow, the lift L (per unit length perpendicular to the plane of the flow pattern) is given by*

$$L = \rho v_0 C, \qquad (12.63)$$

where ρ is the fluid density, v_0 the undisturbed stream velocity, and C the circulation. To prove this, consider a circle of radius r so large that the *circulation velocity* v round it can be written as $v = C/2\pi r$, where $v \ll v_0$. The total velocity at a point r,θ (cf. Fig. 12.22) on the circle is thus given by $v_1 = v_0 + v \sin \theta$, to the first approximation in v. From Bernouilli's theorem, with $U = $ constant, the pressure at this point is given by $p =$ constant $- \rho v_0 v \sin \theta$, again neglecting powers of v. Since the constant cannot give lift, we can neglect it. The term in $\sin \theta$ gives a lift, when integrated round the circle, but not a horizontal resultant force. The lifting force applied from outside on an element ds ($= r \, d\theta$) of the circle at r,θ is $-p \sin \theta \, ds$, i.e. $\rho v_0 v \sin^2 \theta \, r d\theta$. We must add to this the upthrust due to the flow of momentum across this element. The horizontal fluid velocities are equal at θ and $\pi - \theta$ and so do not contribute to the change of momentum of material inside the circle. The vertical velocity $v \cos \theta$ does, however, contribute. The momentum per unit volume due to this is $\rho v \cos \theta$. The rate of flow through ds is $v_0 \cos \theta \, ds$, i.e. $v_0 r \cos \theta \, d\theta$ (neglecting v by comparison with v_0). The upward force through ds due to the momentum flow is $(\rho v \cos \theta)(v_0 r \cos \theta \, d\theta)$, i.e. $\rho v_0 v \cos^2 \theta \, r d\theta$. Adding the upward force due to p, we obtain $\rho v_0 v r \, d\theta$ for the total upthrust. Integrated round the circle, this gives the total lift on the material inside as $2\pi \rho v_0 v r$, i.e. $\rho v_0 C$.

12.9 VISCOSITY AND TURBULENCE

Fluid flow is strikingly altered by viscosity. In the presence of solid surfaces, streamline flow tends to be unstable and to break down to *turbulent* flow if there is the *slightest* degree of viscosity; but if the viscosity is very high, streamline flow is stabilized again and the simple flows of Sections 7.8–7.11 are then observed. In his classical experiments,

(a)   (b)

Fig. 12.26. Flow revealed by a central thread of coloured fluid: (a) laminar; (b) turbulent.

Reynolds allowed water to flow along a horizontal glass tube from a still reservoir, and along the axis he injected a thin thread of coloured water, as in Fig. 12.26. Two quite different types of flow were seen; *laminar* (or *streamline*) at low speeds and *turbulent* (or *eddying*) at high speeds. He also concluded that the critical quantity determining which occurs is the parameter R, now called the *Reynolds number*,

$$R = vr/v, \tag{12.64}$$

this being the only dimensionless combination of the obviously important variables, fluid velocity v, tube radius r and kinematic viscosity v. Reynolds number appears naturally (e.g. as $av\rho/\eta$ in Equation (7.55)) when viscous flow is analysed dimensionally, as in Section 7.11, and it measures the relative strengths of the inertial ($\simeq \rho v^2/r$) and viscous ($\simeq \eta v/r^2$) forces in the fluid, with r as a characteristic dimension in the problem, e.g. radius of a tube or width of an obstacle.

In a tube the laminar–turbulent transition occurs at $R \simeq 10^3$; e.g. at $v \simeq 18$ cm sec^{-1} for water ($v = 0.0178$ cm^2 sec^{-1}) at $0°C$ in a tube of 1 cm radius. Air, because of its higher kinematic viscosity (0.133 cm^2 sec^{-1} at $0°C$), remains laminar to much higher speeds. With careful avoidance of disturbances, laminar flow can be maintained to much higher Reynolds numbers ($\simeq 10^4$), but turbulent flow never continues below $R \simeq 10^3$. When the flow is turbulent, the parabolic velocity profile of Poiseuille flow (cf. Section 7.8) is destroyed by the irregular mixing of fast and slow streams and the velocity is fairly equal over the cross-section of the tube. At high velocities the resistance to flow becomes independent of viscosity, being controlled completely by the forces of inertia. The viscosity η disappears from Equation (7.55) if we assume $q = 2$. The resistance f then becomes proportional to the kinetic energy ρv^2. In highly turbulent flow the forces of resistance (e.g. pressure drop along the tube) thus vary approximately as the square of the velocity.

To estimate roughly the resistance on an obstacle in a stream under these conditions, we suppose that no pressure is exerted in the wake (in practice there may be an *underpressure* in a turbulent wake) and that the dynamic pressure $\frac{1}{2}\rho v^2$ acts on the front side of the obstacle (cf. Equation (12.11)). The total force F on a projected area A presented to the impinging stream is thus

$$F = \tfrac{1}{2}\rho C_D A v^2, \tag{12.65}$$

where C_D is the *drag coefficient*. For rapid flow ($R \gg 10^3$) against a disk (cf. Fig. 12.13), $C_D \simeq 1$ in agreement with our rough estimate. For slow viscous flow, where the argument is no longer applicable, C_D varies strongly and systematically with R. In rapid flow against spheres and "streamlined" bodies the fluid retains much of its initial momentum as it divides round the obstacle; C_D is then small. Figure 12.27 gives observed values. For slow flows ($R < 1$) the sphere obeys Stokes' law (Equation (7.52)). In the turbulent range C_D varies in a complicated way (see below) but generally lies between 1.0 and 0.1. In practical problems of aerodynamics the Reynolds number is usually very large, running into many millions for flying aircraft.

The drag on a body in a fluid can be divided into *form resistance* and *skin friction.* On flying aircraft there is also an *induced drag* due to the energy carried away in the vortices trailing from the wing tips. Form resistance is due to the effect described above, a dynamic pressure on the nose which, because of the wake, is not balanced by a similar pressure on the tail. It can be minimized by "streamlining" the body, i.e. shaping

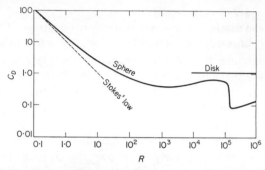

Fig. 12.27. Drag coefficients for flow at various Reynolds numbers.

it so as to minimize the wake. This leads to the familiar airship and fish shapes with smoothly rounded blunt noses (to avoid sharp curvatures which might separate the flow there) and long tapering tails. The form resistance of a fully streamlined body can be as little as 0·02 of that of a disk of the same cross-section. The skin friction is due to the fluid brushing over the surface, and can be reduced by making the surface smooth. It is unimportant compared with form resistance unless the body is streamlined, but in aircraft it consumes most of the engine power in level flight.

A major advance in the theory of fluid friction at solid surfaces was made by *Prandtl*. He showed that a body moving through a fluid drags with

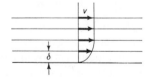

Fig. 12.28. Velocity profile near a flat surface.

itself a thin envelope of fluid and that outside this *boundary layer* the fluid behaves almost as if frictionless and irrotational. As shown in Fig. 12.28, the stream velocity v is fairly uniform except in the thin boundary layer itself, of thickness δ, across which the velocity changes from that of the stream to that of the solid. In an ideal fluid, which can allow an infinite rate of shear, the slip all takes place at the exact interface and the boundary layer is infinitesimal. Prandtl argued that in a real fluid the boundary layer grows to such a thickness that the viscous forces in it are of the same order of magnitude as the inertial forces. The infinitesimally thick vortex sheet at the interface of a solid with an *ideal* fluid thus becomes a thin layer of parallel vortex sheets in the *real* fluid. To estimate δ, consider in Fig. 12.29 flow along a thin plate of length l and width w. All fluid in a

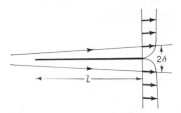

Fig. 12.29. Velocity profile at a distance l from the leading edge of a thin plate parallel to a stream of flow.

band of width 2δ loses about one-half of its initial velocity v by gliding past the plate. The mass m of this fluid which flows past in unit time is $2\rho wv\delta$. Hence, the momentum lost in skin friction is of order $m(v/2) \simeq \rho wv^2\delta$ per second. The frictional force on the plate is of order $(\eta v/\delta)wl$. Equating these,

$$\delta \simeq \sqrt{\frac{\eta l}{\rho v}} \simeq \sqrt{\frac{\nu l}{v}} \qquad (12.66)$$

This is often very small; e.g. if air ($\nu \simeq 0\cdot 15$ cm^2 sec^{-1}) flows over a surface at 10 mile/hr ($\simeq 440$ cm sec^{-1}), then $\delta \simeq 0\cdot 02\, l^{1/2}$ (with l in cm). It increases as $l^{1/2}$, however, so that, for example, down a long pipe the boundary layer eventually spans the whole cross-section.

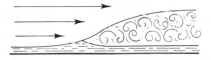

Fig. 12.30. Formation of turbulent boundary layer over a laminar sub-layer.

As the boundary layer becomes thick, it also becomes turbulent. The Reynolds number $v\delta/\nu$ in it increases, with δ and l, and eventually reaches values which allow turbulence. On the basis of Reynolds' experiments, we expect this to happen at about $v\delta/\nu \simeq 10^3$, i.e. at $l \simeq 10^6\nu/v$. The observed critical value of l is about one-half of this. Once the layer becomes turbulent, its thickness rapidly increases, owing to turbulent mixing, as shown in Fig. 12.30.

The size of the wake behind a body depends on where, along the body, the boundary layer separates from it. On the trailing side of the body the streamlines begin to spread apart, i.e. the fluid velocity begins to decrease and the pressure to increase. Under these adverse conditions the fluid in the laminar part of the boundary layer next to the surface of the body, which at best has only a slight downstream motion, may actually begin to flow *upstream* along the surface. In this way a "pad" of almost stagnant fluid may gradually build up under the original boundary layer, so pushing the latter away from the surface and out into the main stream. Separation has then occurred, and the vortex sheets of the boundary layer are injected into the main stream to form a turbulent wake. In an ideal fluid the boundary layer (i.e. an infinitesimal layer at the interface itself) could be separated only at sharp edges, but in a real fluid it can separate even at rounded edges. As an example, Fig. 12.31 shows the flow round an airfoil at normal and stalling incidences.

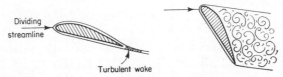

Fig. 12.31.   Separation of boundary layer and size of wake behind an airfoil in normal and stalling incidences.

In addition to streamlining, there are several ways in which the form resistance of a body may be reduced. Turbulence in the boundary layer helps because, through an effect now to be described, it causes the point of separation to move further towards the tail, so reducing the size of wake. The sharp drop in resistance of a sphere, when R exceeds about 10^5 (cf. Fig. 12.27), is due to this. We have seen that separation is produced by backward flow in the boundary layer, which causes a pad of fluid to build up beneath it. This build-up can be opposed by the forward momentum which enters the boundary layer from the main stream of forward flow impinging on it. Turbulent mixing in the boundary layer helps to carry this forward momentum into the inner levels of the layer. The critical Reynolds number at which the drag is reduced by the onset of turbulence in the boundary layer can be reduced by roughening the surface to promote fine-scale turbulence there. This seems to be why golf balls with dimpled surfaces can be driven farther than smooth ones.

Another method of reducing form resistance is to suck fluid into the body, through small holes at points where the backflow occurs. In this way the pad of stagnant or eddying fluid can be drained off and the boundary layer pulled back on to the surface.

At high speeds, as in jet aircraft, skin friction becomes much more important and airfoils have to be designed in which the boundary layers remain *laminar* but nevertheless do not separate. Such designs have been achieved, but the problems of high-speed flight are dominated by various additional effects which we shall discuss in Section 12.11.

Turbulence is important in connexion with the *mixing* of fluids. Molecular diffusion is a macroscopically slow process, and large-scale mixing, for example in the atmosphere, is brought about by turbulent stirring through the movement of eddies. These eddies behave almost as "supermolecules", i.e. as large independent globules of fluid. As well as drifting with the general stream, they have some independent motion of their own, and so may pass from one layer of the stream to another, in this way transporting momentum, energy and matter between layers. In fact, they act macroscopically like the molecules of the kinetic theory of gases and produce *eddy viscosity, conductivity* and *diffusion* in a manner similar to that described in Section 1.10.

They have no permanent existence of course, being merely regions of spinning fluid created in turbulent boundary layers, and as they drift through the fluid they gradually disintegrate under the viscous forces at work in them. Prandtl introduced the idea of a *mixing length L*, like a molecular mean free path, which is the average distance travelled by an eddy before it disappears. Let C be the mean speed of an eddy. Then we can replace Equations (1.41), (1.45) and (1.47) by the corresponding formulae,

$$\kappa \simeq \tfrac{1}{3}\rho c_p CL, \qquad \eta \simeq \tfrac{1}{3}\rho CL, \qquad D \simeq \tfrac{1}{3}CL, \qquad (12.67)$$

for eddy conductivity, viscosity and diffusion. In the atmosphere these are commonly some 10^5 or more times larger than their molecular counterparts, owing to the much larger value of L, so that turbulent mixing is vastly more effective. For turbulent air over solid surfaces, $L \simeq 0\cdot4\,h$, where h is distance from the surface.

12.10 SURFACE WAVES ON LIQUIDS

The motion of water in surface waves can be seen in the movements of small floating objects. These rise and advance as a wave crest passes through them, then fall and retreat in the next trough, moving irrotationally in approximately circular paths, as in Fig. 12.32. To treat this as a problem of steady motion, we allow for the wave velocity c by supposing the wave to be stationary and the water to be flowing beneath it with average velocity $-c$. Let x be the horizontal co-ordinate and h be the height as measured from the average surface level. We assume a stream function of the form

$$\psi = ch + \psi_1, \tag{12.68}$$

the first term giving, by Equations (12.42) and (12.55), the average flow $u_x = -\partial\phi/\partial x = -\partial\psi/\partial h = -c$, the *small* second term ψ_1 giving the wave

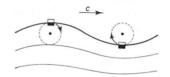

Fig. 12.32. Surface waves on a liquid.

motion. We consider *surface waves on deep water*, so that $\psi_1 \to 0$ as $h \to -\infty$. The streamlines then appear as in Fig. 12.32. We use Laplace's equation,

$$\frac{\partial^2\psi_1}{\partial x^2} + \frac{\partial^2\psi_1}{\partial h^2} = 0, \tag{12.69}$$

and seek a solution that is periodic in x and vanishes at $h = -\infty$. The simplest is

$$\psi_1 = A\, e^{kh} \sin kx, \tag{12.70}$$

and this defines the *surface profile* (i.e. streamline for which $\psi \simeq 0$) approximately as

$$h = -(A/c)\, e^{kh} \sin kx = -(A/c) \sin kx, \tag{12.71}$$

since $e^{kh} \simeq 1$ at $h \simeq 0$. We apply Bernouilli's equation, i.e.

$$\text{constant} = p + \rho g h + \tfrac{1}{2}\rho\left[\left(c + \frac{\partial\psi_1}{\partial h}\right)^2 + \left(\frac{\partial\psi_1}{\partial x}\right)^2\right]$$

$$= p + \rho g h + \tfrac{1}{2}\rho c^2 + \rho c\left(\frac{\partial\psi_1}{\partial h}\right) \tag{12.72}$$

at the surface, where $p = \text{constant}$. Substituting for h and ψ_1 we obtain

$$-(\rho g A/c) \sin kx + \rho c A k \sin kx = \text{constant}, \tag{12.73}$$

which is obeyed if

$$c = \sqrt{\frac{g}{k}} = \sqrt{\frac{g\lambda}{2\pi}}, \tag{12.74}$$

where $\lambda\ (=2\pi/k)$ is the wavelength.

The wave velocity thus increases with wavelength; for example, the approach of a distant storm at sea is first indicated by a *swell* of long wavelength. Such waves are *dispersive* (cf. Section 6.1) and c is their *phase velocity*. A packet of such waves, all with wavelengths near λ_0, travels with the *group velocity* (cf. Equation (6.19))

$$c_g = c - \lambda_0 \frac{dc}{d\lambda} = \frac{c}{2}. \qquad (12.75)$$

Individual wave crests thus move twice as fast as a packet, so that when, for example, a pulse travels over a smooth pond, new waves can be seen continually forming behind it, passing through it, and then disappearing in front. The V-shaped wave of a ship is a pulse of individual waves from the ship which interfere constructively along the line of the wave. The ship creates individual waves continuously at its bow (and stern) with velocities up to that of its own velocity, but their wave packets move slowly and the ship is "supersonic" to them. The kinetic energy radiated into these waves (cf. Section 6.3) sets up a *wave-making resistance* to the ship. A sharp bow minimizes this.

Despite Equation (12.74), c does not go to zero with λ, because very short waves, called *ripples*, are controlled by the surface tension γ. Consider a point on the surface at a height h. The approximately vertical force due to γ is γ/R, where R is the radius of curvature of the surface, i.e. $R^{-1} \simeq d^2h/dx^2$. Equation (12.71) gives $R^{-1} \simeq k^2h$. The forces $k^2\gamma h$ and $\rho g h$ act in the same direction, so that the effect is as if the force of gravity had been increased to $g + (k^2\gamma/\rho)$. Equation (12.74) is then changed to

$$c = \sqrt{\frac{g + (k^2\gamma/\rho)}{k}} = \sqrt{\frac{\lambda}{2\pi}\left(g + \frac{4\pi^2}{\lambda^2}\frac{\gamma}{\rho}\right)}. \qquad (12.76)$$

This gives a *minimum* wave velocity

$$c_{min.} = \left(\frac{4\gamma g}{\rho}\right)^{1/4} \qquad (12.77)$$

at the wavelength

$$\lambda_c = 2\pi\sqrt{\frac{\gamma}{g\rho}}. \qquad (12.78)$$

For water ($\gamma = 85$ dyn cm^{-1}) the values are $c_{min.} = 23$ cm sec^{-1} and $\lambda_c = 1 \cdot 7$ cm. For shorter wavelengths the velocity increases, eventually becoming

$$c = \sqrt{\frac{2\pi\gamma}{\lambda\rho}} \qquad (12.79)$$

when $\lambda \ll \lambda_c$. Thus, if an obstacle such as a vertical fishing line is held at rest in a smooth stream flowing faster than 23 cm sec^{-1}, a stationary pattern of ripples (or *capillary waves*) is formed upstream and a separate set of *gravity waves* ($\lambda > \lambda_c$) downstream.

Equation (12.74) suggests that waves of infinite lengths could cross the sea at infinite speed, but in practice the speeds of very long waves are limited by the finite depth of the water. In these so-called *tidal* or *canal* waves the wavelength is large compared with depth; the velocity of the particles is nearly the same at all depths. The circular motions of Fig. 12.32 are replaced by flat ellipses, the particles moving mainly forwards and backwards. To calculate the wave velocity, we consider in Fig. 12.33 a rise in level, from h_1 to h_2, propagating with speed c along the surface,

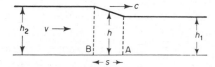

Fig. 12.33. A wave-front moving across shallow water.

the water velocity being zero in front of A and v behind B, where $v \ll c$. In a time t an amount of water $h_2 vt$ crosses the fixed point B from the left. To accommodate this water, the level to the right of B must rise from h_1 to h_2 over a distance ct, where $c(h_2 - h_1)t = h_2 vt$. Thus,

$$h_2 v = c(h_2 - h_1) \tag{12.80}$$

independently of the thickness s of the front. The momentum ρv, per unit volume, of the water increases at the rate ρvch_2 per unit time. This must be equated to the horizontal force acting, which is the difference in pressure $\rho g(h_2 - h_1)$ at any point across the wave-front multiplied by the total height of the front, which we can take as h_2 when $h_2 \simeq h_1$. Hence,

$$vc = g(h_2 - h_1). \tag{12.81}$$

We eliminate $h_2 - h_1$ from Equations (12.80) and (12.81), and set $h_2 = h$ to obtain

$$c = \sqrt{gh} \tag{12.82}$$

as the velocity of the wave. This velocity is independent of the form of the wave (provided $h_2 \simeq h_1$), so that shallow waves are *non-dispersive*; the sinusoidal components of a pulse such as that of Fig. 12.33 all travel with the same speed.

The wave velocity increases with h. This has several effects. If waves of increasing height follow one another, the later ones travel faster and catch their predecessors up. Thus, a continuous wave-front sharpens up into a *shock-front* as it progresses, in the manner of Fig. 6.11. In *tidal bores* and in waves advancing on the sea shore this effect is enhanced by the sloping bottom, which slows down the shallow water waves until they are overtaken by faster, deeper waves following behind. The height of the bore is also increased if the waves are advancing up a narrowing channel such as an estuary. A wave approaching a sloping beach at an angle turns parallel to the beach as it runs into shallow water, since those parts nearest the shore are slowed down first. When the decreasing depth h_1 becomes comparable with the height $h_2 - h_1$ of the wave, the wave-form is distorted by various non-linear effects. The wave, in fact, rolls over and breaks into surf.

The uniformity of particle motion at all levels in a shallow wave implies that the water slides to and fro along the bottom on an oscillating sheet of vortices. We can see this in the agitation of suspended matter near the bottom of shallow sea-water, and in the corrugations left on sandy beaches by the tide.

12.11 SUPERSONIC FLOW

We have already seen, in Sections 6.3 and 12.10, examples of waves which sharpen into shock-fronts as they propagate. Similar effects occur when strong sound waves, e.g. detonations, propagate through fluids. The wave-fronts sharpen up into thin zones, only a few molecular paths thick ($\simeq 10^{-4}$ cm in air), which can be regarded macroscopically as surfaces of discontinuity across which the state of the fluid changes *finitely*. The sonic boom of an aircraft and the crack of a whip are the sounds of shock waves radiated from bodies moving supersonically in air.

In an incompressible fluid all pressure disturbances are propagated instantly with infinite velocity. Supersonic flow is then impossible. Consider, however, a compressible gas. A small disturbance started at some source point spreads through the gas at the speed of sound $c \ (= \sqrt{dp/d\rho})$. At a time t later, all points beyond a distance ct from the source are still unaffected. If the source is itself moving, with speed v relative to the undisturbed gas, two distinct wave patterns are then possible, as shown in Fig. 12.34. The same patterns are clearly also possible if the source is at rest and the fluid is flowing past it. In the subsonic case ($v < c$) the waves from the source travel faster than the source itself and regions to a distance $(c - v)t$ upstream are disturbed. In the supersonic case ($v > c$) the source travels faster than its waves, so that the disturbance all lies *downstream*, inside a cone (three dimensions) or a V (two dimensions), called the *Mach surface* or *characteristic surface*. Fluid upstream of this surface has no warning of its approach. The sudden, almost discontinuous, jump in pressure, density and other properties in an element of the fluid as the surface sweeps through it identifies the surface as a *shock wave*. The *Mach angle* α is easily found from the fact that $AO = vt$, $AB = ct$; thus

$$\sin \alpha = \frac{c}{v} = \frac{1}{M}, \tag{12.83}$$

where M is the *Mach number*. In high-speed compressible flows c is not generally constant, since it depends on pressure and density. In calculations of Mach numbers and angles it is therefore necessary to use the

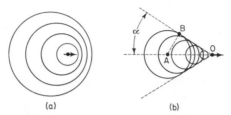

(a)   (b)

Fig. 12.34. Subsonic (a) and supersonic (b) wave patterns round a moving source.

local value of c for the element of fluid considered. The Mach number is a dimensionless quantity and plays as important a role in high-speed compressible flow as Reynolds number does in viscous flow. It measures the relative velocities of the mass motion of the fluid and thermal motion of the molecules.

Shock waves form at the nose, tail and sharp edges of bodies moving supersonically. They can be photographed by spark illumination and can

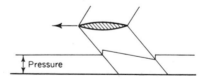

Fig. 12.35. Shock waves and pressure distribution produced by a tapered supersonic body.

occasionally be seen by eye. Figure 12.35 shows the typical pattern formed from a streamlined object such as a supersonic aircraft. In a flying aircraft, lift and control both decrease at speeds $M > 0.75$ (although they are partly regained at speeds $M > 1$), and there is then a danger of stalling (*shock stall*). If *swept-back* wings are used, however, with leading edges turned back through an angle θ, the air velocity *perpendicular* to the leading edge is reduced to $v \cos \theta$, and this can be subsonic even when the speed

v of the aircraft is supersonic. The onset of shock stall is then delayed to higher speeds.

Compression pressure waves of *finite* amplitude sharpen into shock waves. To see this, let us use Fig. 12.33 now to represent a pressure wave between A and B, with h now representing pressure in the fluid, which is advancing into stationary fluid to its right. Fluid behind the wave is then moving forward, in the same direction as the wave. Each increment of pressure between A and B moves forward at the speed c *relative to its own region of fluid*. Its *absolute* speed is then greater than c because its fluid is also moving forward. Thus, the later parts of the pressure wave overtake their predecessors and a shock-front is formed. Since the later parts of the pressure wave help push the earlier parts forward, through their greater fluid velocities, the total velocity of the shock wave exceeds c, i.e. is *supersonic* relative to the undisturbed fluid ahead. Conversely, since the earlier parts of the wave tend to retard the later parts, the shock wave is *subsonic* relative to the fluid behind the shock-front. In a finite *rarefaction* wave, by contrast, the fluid moves in the opposite direction to the wave-front and the later parts of the wave are thereby slowed down relative to the earlier parts. The wave thus becomes more diffuse. Shock waves are therefore necessarily compressive.

To consider this more precisely, we show in Fig. 12.36 a plane shock-front AB in a tube of unit cross-section, and use subscripts 1 and 2 to

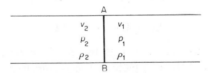

Fig. 12.36. A shock wave in a tube.

distinguish between opposite sides of the shock. We also denote the *finite* change in these properties across the shock from 1 to 2 by Δ; e.g. $\Delta v = v_2 - v_1$. It is convenient to regard the shock as stationary and the fluid as flowing through it with velocities v_1 and v_2 on the two sides. The relation between the states of the fluid on the two sides depends only on the conservation laws of mass, momentum and energy. These laws give, respectively,

$$\Delta(\rho v) = 0, \tag{12.84}$$

$$\Delta(p + \rho v^2) = 0, \tag{12.85}$$

$$\Delta[\rho v(H + \tfrac{1}{2}v^2)] = 0, \tag{12.86}$$

where H is the *enthalpy* (cf. section 2.3) of unit mass of fluid, i.e.

$$H = E + pV, \tag{12.87}$$

$$V = \rho^{-1}. \tag{12.88}$$

To derive these relations, we note first that the masses of fluid entering the shock on one side and leaving on the other must be equal; hence, $\rho_1 v_1 = \rho_2 v_2$. Equation (12.85) similarly expresses the flow of momentum through the shock, since in unit time a mass of fluid $\rho_1 v_1$ enters with momentum $\rho_1 v_1 v_1$ and leaves with momentum $\rho_1 v_1 v_2$; by Equation (6.1), the change in momentum must be equated to the net force $p_1 - p_2$ on the fluid. Hence, $\rho_1 v_1 (v_2 - v_1) = p_1 - p_2$. Combining this with $\rho_1 v_1 = \rho_2 v_2$, we obtain $\rho_1 v_1^2 + p_1 = \rho_2 v_2^2 + p_2$, which is Equation (12.85). The mass $\rho_1 v_1$ flowing through per unit time also changes its kinetic energy by $\rho_1 v_1 [\frac{1}{2}(v_2^2 - v_1^2)]$ and its internal energy by $\rho_1 v_1 (E_2 - E_1)$ as it passes through the shock, and it does work $p_2 v_2 - p_1 v_1$ through the displacement of the forces on its faces. The sum of these energy changes is zero. Again using $\rho_1 v_1 = \rho_2 v_2$, we obtain Equation (12.86).

Equations (12.84), (12.85) and (12.88) give us

$$v_1 = V_1 \sqrt{-\frac{\Delta p}{\Delta V}} \quad \text{and} \quad v_2 = V_2 \sqrt{-\frac{\Delta p}{\Delta V}}. \tag{12.89}$$

We also obtain

$$\Delta H = \tfrac{1}{2}(V_1 + V_2)\,\Delta p \tag{12.90}$$

and

$$\Delta E = \tfrac{1}{2}(p_1 + p_2)\,\Delta V. \tag{12.91}$$

These four equations are known as the *Rankine–Hugoniot* equations of the shock. We notice the complete symmetry upstream and downstream. If we assume that fluid enters the shock from the side with subscript 1, then v_1 is the velocity of the shock relative to the fluid in front and v_2 is its velocity relative to the fluid behind. In the limit of an infinitesimal shock we can reduce Equations (12.89) to

$$v_1 = v_2 = \sqrt{dp/d\rho}, \tag{12.92}$$

which represents an ordinary infinitesimal sound wave.

To distinguish between compression and rarefaction waves, we have to consider the *entropy* S of the fluid (cf. Section 2.3), because, by the general principles of thermodynamics, the changes which occur in an element of fluid when a shock wave passes through it *cannot decrease* its entropy; thus, $\Delta S = S_2 - S_1 \geqslant 0$. To connect ΔS with Δp we consider a weak shock and, taking the values of the physical parameters belonging to side 1 as references, we express the changes across the shock-front as powers of

ΔS and Δp. We shall need to go to the third power in Δp but only the first in ΔS. Then,

$$\Delta H = \left(\frac{\partial H}{\partial S}\right)_p \Delta S + \left(\frac{\partial H}{\partial p}\right)_S \Delta p + \frac{1}{2}\left(\frac{\partial^2 H}{\partial p^2}\right)_S \Delta^2 p + \frac{1}{6}\left(\frac{\partial^3 H}{\partial p^3}\right)_S \Delta^3 p \quad (12.93)$$

and

$$V_1 + V_2 = 2V + \left(\frac{\partial V}{\partial p}\right)_S \Delta p + \frac{1}{2}\left(\frac{\partial^2 V}{\partial p^2}\right)_S \Delta^2 p, \quad (12.94)$$

where the subscripts p and S denote variables held constant when evaluating the differentials; and $V = V_1$. Equation (12.87) gives $dH = dE + p\,dV + V\,dp$. The change dE in internal energy is equal to the heat dQ received minus the work $p\,dV$ done by expansion; hence, $dH = dQ + V\,dp$. From Equation (2.8), this gives $dH = T\,dS + V\,dp$. Hence,

$$\left(\frac{\partial H}{\partial S}\right)_p = T \quad \text{and} \quad \left(\frac{\partial H}{\partial p}\right)_S = V. \quad (12.95)$$

With the help of these standard thermodynamic formulae, Equation (12.93) becomes

$$\Delta H = T\,\Delta S + V\,\Delta p + \frac{1}{2}\left(\frac{\partial V}{\partial p}\right)\Delta^2 p + \frac{1}{6}\left(\frac{\partial^2 V}{\partial p^2}\right)\Delta^3 p. \quad (12.96)$$

Hence, substituting Equations (12.94) and (12.96) in Equation (12.90), we obtain

$$\Delta S = \frac{1}{12T}\left(\frac{\partial^2 V}{\partial p^2}\right)\Delta^3 p, \quad (12.97)$$

where we take T and $(\partial^2 V/\partial p^2)$ as evaluated in front of the shock. We see that the entropy changes across the shock by an amount proportional to the cube of the *strength* (i.e. Δp) of the shock. Ordinarily, fluids become less compressible the more they are compressed, in which case $\partial^2 V/\partial p^2$ is positive. It then follows that Δp has the same sign as ΔS, i.e. that $p_2 > p_1$, i.e. the pressure behind the shock is higher than in front. It also then follows, since $p_2 - p_1 = \rho_1 v_1(v_1 - v_2) = \rho_2 v_2(v_1 - v_2)$, that $v_1 > v_2$, i.e. the fluid velocity is greater in front of the shock than behind. Equations (12.84) and (12.88) give other relations. To conclude, when region 1 is upstream,

$$v_1,\, V_1,\, S_2,\, p_2,\, \rho_2 > v_2,\, V_2,\, S_1,\, p_1,\, \rho_1, \quad (12.98)$$

i.e. the shock is compressive. The p,V relation is usually of the general form shown in Fig. 12.37. Clearly, at the shock-front, i.e. point 1, the slope $(p_2 - p_1)/(V_1 - V_2)$ is steeper than the gradient dp/dV. But $V\sqrt{-dp/dV}$ is the velocity of sound (cf. Equations (12.27) and (12.88)).

Hence, from Equation (12.89), $v_1 > c_1$, so that the shock is supersonic relative to the fluid in front. The same argument applied at the back of the shock, point 2 in Fig. 12.37, shows that $v_2 < c_2$, i.e. the shock is subsonic relative to the fluid behind.

The increase of entropy implies that energy is *always* dissipated in a shock wave (cf. Section 6.3). Thus, D'Alembert's paradox does not arise for bodies moving supersonically through fluids. There is necessarily a

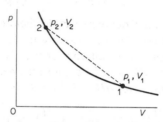

Fig. 12.37.   Points 1 and 2 on an adiabatic curve, representing states at the front and back of a shock.

form resistance, even when there is no wake. To reduce this form resistance in supersonic aircraft, the shock waves are made as weak as possible by using a tapered, pointed nose, by using wings with sharp leading edges, and by avoiding projections from which shock waves could start.

We notice that the creation of the entropy ΔS is *demanded* by the conservation laws, without regard to the frictional properties (viscosity and thermal conductivity) of the fluid. As fluid passes through the shock, some of its mechanical energy has to be turned into heat and, to enable the viscosity and heat flow to do this at the required rate, the shock-front reduces its thickness to the appropriate value. For shocks in the atmosphere this is usually about 10^{-4} cm. The temperature rise across a shock can be considerable. For an ideal gas (cf. Equation (1.6)) we have

$$\frac{T_2}{T_1} = \frac{p_2 \rho_1}{p_1 \rho_2}. \tag{12.99}$$

When $M_1 \ (= v_1/c_1)$ is *large*, it can be shown that this becomes, approximately

$$\frac{T_2}{T_1} \simeq \frac{2\gamma(\gamma - 1)}{(\gamma + 1)^2} M_1^{\,2}, \tag{12.100}$$

where γ is defined by Equation (1.30). For common gases this gives $T_2/T_1 \simeq 0\cdot2 \ M_1^{\,2}$.

A well-known effect of the temperature rise is the *kinetic heating* of meteorites, space vehicles and high-speed aircraft. Figure 12.38 shows the streamlines and shock wave at the nose of a body in a supersonic stream. The temperature T_2 at the stagnation point is given ideally by

$$T_2 - T_1 = v^2/2c_p. \qquad (12.101)$$

Since the specific heat c_p is proportional to the mean-square thermal speed of the molecules, divided by temperature, $T_2 - T_1$ is proportional to $M_1{}^2 T_1$. On leading edges of aircraft the rise in air temperature (in °C) is

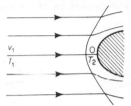

Fig. 12.38. Streamlines and shock wave at the nose of a body in supersonic flow.

roughly equal numerically to the square of the airspeed (in 100 mile/hr). Below the *hypersonic range* (i.e. below $M = 5$) this heating problem can be solved in aircraft by using refractory metals such as stainless steel, but at the speeds of re-entry of space vehicles ($M \simeq 20$) temperatures far above 5000°C become possible and more drastic measures are necessary. Some heat can be removed by radiation, if materials of high reflectivity are used. Heat shields on space vehicles make use of *ablation*, i.e. the removal of heat by evaporation of the outer layers of the material itself. This requires materials of low thermal conductivity, to prevent the heat from penetrating the interior, and of high evaporation energy (light atoms, strong bonds) to carry away as much heat per unit mass as possible. In addition, space vehicles are given large flat noses, to produce very strong shock waves and so transfer as much mechanical energy as possible to these waves rather than convert it to heat by skin friction on the surface of the vehicle.

Figure 12.38 shows the streamlines as *refracted* where they pass through an inclined shock wave. This is a general effect in supersonic flow. Starting from the perpendicular shock wave of Fig. 12.36, we can superpose a uniform fluid velocity u, *parallel* to the shock-front, and so obtain an *oblique*

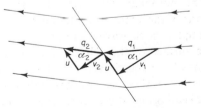

Fig. 12.39. Refraction of streamlines at an oblique shock.

shock as in Fig. 12.39. The resulting uniform flows due to the perpendicular components u and v_1 upstream, and u and v_2 downstream, are indicated by q_1 and q_2, respectively. Since $v_1 > v_2$, then $\alpha_2 > \alpha_1$, i.e. the streamlines change direction as they cross the shock-front. Reversing the argument, if a family of supersonic streamlines is deflected by an obstacle, as in Fig. 12.40(a), a shock wave is formed with a strength that increases with the angle of deflexion. Conversely, if the stream flows over an edge, as in diagram (b), refraction occurs in the opposite sense. In this case,

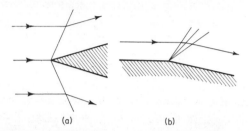

(a)   (b)

Fig. 12.40. Deflexion of supersonic streamlines at obstacles.

however, a rarefaction shock wave is produced, which is unstable. As a result, the Mach lines of the shock-front open out to form a *Prandtl–Meyer expansion fan*, across which the streamlines bend continuously. By combining such shock-fronts and expansion fans, various patterns of supersonic flow can be described. Figure 12.41 shows such a pattern round a primitive airfoil.

Fig. 12.41. Supersonic flow about an inclined plate.

At very high speeds and high altitudes various new effects appear. The gas is largely ionized and its movements can then respond to electromagnetic fields. The molecular mean free path becomes large relative to the linear dimensions of the vehicle, and this makes the flow pattern quite different. For example, shock waves cannot form if the molecules, after reflection from the surface of the vehicle, have to travel distances larger than the dimensions of the vehicle before colliding with molecules of the oncoming gas. Absence of shock waves means that the vehicle is less protected from kinetic heating; most meteors, in fact, burn up owing to heating from free molecular collisions in the upper atmosphere. The condition for this *super-aerodynamic behaviour* is that the *Knudsen number*, i.e. the molecular mean free path l divided by the linear dimension L of the body, should be large. From the Mach number $M = v/c$, the Reynolds number $R = vL/v$, and the relations connecting the kinematic viscosity with the mean free path (Equations (1.44) and (1.45)), we see that the Knudsen number is equal to M/R. When $M > R$, we can no longer pretend that the gas is a continuous fluid but must recognize it as an assembly of separate molecules. Under these conditions fluid mechanics turns into the kinetic theory of gases, which completes the circle in our survey of the mechanical properties of matter.

FURTHER READING

LAMB, SIR HORACE, *Hydrodynamics*, Cambridge University Press, London, 1932.

PRANDTL, L., *Essentials of Fluid Dynamics*, Hafner, New York, 1952.

RAYLEIGH, LORD, *Theory of Sound*, Cambridge University Press, London, 1896.

LANDAU, L. and LIFSHITZ, E., *Fluid Mechanics*, Pergamon, Oxford, 1959.

GOLDSTEIN, S. (ed.), *Modern Developments in Fluid Dynamics*, Oxford University Press, London, 1948.

BIRKHOFF, G., *Hydrodynamics*, Dover, New York, 1955.

SUTTON, O. G., *Atmospheric Turbulence*, Methuen, London, 1949.

RUTHERFORD, D. E., *Fluid Dynamics*, Oliver and Boyd, Edinburgh, 1959.

COLE, G. H. A., *Fluid Dynamics*, Methuen, London, 1962.

MILNE-THOMSON, L. M., *Theoretical Hydrodynamics*, Macmillan, London, 1949.

COURANT, R. and FRIEDRICHS, K. O., *Supersonic Flow and Shock Waves*, Interscience, New York, 1948.

RICHARDSON, E. G., *Dynamics of Real Fluids*, Arnold, London, 1950.

PROBLEMS

12.1 A liquid in a cylindrical vessel is rotated with constant angular velocity ω about the cylindrical axis, which is vertical. By considering the equilibrium of forces on a small element of liquid in the free surface, show that this surface is parabolic and has the form $h = r^2\omega^2/2g$, where h is the height, above the lowest point of the surface, of a point in the surface at a distance r from the cylindrical axis, and g is the acceleration due to gravity.

12.2 Water is flowing down a pipe which tapers from 10 cm diameter to 8 cm diameter. The pressure difference between the ends of the pipe amounts to 15 cm head of water. What is the rate of flow?

[$3\cdot6$ l sec^{-1}]

12.3   Explain why a ping-pong ball can be supported stably on a vertical current of air issuing from a tube.

12.4   The velocity of water flowing in streamlines past a fixed reference point remains steady at 100 cm sec^{-1}. A small floating object, moving freely with the water, increases its velocity by 0·02 cm sec^{-1} per cm travelled. What is the acceleration of the object 1 sec after it has passed the reference point?   [2·04 cm sec^{-2}]

12.5   A gas which obeys Boyle's law, $p/\rho =$ constant, is contained in a vessel at pressure p_1, the pressure outside being p_0. Neglecting gravity, show that the velocity v of fluid flow of the gas out of a hole in the vessel is given by

$$v^2 = \frac{2p_1}{\rho_1} \ln\left(\frac{p_1}{p_0}\right).$$

12.6   Show, by constructing the streamlines $\psi =$ constant, that the complex function $w = \phi + i\psi$ defined by

$$x + iy = w + \exp w$$

represents the flow of a fluid from a large reservoir into a narrow parallel canal.

12.7   Show that for steady, irrotational, non-viscous adiabatic flow Bernouilli's equation, with $U =$ constant, takes the form

$$\frac{v^2}{2} = \frac{c_0{}^2 - c^2}{\gamma - 1},$$

where γ is defined by the adiabatic equation of state and c_0 is the velocity of sound at a stagnation point.

12.8   Continuing with the previous problem, show that the pressure at Mach number $M = 1$ is given by

$$p_0\left(\frac{2}{\gamma + 1}\right)^{\gamma/\gamma - 1}$$

where p_0 is the stagnation pressure.

12.9   A spherical bubble of gas under high pressure is created by an under-water explosion. The gas expands adiabatically, pushing out the (in-compressible) water radially. Assuming that the sum of the potential energy of the gas (i.e. work of adiabatic compression) and kinetic energy of the water remains constant during the expansion and that γ (cf. Problem 12.7) equals 4/3, show that the rate of expansion of the bubble is given by

$$\frac{t}{R_0}\left(\frac{p_0}{\rho}\right)^{1/2} = (2\theta)^{1/2}\left(1 + \frac{2}{3}\theta + \frac{1}{5}\theta^2\right),$$

where R_0 is the initial radius, R is the radius at time t, $\theta = (R - R_0)/R_0$, p_0 is the initial gas pressure, and ρ is the density of the water. If $R_0 = 1$ m and $p_0 = 1000$ atm, how long does it take for the radius to double?

[0·004 sec]

12.10 Show that for a compressible fluid, in which the density depends only on the pressure, Bernouilli's equation can be generalized to

$$\int \frac{dp}{\rho} + \tfrac{1}{2}v^2 + U = \text{constant},$$

where the integral is taken along a streamline. Consider as a one-dimensional problem the flow of such a fluid along a gradually tapering nozzle, under conditions where $U = \text{constant}$. Using this equation and also Equations (12.7) and (12.27), prove that the velocity *decreases* as the cross-section *increases*, in subsonic flow, but *increases* in supersonic flow.

Index

423